PROFESSIONAL AND TECHNICAL WRITING STRATEGIES

Communicating in Technology and Science

FIFTH EDITION

Judith S. VanAlstyne

with

Merrill D. Tritt

Prentice Hall

Upper Saddle River, New Jersey 07458

Library of Congress Cataloging-in-Publication Data

VanAlstyne, Judith S.,
  Professional and technical writing strategies : communicating in technology and science
/ Judith S. VanAlstyne with Merrill D. Tritt.—5th ed.
    p. cm.
  Includes bibliographical references (p.) and index.
  ISBN 0-13-041279-1
    1. English language—Technical English.   2. Communication of technical
  information—Problems, exercises, etc.   3. Communication in science—Problems,
  exercises, etc.   4. Technical writing—Problems, exercises, etc.   5. English
  language—Rhetoric.   I. Tritt, Merrill D.   II. Title.

PE1475 .V36 2002
808'.0666—dc21

00-140084

AVP, Editor in Chief: Leah Jewell
Acquisitions Editor: Craig Campanella
Senior Managing Editor: Mary Rottino
Production Liaison: Fran Russello
Editorial/Production Supervision: Marianne Hutchinson (Pine Tree Composition, Inc.)
Prepress and Manufacturing Buyer: Mary Ann Gloriande
Art Director: Jayne Conte
Cover Designer: Bruce Kenselaar
Marketing Manager: Rachel Falk

This book was set in 10/12 New Century Schoolbook by Pine Tree Composition, Inc.,
and was printed and bound by RR Donnelley & Sons Company.
The cover was printed by Phoenix Color Corp.

Prentice Hall ©2002, 1999, 1994, 1990, 1986 by Pearson Education, Inc.
Upper Saddle River, New Jersey 07458

Printed in the United States of America
10 9 8 7 6 5 4 3 2 1

ISBN 0-13-041279-1

Pearson Education LTD., *London*
Pearson Education Australia PTY. Limited, *Sydney*
Pearson Education Singapore, Pte. Ltd
Pearson Education North Asia Ltd., *Hong Kong*
Pearson Education Canada, Ltd., *Toronto*
Pearson Educación de Mexico, S.A. de C.V.
Pearson Education—Japan, *Tokyo*
Pearson Education Malaysia, Pte. Ltd
Pearson Education, Upper Saddle River, New Jersey

Melvin J. Hoffman KH308 · 826-851

CONTENTS

PART ONE
General Technical and Professional Communication Strategies

CHAPTER 3
Utilizing Graphics and Other Visuals   **71**

CHAPTER 4
Designing and Producing Documents and Presentations   **113**

PART TWO
The Research Strategies

PART THREE
The Technical Strategies

PART FOUR
The Professional Strategies

CHAPTER 12
Composing Correspondence   361

CHAPTER 13
Preparing Resumes, Cover Letters, and Interviews   414

CHAPTER 14
Writing Brief Reports   452

CHAPTER 15
Devising Longer Reports, Proposals   491

CHAPTER 16
Producing Professional Papers   533

PART FIVE
The Presentation Strategies

CHAPTER 17
Performing Verbal Communications   575

PREFACE

The fifth edition of *Professional and Technical Writing Strategies: Communicating in Technology and Science* incorporates major revisions and additions. It is designed for students majoring in technical and scientific fields, but is helpful for all students who will share the writing responsibilities in any field of endeavor. Comprehensive and flexible, the text is suitable for college and university students at any level, technical school students, professional and technical writers, and others in technical/scientific employment seeking a guideline and model text. The materials have been tested in academic classes and training workshops in a variety of businesses and technical industries. The book has won two professional awards: the Award of Distinction from the Everglades Chapter of the Society for Technical Communication and the Award of Achievement from the International Society of Technical Communication.

College students of the twenty-first century are a heterogeneous group with a myriad of interests, needs, and skills. More often than not, they are computer literate and adept at word processing, graphics generation, document design, Internet research, and electronic mail. Their ages range from 17 to over 70 (average age, 28). They major in every field from aviation to zoology. The text includes writing samples that illustrate actual writing demands in a cross-section of career fields: architecture, aviation, computer sciences, ecology, engineering, fire science, insurance and real estate, landscape technology, manufacturing, medicine and dentistry, nursing and paramedical fields, pest control technology, and all of the sciences. The samples in the text have been culled from students and professionals in all of these fields and more.

MAJOR REVISIONS

Major revisions in organization, expansions of previous subjects, and many new additions mark this fifth edition. It covers the expanded role of the technical writer, with extensive coverage of ethical and legal concerns in communications. Because collaborative writing is now more the norm than the exception, this edition expands discussion of collaborative writing and offers more collaborative writing projects than in the past. The issue of international/multicultural aspects in technical communications is also extended, as is the entire writing process coverage.

Some 38 new illustrative graphics and page design options pave the way to enhanced documents. Ten new sample letters, reports, and professional papers amplify the writing instructions in five chapters. Part Five, "The Presentation Strategies," elaborates on verbal and visual communications, and covers the uses of employment and job performance videos and interactive videoconferenced employment interviews. This section also contains a new chapter on Web site design and management.

You will find extensive new computer technology coverage (expanded library and personal computer research techniques, with an emphasis on accessing and evaluating sources, new research Web sites, up-to-date electronic source documentation styles, electronic mail guidelines (format, privacy, and e-mail etiquette), expanded career Web sites, fleshed-out coverage of posted online resumes, and more.

Each chapter has an expanded list of skills to be obtained, new and revised strategy guidelines and instruction, extensive samples, a writing checklist, updated exercises, and writing projects, with collaborative options for each chapter. Updated exercises are included for most chapters.

ORGANIZATION

Part One, General Communication Strategies. Chapters 1 through 4 explain professional and technical communications, discussing the role of the technical writer and communication ethics and law. Part One covers the writing process—theory and practice—with an emphasis on collaborative writing and international and multicultural concerns; organizing, revising, and editing documents; and techniques for the design and presentation of graphics, visuals, and documents.

Part Two, The Research Strategies. Chapters 5 and 6, which focus on accessing information and documenting research, have been greatly expanded and moved forward from the fourth edition, inasmuch as all professional writing and the writing projects assigned throughout this book require some library, Internet, or primary research and documentation. These chapters include extensive updating on library computer retrieval systems, specialized databases and search engines for accessing online materials, methods of evaluating sources, and 16 actual professional Web sites. Chapter 6, "Documenting Research," investigates four major systems for documenting sources (APA, MLA, CBE and variations of numbering systems, and the Chicago style) with coverage of plagiarism and updated Internet source in-text citations and final bibliography styles.

Part Three, The Technical Strategies. Chapters 7 through 11, with revisions, cover user manuals production plus the integral skills of defining terms, explaining mechanisms, giving instructions, and analyzing processes. New samples and graphics enhance this section.

Part Four, The Professional Strategies. Chapters 12 through 16 cover composing professional correspondence, preparing resumes and cover letters, plus interacting in employment interviews, writing a variety of brief reports, devising longer reports and proposals, and producing professional papers. Chapter 12, "Composing Correspondence," addresses electronic mail concerns as well as offers instruction on memos, faxes, and letters, with updated samples throughout. New report samples, abstracts, and summaries illustrate both brief and lenghty reports. The revised Chapter 16, "Producing Professional Papers," explains how to write papers for publication and academic assignments, with samples of each.

Part Five, The Presentation Strategies. Chapters 17 through 19 expand on verbal and visual communications, plus Web site design and management. Chapter 17, "Performing Verbal Communications," includes information-gathering interviews, telephone conversations, group discussions, appraisals and reprimands, and extemporaneous oral reports. Chapter 18, "Managing Visual Communications," discusses twelve types of oral report visuals, along with effective design tips. This chapter also expands on videotaped and videoconferenced employment interviews. The final new chapter, "Designing and Managing Web Sites," provides nine steps for creating and maintaining an effective personal or business Web site.

Appendixes. Appendix A, "Conventions of Construction, Grammar, and Usage," provides explanations and exercises on sentence elements and construction, basic sentence errors, and agreement problems, and provides a usage glossary of often confused terms to aid in document writing, revision, and editing. Appendix B, "Punctuation and Mechanical Conventions," covers all the marks of punctuation, mechanical conventions of abbreviation, capitalization, hyphenation, italics, numbers, and spelling, and includes tables of common technical abbreviations and frequently misspelled words.

Although the organization of the text is intended to offer cumulative skills, the instructor and the student may move about the text as freely as their purposes dictate.

An available *Instructor's Manual* offers general notes to the instructor, student preparation guidelines, and sample syllabi for a variety of class structures. In addition, the manual discusses approaches to each chapter, provides exercise solutions, and exhibits models for the writing projects than can serve as transparency masters. Questions to the author may be sent by e-mail to **JudyVanAlstyne@aol.com**.

ACKNOWLEDGMENTS

I am indebted to the students in my academic classes and professional workshops and seminars and also to the business, technical, and industrial professionals who gave me advice and permissions to reproduce their materials in all the revisions of the original text.

Without my associate, Merrill Tritt, who is a wizard of the complexities of sophisticated computer applications and a reliable advisor, this edition would not have evolved. Neil Linger of Broward Community College/Central offered superb assistance to update library computer retrieval systems and Internet research techniques. Laquita Pimm, educator par excellence, graciously copyread the fourth edition and offered excellent style revision advice. I gratefully acknowledge the evaluations and suggestions from the following reviewers: Gary Christenson, Elgin Community College; John E. Moscowitz, Broward Community College; Jo Ellen Locher, Monroe County Community College; and William Reich, C.S. Mott Community College. I have conscientiously tried to incorporate their suggestions, revisions, additions, and deletions whenever feasible.

I particularly thank Marianne Hutchinson, Production Editor, whose professional assistance throughout the editing and production process made this edition possible. I am also grateful to Craig Campenella, Acquisitions Editor, English, and his Editorial Assistant Joan Polk, who have encouraged me and answered all of my questions throughout the revision process. Thanks also to Mary Dalton-Hoffman who organized and obtained all of the permissions for reprints and to Carol J. Lallier who rigorously and conscientiously copyedited this edition. Finally, I proffer sincere words of appreciation to the marketing department personnel, artists, editors, book representatives, and all the good people at Prentice Hall who have had a hand in this edition.

Judith S. VanAlstyne

ABOUT THE AUTHOR AND THE ASSOCIATE

Judith S. VanAlstyne taught English at Broward Community College for thirty years and was an adjunct professor at Florida Atlantic University. She has served as an industry technical communications consultant and conducted technical writing workshops and seminars throughout three South Florida counties. Ms. VanAlstyne initiated the Technical Writing Curriculum and Certificate Program at Broward Community College. She has authored numerous graphic design texts, instructional videos for technical writing, and technical workshop/seminar workbooks. In addition to teaching expository and creative writing and literature courses, she introduced the technical writing course curriculum to a private college in Malaysia. She has also published literary criticism, travel articles, and poetry.

Merrill D. Tritt has been in the information technology and human resources fields for five years. He is currently a Human Resource Information Systems Analyst for Spherion Corporation. He is also a personal computer consultant in the South Florida tri-county area. In the past, he was an adjunct professor at Broward Community College, teaching computer-related courses. He is responsible for updating most of the graphics and artwork and has also submitted numerous professional papers for this fifth edition.

PART one

General Technical and Professional Communication Strategies

CHAPTER 1

Explaining Professional and Technical Communications

DUFFY   by Bruce Hammond

S K I L L S

After studying this chapter, you should be able to

1. Define *technical writing.*
2. Name three advantages of studying technical writing.
3. Name at least fourteen types of communication strategies that a technical writer needs to master.
4. Name at least five journals to which a technical writer might subscribe.
5. Define *ethics* and explain three theories that guide society's notions of ethics.
6. Name three motives that may result in unethical organizational, corporate, or personal actions.
7. Name at least twelve on-the-job personal activities that are unethical.
8. Name three ethical obligations of manufacturers.
9. Define the ethical problem of whistleblowing.
10. Explain the six principles of ethical technical communications as delineated by the Society for Technical Communications.
11. Name at least five ethical commitments for technical communicators.
12. Name at least four behaviors that ethical communicators should avoid.
13. Name and briefly explain the four basic bodies of law that concern technical writers.

INTRODUCTION

Breakthrough discoveries are occuring in gigabytes at the speed of milliseconds. The twenty-first century will see more innovations than we can predict or imagine. These innovations are occurring in biotechnology, information technology, transportation and energy, biodiversity studies, and exploration, to name only a few categories. Each discovery, invention, manufactured product, and procedure will demand written and oral definitions, explanations, analyses, and instructions for understanding and usage.

Biotechnology breakthroughs will include the cloning and growing of hearts, livers, kidneys, artificial eyes and ears, and stem cell production of muscles, nerves, and blood. The human genome projects of mapping the 100,000 or more genes in the human body will open the doors to diagnostic testing for hereditary diseases and more. Vaccines will neutralize sperm's fertilizing capacity, and others will prevent AIDS and control such diseases as Parkinson's and Alzheimer's.

Information technology has allowed us to embed tiny, digital microchips into all of our machines so that we can chat with our automobile navigation systems, browse the Web, and send and receive e-mail through our pocket-size cell phones and palm computers. Robotic lawn mowers can be operated from the comfort of a hammock, once you have established a few onboard computer guidelines, set up hidden perimeter wires, and recharged the batteries.

Engineers are seeking solutions to our depleting natural resources by inventing and perfecting direct and indirect solar systems. Electric cars will be perfected. Studies are underway to preserve our ecosystems, protect endangered species, and meet the threat of global warming. Exploration of the earth and space continues seeking new organisms, new solar systems, and extraterrestrial resources and intelligence.

State-of-the-art computers, scanners, and printers already offer sophisticated software to enhance word processing, videoconferencing, slide program capabilities, and more, bombarding us with communication materials. E-mail, Internet documents, telephone voice mail, cellular phones, beepers, digital videos, and compact discs continually demand our attention. Without the scientific and technical writer to provide the explanations and operations of these breakthroughs, however, the twenty-first century technology would overwhelm us.

Scientific and technical communicators must decode for us by correspondence, directions, instructions, analyses of operating processes, descriptions, researched reports and articles enhanced by visuals and document design features, and oral presentations the breakthroughs and inventions occurring daily. Such communicators must be trained in computers, word processing, desktop publishing, and design principles and production; further, they need to study and practice ethical communications, international and intercultural differences, and a lengthy but manageable number of writing strategies.

THE TECHNICAL WRITER

Industry is pressed for people with both technical and communications training. A new technical product hits the market every 17 seconds. According to the U.S. Department of Labor's *Occupational Outlook Handbook* (2000–2001), the employment of writers and editors, including technical writers, is expected to increase faster than the average of all occupations through the year 2008 because of the continuing expansion of scientific and technical information and the need to communicate it to the public. To meet the challenge, colleges and universities across the country have developed technical communications courses at all levels. You may take an introductory course such as this one, or continue a degree program, such as a bachelor of science in technical communications, a master of science in technical information design and management, or even a doctor of philosophy in rhetoric and scientific and professional communications, to name a few.

The median of technical writer earnings was $36,480 in 1998, with the highest 10 percent of such writers earning more than $76,660. These earnings have been rising steadily and are expected to continue to rise. Not every technical writer is an employee of a company. Some people establish themselves as freelance writers, marketing themselves to technological and manufacturing companies and charging very respectable fees. Some freelance writers may earn up to $100,000 a year.

Not only the professional technical writer, the engineer, and the scientist, but also the entry-level employee, must master the skills of clear technical communication. The clerk, the firefighter, the pest control technician, the health care worker, anyone in the technical, scientific, and industrial disciplines will contribute to the written records of an organization. Despite the diversity of careers, the writing skills required are more similar than dissimilar.

Countless surveys reveal that employees at all levels, not just professional technical writers, spend from 25 percent to 80 percent of their time on the job writing. That represents slightly more than two to six hours a day, or ten to thirty hours each week. Another significant percentage of time is spent dealing with the writing of others. Consider the time spent reading letters, e-mail, memos, reports, operation manuals, professional journals, and so forth. If such documents are poorly written, the reader wastes time in having to reread material, and, worse, may misinterpret the meaning—which could result in costly errors.

THE VALUE OF TECHNICAL WRITING

The employee who writes well is the one who is noticed by management and marked for promotion. Your ability to write well reveals not only your command of language and grammar, but also your organizational skills, your attention to detail, your persuasiveness, and your logic. Your ability to recognize and execute various strategies of technical writing and to work collaboratively with print design variety, graphic art, and other visuals to enhance your documents will be in great demand. Your writing may even suggest a solid commitment to your employer and profession. It is certain that the sloppy writer will be overlooked for promotion at every level.

DEFINITION

Technical writing informs, explains, instructs, or persuades a specific audience through mastery of specific strategies and sometimes a special language, so that the readers gain new knowledge or the ability to perform their own complex and specialized jobs more effectively. The writer or collaborative writing team may range from entry-level employees to highly trained technicians and communicators. Increasingly, professional technical writers provide written manuals and product support materials. They also assume the tasks of other information channels, such as the development of Web instruction, Internet instructional movies for downloading, development of materials to be imbedded into the product similar to the "Help" sections on Windows and computer software, CD-ROM production, and video and audiotape materials. The technical writer also requires skilled interviewing techniques to question product engineers, marketing specialists, and others involved in the development of a product.

Technical and scientific writing differ somewhat in that technical writing serves technology whereas scientific writing serves science investigating natural phenomena. Science writing, however, is taught in journalism courses and sometimes in technical writing courses. Science writing is more like technical writing in that it informs and explains discoveries, theories, research, and systems to lay people. Technical and scientific writers need command of communication strategies. They will routinely be asked to

- Write clear memos, e-mails, and letters to seek action, provide information, or respond to requests.
- Write progress, laboratory, and test reports on ongoing work.
- Prepare work logs, expense reports, and request-for-leave reports.
- Write field, incident, and feasibility reports.
- Persuade management through proposals to approve new projects, procedures, funding, personnel increases, or duty changes.
- Describe a product or procedure to employees or customers.
- Instruct employees or clients on how to use and maintain a product.
- Explain or analyze a product or procedure for better understanding or improvement.
- Contribute articles or other materials for the employees' newsletter, catalogue, or other publications.
- Develop Web pages and be competent to take advantage of all Internet potential.
- Interview all participants in a manufacturing or scientific endeavor.
- Design documents and enhance them with illustrations, such as graphs, photographs, and other visuals.
- Edit and review documents written by other colleagues.
- Make oral or visual presentations, or both.

As writers improve their communication skills, they will move from writing and editing strategies to the more sophisticated skills of team-produced documents, critical and ethical thinking and problem solving, research, and project management.

Textbooks on technical and scientific writing offer valuable information, and a number of journals provide up-to-date studies in professional writing. They include

ATTW Bulletin
Business Communications Quarterly
IEEE Transactions on Professional Communications
Journal of Computer Documentation
Journal of Technical Writing and Communication
Science, Technology, and Human Values
Technical Communications
Technical Communication Quarterly

Serious students of technical communication will examine these journals carefully and subscribe to the one or two that best supplement their studies.

ETHICAL CONSIDERATIONS

Technical communicators are faced daily with ethical and legal issues, both personally and professionally, in their actions and in their writing or other communications. Ethical and legal considerations are often intertwined. Ethics are the system of principles, based on the values of society, that establish the guidelines for correct behavior. While some motives that may be deemed unethical are not necessarily illegal, there are many laws to protect society against the results of unethical behavior. All communicators (virtually all employers and employees) need to know the principles of ethics and the basics of different bodies of law.

Ethical Issues

Three ethical principles should guide our actions and communications:

1. The **utilitarian theory** asserts that people should base their actions on the degree to which they promote the general welfare or the greatest good for the greatest number of people.
2. **Deontological theories** shift the focus from the consequences of actions to the essential rightness or wrongness of motives.
3. **Justice theories** assert that like situations should be treated alike and should not violate basic human rights.

The very language of these ethical theories poses problems in that definitions of "the greatest good," "essential rightness or wrongness of motives," and "basic human rights" are subject to interpretation, yet all of these principles should be considered when examining our personal and professional ethics.

Personal Employment Ethics

You are or will be competing in a political, bureaucratic, and market-driven environment. Personal ambition, competition, and your need for sales and self-promotion often strain personal and company ethics. Personal ethics become an issue the moment you seek a job. It is unethical to exaggerate or misstate your credentials, experience, or expertise. It is also illegal to claim you have earned a degree at a college or university when, in fact, you have not. It is just as unethical to write an unfactual or glowing recommendation for a friend when, in fact, you know of that person's shortcomings or the superiority of another applicant for the same position. Eventually, your distortions will catch up with you and reflect negatively on your good name.

Equally, it is unethical to suppress knowledge or give false information that would sabotage a colleague, such as failure to admit that another person is more responsible for work for which you are being credited. Understatement may be as unethical as overstatement. For example, if you only faintly acknowledge the contribution of another instead of being explicit, you may undermine that person's expertise. Acts of nepotism (hiring and favoring of family members) are usually unethical.

You must guard against championing a position just because the majority of others in your group or organization do. Avoid *groupthink,* the conformist position due to the need for individual acceptance, the fear of "making waves," or the pressure to "go along with the crowd." Ask yourself if you are being fair and completely honest. If you verbally attack or maliciously ridicule someone, you may be subject to libel laws, which can hold you to verbal or monetary restitution. Racist or sexist remarks or actions are unethical (and can lead to harassment charges and dismissal). You must respect everyone's legitimate rights to privacy and confidentiality.

It should be obvious that to use company supplies (stationery, envelopes, staples, stamps, and the like) or to use your company-owned computer for personal e-mail and Internet capabilities is unethical. It is also unethical to divulge company plans, formulas, and other company secrets. As an employee, you must be loyal to your company as long as it acts within ethical boundaries. Some employees have been induced by rival companies to disclose confidential information, specific technologies, and other confidential material. It is unethical to receive kickbacks or bribes, or to engage in excessive moonlighting, which may be seen as a conflict of interest or may simply make you less productive in your main employment.

Employer Ethics

Employers have ethical obligations, too. Management must adhere to three ethical theories:

1. The **contract theory** that suggests that the manufacturer or provider is entering a contract with the person who purchases a product or service. The contract suggests that the product or service does what it is said to do, is safe, and will operate for a specified length of time before needing service.

2. The **due-care theory** that guarantees that a product will be designed, manufactured, and tested for safety as well as be advertised honestly and supported by operation manuals, warranties, and other written materials.

3. The **strict liability theory** that puts the manufacturer or provider at fault if a product has a defect that causes harm or injury, even if due care was exercised and proven.

The U.S. government and other agencies provide online training and games to sharpen employer and employee awareness of ethics. Many companies are devising statements of ethical behavior and communications. These statements concern conflicts of interest, confidentiality, impartiality in carrying out policies, abuse of power, financial disclosures, and even just the appearance of impropriety.

Manufacturers have ethical obligations to conserve our natural environment by avoiding pollution, unethical and illegal disposal of hazardous wastes, depletion of our natural resources, and destruction of our ecosystems, to name a few.

Whistleblowing is another important ethical issue. We may encounter information or activities within an organization that cause us alarm or concern over their ethics and legality. The question, of course, is whether to "blow the whistle" by bringing such concerns to the attention of higher management or external authorities. Extensive federal laws and regulations have been enacted to protect the rights of whistleblowers. These measures help us to protect the rights, safety, and interests of society and, of course, to protect the individual whistleblower from being fired and from stigmatization, such as loss of reputation, income, security, and even health. Check your organization to determine if there is an ethics officer and learn to consult this individual when you encounter ethical concerns.

Unethical personal and company behavior makes news headlines daily. Consider these incidents:

Incident 1. In 2000 a Florida owner of a paint factory paid two men with a truck just $2,000 to dispose of metal vats of toxic wastes. Normally, it would cost $20,000 to $25,000 to handle such disposal legally and ethically. The two men hid the toxic waste vats on the edges of Everglades National Park, deep in South Florida. Luckily, the vats were discovered before they disintegrated and the waste polluted the Everglades, the Miami River, the Port of Miami, and Biscayne Bay. The paint factory owner's actions were both illegal and unethical.

Incident 2. In 1996 Newt Gingrich, the Speaker of the U.S. House of Representatives, was investigated by the House Ethics Committee for using three tax-exempt foundations to finance his partisan, televised college course. Initially, he denied this to the committee, but months later admitted he had given it wrong information—inadvertently, he said. He agreed not to refute the charge. However, an equally unethical and illegal tape of Gingrich's cell phone calls, recorded by a Florida couple, revealed that he intended to bend that promise. It can also be argued that the Ethics Committee was unethical by postponing the full hearing until after Gingrich was reelected Speaker. The hearing resulted in a $600,000 fine but only a verbal slap on the wrist. This incident underscores the questionable ethics of many participants in government.

Incident 3. In 1999 Governor Jeb Bush of Florida and David Struhs, chief of the Department of Environmental Protection (DEP), stated publicly that they would deny a permit to the Suwannee American Cement Company based on the potential mercury threat to three Florida rivers. Less than a year later in the spring of 2000, the Florida Cabinet voted to allow the cement company to build a plant. Steven MacNamara, a Florida State University faculty member and chief of staff for the House Speaker, also moonlighted as an attorney for the Suwannee American Cement Company—at the same time that he was paid $255,000 for 11 months as chief of staff. He was being compensated by taxpayers to work for a state legislator while privately providing services to a company with a $130 billion project in dire need of state approval. Despite MacNamara's denial, state records show that work for both the company and the legislator overlapped. After investigation, he was forced to acknowledge that he participated in the negotiations that resulted in the DEP's reversal. These actions were flagrantly unethical.

Communication Ethics

Your communication (writing, graphics production, document designs) can easily pose ethical problems. Such communications affect the values and judgments of others. Therefore, you should always adhere to the strictest code of conduct in communicating. While you may aim for persuasion, you have a strong ethical obligation to present the unadorned truth in its full content.

The Society for Technical Communications (STC), the largest organization of its type in the world, devised ethical guidelines (see Figure 1.1 containing the STC Ethical Guidelines) for technical communicators in 1998. These guidelines, although sufficiently broad to apply to widely

STC Ethical Principles for Technical Communicators

As technical communicators, we observe the following ethical principles in our professional activities.

Legality

We observe the laws and regulations governing our profession. We meet the terms of contracts we undertake. We ensure that all terms are consistent with laws and regulations locally and globally, as applicable, and with STC ethical principles.

Honesty

We seek to promote the public good in our activities. To the best of our ability, we provide truthful and accurate communications. We also dedicate ourselves to conciseness, clarity, coherence, and creativity, striving to meet the needs of those who use our products and services. We alert our clients and employers when we believe that material is ambiguous. Before using another person's work, we obtain permission. We attribute authorship of material and ideas only to those who make an original and substantive contribution. We do not perform work outside our job scope during hours compensated by clients or employers, except with their permission; nor do we use their facilities, equipment, or supplies without their approval. When we advertise our services, we do so truthfully.

Confidentiality

We respect the confidentiality of our clients, employers, and professional organizations. We disclose business-sensitive information only with their consent or when legally required to do so. We obtain releases from clients and employers before including any business-sensitive materials in our portfolios or commercial demonstrations or before using such materials for another client or employer.

Quality

We endeavor to produce excellence in our communication products. We negotiate realistic agreements with clients and employers on schedules, budgets, and deliverables during project planning. Then we strive to fulfill our obligations in a timely, responsible manner.

Fairness

We respect cultural variety and other aspects of diversity in our clients, employers, development teams, and audiences. We serve the business interest of our clients and employers as long as they are consistent with the public good. Whenever possible, we avoid conflicts of interest in fulfilling our professional responsibilities and activities. If we discern a conflict of interest, we disclose it to those concerned and obtain their approval before proceeding.

Professionalism

We evaluate communication products and services constructively and tactfully, and seek definitive assessments of our own professional performance. We advance technical communication through our integrity and excellence in performing each task we undertake. Additionally, we assist other persons in our profession through mentoring, networking, and instruction. We also pursue professional self-improvement, especially through courses and conferences.

Adopted by the STC Board of Directors
September 1998

FIGURE 1.1   *STC Ethical Principles for Technical Communicators*   (Used with permission from the Society for Technical Communication, Arlington, Virginia.)

divergent working environments, aim to guide all writers in business, industry, and government, as well as editors, graphic designers, multimedia artists, Web and intranet page information designers, translators, and others whose work involves making technical information available to those who need it.

The following discussion will explain the implications of the ethical principles of legality, honesty, confidentiality, quality, fairness, and professionalism, as set forth by the STC.

Ethical Shoulds and Should Nots

The STC Code for Communicators asks for a personal commitment to professional, ethical behavior. It states that writers **should**

- Prefer simple, precise expression of ideas and avoid cliches (*user-friendly, state of the art,* and so on), euphemisms (this ladder may be *horizontally challenged*), and other vagaries of slang, shoptalk, and gobbledegook (see Chapter 2, "Executing the Writing Process: Theory and Practice").
- Use graphics and other visuals that do not distort the truth (see discussion in this chapter and in Chapter 3, "Utilizing Graphics and Visuals," and Chapter 4, "Document Design").
- Satisfy the audience's need for information, not the writer's own need for self- expression.
- Hold themselves responsible for how well the audience understands the message.
- Respect the work of colleagues, knowing that a communication problem may have more than one solution.
- Strive continually to improve their professional competence.
- Promote a climate that encourages the exercise of professional judgment and that attracts talented individuals to careers in technical communication.

To these we may add that the writer **should not**

- Suppress knowledge by omission or distortion.
- Exaggerate claims about products and services.
- Fail to acknowledge the work of collaborative team writers (see Chapter 2).
- Exploit cultural differences (see Chapter 2).

Brenda R. Syms, in her article "Linking Ethics and Language in the Technical Communication Classroom," suggests that students and employees should ask these questions about their own actions and technical writing:

1. Is the communication honest and truthful?

2. Am I acting in the company's best interest? in the public's best interest? in my own best interest?

3. What if everybody communicated in this way? If the action or communication is right now, is it right for everyone in the same situation?

4. Am I willing to take responsibility for the action or communication publically and privately?

5. Does the action or the communication violate the rights of any of the people involved?

6. Am I treating similar situations similarly?

Careful consideration of these questions should guide your communications and actions into ethical channels.

Visuals Ethics

Consider carefully all illustrations, which can easily unethically distort the truth. Several studies point out that skim readers will see visuals even if they don't read the text and that readers remember visuals longer. Document designs, photographs, font size, and other illustrative elements may all be subject to ethical examination.

Photographs can easily be manipulated with today's photo software tools, which can be used to delete any portion of a picture, zoom in on close-ups to distort the entire picture, lighten or dim features, put two or more images together (such as Oprah's head on Ann Margaret's body, as *TV Guide* did, or to make countless other alterations. *Time* magazine unethically altered a Los Angeles Police Department mug shot of O. J. Simpson, the popular football player who was charged with murdering his wife. The alteration, used on the June 27, 1994 *Time* cover, made him look darker and more sinister than the mug shot portrayed. Critics charged that the alteration was racist and prejudicial. If your alterations overemphasize or under emphasize reality, the use is unethical.

Using small fonts to present product warnings, notifications of upcoming rate increases in credit cards, and other such information de-emphasizes potential problems. Graphs and charts can easily distort information by using too many shadows and lines or cluttered three-dimensional depictions. Pictographs that increase in size but do not accurately reflect numbers may distort reality. Figures 1.2 and 1.3 show two line charts of a company's earning performance over seven years. However, Figure 1.2 ethically includes every year, clearly depicting the drops and increases of the performance, while Figure 1.3 omits the down years and unethically suggests an uninterrupted earnings increase over the seven year period.

Ask yourself these questions about graphics and visuals:

- Does the visual actually do what it seems to promise to do?
- Is it truthful; does it avoid implying lies?
- Does it avoid exploiting or cheating its audience?
- Is it helpful or just decorative?
- Where appropriate, does it clarify text?
- Does it avoid depriving others of a full understanding?

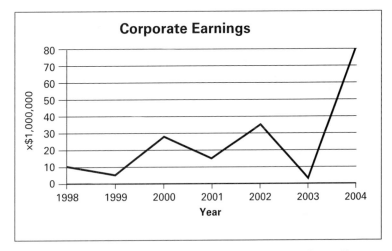

FIGURE 1.2   *An ethical line chart showing earnings dips and rises over a seven-year span*

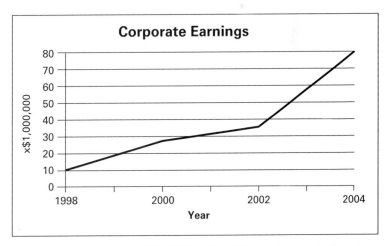

FIGURE 1.3   *An unethical line chart depicting only the rises in earnings over the same seven years and ignoring the odd year drops*

Following are some incidents involving unethical communications in which information was distorted, misrepresented, or suppressed.

Incident 1. A devastating result of unethical communications arose when despite written engineering evidence that the O-rings in the Challenger space shuttle were not safe for a launch in temperatures below 75 degrees and that exhaust leakage had occurred in testing conducted in temperatures less than 53 degrees, Thikol Company top managers neither released the engineering information to NASA (National Aeronautic Space Administration) nor canceled the launch, even though the temperature was barely above 30 degrees. All that the NASA personnel received was the written statement, "The conclusion is that secondary sealing capability in the SRM field joints cannot be guaranteed." The result was the explosion of the shuttle and the deaths of all seven crew members 43 seconds after launch.

Incident 2. Serious consequences have resulted from scientific and pharmaceutical concerns that have suppressed information and documentation on the toxic results of products. In the late 1950s a German manufacturer withheld information on research regarding birth defects in rabbits given K-27, Thalidomide. The drug, already on the market, resulted in more than 8,000 drastically deformed human births.

Incident 3. The makers of silicone implants for breast augmentation issued written documents on the value of their product but withheld information indicating a tie-in to cancer. After silicone implants became the number one surgery in the United States, problems became too numerous to ignore, and huge lawsuits against the manufacturers occurred.

Ethical Dilemmas

Not all disputable issues indicate clear action. Ethical dilemmas arise, according to Vincent R. Ruggiero in his book *The Art of Thinking: A Guide to Critical and Creative Thought*, whenever the conflicting obligations, ideals, and consequences are so very nearly equal in their importance that we believe we cannot choose among them, even though we must. In particular, the medical community offers many ethical dilemmas. Because we can detect birth defects in the womb, is it ethical to alter genes in the uterus before birth? Even though it is legal, is it ethical to abort the defective baby? Is it ethical to terminate the life of a baby whose brain did not develop, and who will die rather quickly without intervention, in order to donate its other organs? Is it ethical to use the medical findings of Nazi doctors who experimented on concentration camp prisoners during World War II? These and other issues will continue to escalate in the twenty-first century due to increasing scientific investigation. Should you write about any of these issues as a technical writer for a laboratory or as an employee of the legal or medical professions, you must present the information in full context without bending or omitting any truth.

Ethics Awareness

In summary, you should question how to develop your own highest ethical behavior. Sam Dagga, a technical communications professor at Texas Tech University, suggests the following four objectives:

1. **Ethical codes.** Look to professional codes of ethics and strive to develop them in your own professional field and personal employment. Familiarize yourself with the STC ethical guidelines. Determine if your employer offers an ethical code or holds discussions about ethics, or both.

2. **Narrative detail.** Examine the narrative details of ethical cases in government, manufacturing, business, and science in order to distinguish between ethical and unethical behavior. Your newspapers report these cases with frequency.

3. **Class and colleague discussions.** Encourage classroom and colleague discussions of ethical questions. Share the details of ethical dilemmas and moral choices that you have faced in your personal and professional life.

4. **Ethical heroes.** Look for the heroes of ethics in society and your profession. Who have been the spokespersons for racial equality, women's rights, homosexuals, and other groups who have often been treated unethically? What have you learned from honest journalists, tobacco company defectors, and others who have taken an ethical or moral stand?

LEGAL CONSIDERATIONS

Persons involved in technical communication need to study four basic bodies of law:

- Copyright law
- Trademark law
- Contract law
- Liability law

1. **Copyright law.** The 1976 Copyright Act provides legal protection to the author, whether an individual or a corporation, of any printed material, software, and photographs, published or unpublished. The author is entitled to a profit from the sale and distribution of the work in exchange for making it accessible. The concept of "fair use" gives to another the right to use material without permission for "purposes of criticism, comment, news reporting, teaching, scholar-

ship, or research." This means that writers and other communicators should avoid excessive use of another's work, seek permission for use in writing (sometimes with payment), cite all sources accurately in their own work, and seek legal counsel if in doubt. Multiple authorship, electronic documents, and electronic storage mediums often pose complicated problems.

2. **Trademark Law.** Companies use trademarks and registered trademarks to ensure that the public recognizes the name or logo of their products. Companies use ^{TM} or $^®$ and other symbols to indicate that a company claims the design or device as a trademark or that the name is a registered trademark. For a company's trademarks to be protected, they should be registered with the U.S. Patent and Trademark Office, be easily identifiable by boldface, italics, type size, or color, referred to with the appropriate name or logo and symbol in written communications, used as adjectives (not as nouns or verbs), and used only in the singular (not as plural or possessive forms of the term).

   Right: a Laserjet® printer
   Wrong: make copies on your Laserjet®
   Right: take some photographs on your Agfa *Agfa* camera
   Wrong: develop your Agfas *Agfa*
   Right: the brightness of Kodacolor^{TM} film
   Wrong: Kodacolor's^{TM} fine quality

3. **Contract law.** This body of law deals with agreements between two parties. The question is whether a product lives up to the manufacturer's claims. The claims are communicated as *express warranties* or *implied warranties*. An *express warranty* is an explicit, written statement that a product has a particular feature or can perform a particular function. An *implied warranty* is not written explicitly but is inferred by the purchaser. If a manual for a camera shows wide-angle photographs, the purchaser has a right to infer that the camera can take wide-angle photographs—even if the manufacturer never explicitly makes that claim. To protect themselves, most companies include disclaimers, a statement that the company hopes will limit its liability for a product. This may read along the lines of "Every effort has been made to supply complete and accurate information. However, ViewSonic assumes no responsibility for the information contained herein or its use."

4. **Liability law.** Under these laws, product documentation in the form of instructions and manuals is treated equally to any other component of the product itself. In cases of personal injury, death,

property damage, or financial loss caused by a defective product, the manufacturer is subject to product-liability action. Therefore, the technical communicators of the company must carefully write their manuals and imbedded Help information to comply with these guidelines, which are based on recent court rulings:

- Knowledge of the product and the likely users through research on the product and the users and their expectations.
- Descriptions of both the product's functions and limitations. An example of a limitation is "This product will not operate in a power outage."
- Instructions for users on all aspects of ownership, including assembly, installation, use and storage, testing, maintenance, first aid emergencies, and disposal.
- Use of appropriate, simple language and graphics in the instructions in terms of the users' education, mechanical ability, manual dexterity, and intelligence. Consideration of the needs of children and audiences from other cultures is also required.
- Warnings about the risks of using or misusing the product (electrical shock, chemical poisoning, danger of fire). Warnings must include the nature of the danger, the uses or misuses which can lead to danger, and the extent of harm possible. Clear and durable warning and caution statements must be appropriately placed in large, easily visible locations, and such language as *must* and *shall,* rather than *might, could,* or *should,* must be used. (See Chapter 10 "Giving Instructions").

Wrong: Plugging too many electrical cords into this socket could result in a fire.

Right: Plugging **more than four** electrical cords into this socket **shall** result in fire.

- Assurance that instructions comply with applicable company standards plus any applicable local, state, or federal statutes.
- Testing of the product and the instructions to ensure that they are accurate, complete, and easy to understand.
- Assurance that the users get the correct information with accompanying instructions. If a problem is discovered later and the entire product design is faulty, the company is obligated to notify the users, using direct mail and/or newspaper advertising to announce a recall of the product.

Think like a lawyer (or even better, a judge or a jury) to protect yourself and your organization. Channel your memos and reports through levels of hierarchy to keep others fully informed, claim authorship of your own ideas, provide a clear written record of your activities in ambiguous situations, and avoid confrontations by always acting in the highest ethical and legal manner.

WRITING PROJECTS

1. **Recognize Technical Writing.** Locate a sample of technical writing (memo, product description, instructions for a product, analysis of a technical process, a magazine article, or a textbook page). Photocopy it and write two or three paragraphs to explain what you think marks it as technical writing.

2. **Future Technical Writing Demands on You.** Write a brief paper to describe the writing you will be expected to do in your chosen field. For sources, interview both an instructor and a professional in the field, look up the profession in the *Dictionary of Occupational Titles* in your library, and examine some writing pertinent to your field. How often will you be expected to write each day? What kinds of writing will be expected? What specific writing skills will be required? How important will writing skills be for your job? How important will writing skills be for a promotion?

3. **Personal Ethics**. Write a brief paper about an ethical (not legal) problem you have encountered. This might be a situation that required your action or an ethical question that a friend or colleague of yours faced. Explain the situation. What choices did you (or the person involved) have? What did you (or the other person) do? What could have been done differently? How can this situation be avoided in the future?

4. **Communications Ethics.** Interview a professor who has published scholarly books or articles. Write a brief report on how the author copyrighted his or her own materials, obtained permission to reprint any other person's or company's materials, and/or documented the words, ideas, and organization of other writers quoted or paraphrased in the project. Organize your notes into a brief report.

5. **Communications Ethics.** Locate in print a product warranty or a product user's manual that gives a false impression, an overstatement on safety, buried or extra small print warnings, or other possible unethical content. Photocopy the material and in a brief paper explain the possible ethical lapses.

NOTES

CHAPTER 2

Executing The Writing Process: Theory and Practice

DILBERT   by Scott Adams

DILBERT reprinted by permission of United Feature Syndicate, Inc.

S K I L L S

After studying this chapter, you should be able to

1. Define the Rhetorical Method and the Bell Telephone Model theories of communication.
2. Define sender, receiver/audience, message, channel, noise, encoding/decoding, interpretation, and feedback
3. Recognize the role and concerns of the sender (individual or collaborative).
4. Name at least five advantages and five disadvantages of collaborative writing.
5. Name the eight strategies for collaborative writing.
6. Recognize the characteristics of the receiver (audience).
7. List five questions the sender must consider about the audience.
8. Name at least seven considerations to be made for an international/intercultural audience.
9. Name the three main steps of preparing a message.
10. Execute the steps of prewriting: analysis of sender and audience, topic list compilation, and logical outlining.
11. Write a rough draft from an outline you have prepared.
12. Write inductive and deductive paragraphs.
13. Avoid an emotional, flowery, judgmental, or pompous tone.
14. Write in a factual, objective, impersonal style.
15. Avoid sexist language.
16. Avoid jargon, shoptalk, and gobbledygook.
17. Avoid wordiness.
18. Avoid redundancies.
19. Avoid nonparallel constructions.
20. Thoroughly execute the postwriting steps of editing and revising a document.
21. Employ copywriting symbols in the editing process.
22. Add signal words where appropriate.
23. Revise word usage ambiguities and mistakes.
24. Recognize and repair fragmentary, run-on, and comma-spliced sentences.
25. Vary sentence construction and sentence length.
26. Avoid unnecessary "There are" constructions.
27. Recognize and repair faulty grammar, including subject/verb disagreements, pronoun/antecedent disagreements, faulty pronoun case usage, misplaced modifiers, and split infinitives.

28. Repair errors in punctuation, abbreviations, capitalization, hyphenation, italics usage, number and symbol usage, and spelling.

29. Employ encoding (document design, graphics and visuals) elements to allow clear interpretation.

30. Choose appropriate channels (letter, manual, e-mail, etc.) and eliminate mechanical problems.

31. Elicit feedback by various methods.

INTRODUCTION

To write effectively, you must understand the basic theories of communication, the concerns of the individual sender or collaborative team, the audience to whom the communication is directed, the process of writing the actual message with concern for planning, executing, and polishing as well as particular concerns (tone, style, grammar, and mechanics), document design options, methods of transmitting, potential barriers to interpretation, and elicitation of feedback methods.

BASIC COMMUNICATION THEORIES

Rhetorical Method

Aristotle, the famous Greek philosopher who lived 300 years before Christ, analyzed a system of communication that orators used to persuade their audiences to adopt their point of view. He called this the *Rhetorical Method* and emphasized that the orator should analyze his audience carefully to adjust the message to the particulars of their biases, desires, education, and so on. Aristotle was concerned with the structure of language, the formation of sentences, and the methods by which one could communicate clearly. Much of the system involved questioning to evoke the desired response. It was particularly relevant for would-be politicians, lawyers, and teachers who needed to command the art of persuasion through words in order to sway the audience to his or her opinion. Aristotle's emphasis on audience analysis is key for modern communicators. More recent research has added new considerations for effective communication in our time.

Bell Telephone Laboratories Model

In the 1940s researchers at the Bell Telephone Laboratories devised a model of the process of human communication with attention not only to audience analysis, but to the sender's capabilities, the message itself, the method of transmitting the message, barriers to clarity, and the strategies used to underscore the message. Modern communicators also pay a great deal of attention to the collaborative nature of the sender, the audience's ability to interpret the message (prior knowledge of the subject, technical level, the hierarchal position of the audience, and international/intercultural differences), and the methods to elicit response.

The Bell Telephone Laboratories model of human communications consists of the following elements:

- **Sender,** the person or team that originates the message
- **Receiver,** the **audience** to the message
- **Message,** the purpose, organization, details, and language of the communique
- **Channel,** the method of transmitting the message (print, e-mail, Internet document, speech, videotape, and so forth)
- **Noise,** mechanical or semantic barriers
- **Encoding/decoding,** the terms, symbols, graphics, and document design which the sender (the encoder) chooses to carry the message and which the receiver/audience (the decoder) accepts.

Two more elements have been added by later researchers:

- **Interpretation,** the audience's understanding of the message
- **Feedback,** the audience's reaction to the message.

Figure 2.1 shows a sender-receiver communications model.

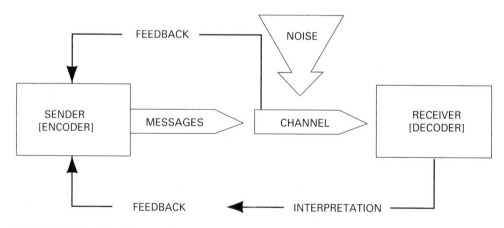

FIGURE 2.1   *Sender-receiver communications model*

Application of Theory

In order to apply the particulars of the Bell Telephone Model, the first step in tackling a professional, technical, or scientific writing project is to sort out answers to the following questions:

- What is my or my team's credibility as the originator? (Sender)
- Who *exactly* is the audience? (Receiver)
- What stated purpose, information, and language are necessary for the communication? (Message)
- What means (print, e-mail, Internet document, speech videos, disks, and so forth) are necessary to transmit the message clearly? (Channel)
- What special considerations (mechanical details and semantics) must be perfect to eliminate misunderstandings? (Noise)
- What devices and supplements (visuals, document design, attachments, and so forth) will make the message more understandable? (Encoding/decoding)
- What else (international/intercultural differences, technical language, idioms, slang, and other stylistic matters) must be considered, included, or eliminated to make the meaning crystal clear? (Interpretation)
- What reaction is desired? (Feedback)
- What considerations, such as overstatement, distortion, omissions, and so forth, must be considered? (Ethics)

Sorting out the answers to these questions is the key to effecting professional, technical, and scientific writing. We will examine each separately, but as you work out your project, you will be working on all of the areas concurrently.

SENDER: INDIVIDUAL OR COLLABORATIVE TEAM

Much of your writing will be individual, but collaborative writing by a group or a team is frequently the route to an effective document. One person seldom produces a major technical document alone. A writing team may be selected to write collaboratively a proposal, a user manual, advertising materials, and the like. The writing team may consult with management, colleagues, engineers, market managers, and others to determine the intentions and the particulars of a document, and then co-produce the actual document. Each member of the team may offer different skills: interviewing expertise, competent research techniques in both print and online materials, organizational skills and style know-how, proficient writing capabilities, acute editing and revising abilities, visuals and document design talents, multimedia materials mastery, final document production efficiency, and so forth.

Advantages and Disadvantages

The benefits of collaborative writing should be obvious:

1. The two heads are better than one principle is foremost here. A team of experts offers a wider base of knowledge than just one writer may have.

2. Collaboration offers wider expertise and skills than just one writer may possess.

3. A group will offer divergent opinions, raise more questions, and point out more problem areas than a single writer will.

4. The responsibility is wider spread, both allowing more members to shine with their accomplishments and removing the burden on just one person to produce a polished document.

5. Collaboration offers the opportunities for co-workers in business and industry to know and respect each other more intimately and thus work more productively. Differences in length of company tenure, gender, and multicultural concerns should be brought to light and worked out democratically toward a shared group and organizational goal.

Despite the obvious advantages of group writing, the members of the collaborative team must be aware of potential problems that may surface in a team effort:

1. There may be disparities in the workload, and some people will inevitably have to work harder than others causing resentments, which must be addressed and mollified.

2. Interpersonal conflicts may arise due to work disparities and disagreements.

3. A collaborative work may take longer to produce than one produced by a single technical writer due to the time it takes to share opinions, edit, and improve the document to everyone's satisfaction.

4. Unless the preliminary planning and closing editing is handled with extreme care, the work may actually result in a document disunified in purpose, style, tone, and design.

5. A group may tend to emphasize getting along together rather than being critical of other's opinions or concentrating clearly on the subject, which may lead to *groupthink*, the negative flip side of team work.

6. Gender and multicultural differences in communications may arise, causing misunderstandings and challenges for accepting other perspectives.

7. Unless the contributions of each are clearly appreciated, members may feel that they are just replaceable nonentities in the scheme of things, and the document will be of low quality.

Figure 2.2 shows the advantages and disadvantages of collaborative team waiting.

FIGURE 2.2   *Collaborative team writing advantages and disadvantages*

These disadvantages can be overcome if all members are aware of the possibilities for misunderstandings and disparities. The wise team leader should be well versed on the potential negative aspects of collaborative writing in order to help each member to exercise clear thinking and to focus on the project.

There may be some differences in the communication behaviors between men and women. Some women tend to use more qualifiers and tag questions (such as "In my opinion, anyway" or "Don't you think?"), which may inaccurately suggest subservience or uncertainty. Women appear to value consensus and relationships, show more empathy, and listen more closely. Perhaps this is because they are raised as the primary family caregivers who nurture, connect, and cooperate. On the other hand, men may value more separateness, competition, debate, and even conflict due to their role as primary breadwinners who are decisive, strong, and informed. Awareness of these perceived differences must be noted and managed in a collaborative effort.

Similarly, people of other cultures may display some differences in communication behavior. Asians and others who value family and community more highly than they value outside individuals may be misunderstood or hesitant to speak their minds if the group is not close knit. In many cultures, it may be rude to respond with a definite "no." Japanese value their privacy and avoid criticizing others or admitting their own confusion. More multicultural differences are discussed later in this chapter.

Collaborative Writing Strategies

Collaborative writing demands that each member understands the strengths and weaknesses of the team effort and understands his or her exact role. A successful collaboration requires at least eight considerations:

1. The **selection of the team members** who can collaborate with common goals and recognize the strengths and differences of others in the group. Sometimes the team will all be members of the same discipline (engineers, computer specialists, technical writers, biochemists, classroom students). Another team may consist of members from diverse fields (an engineer, a computer whiz, a market specialist, a graphic artist). Ideally, each member will have an equal say in the selection of the group leader, the content, particulars, the activity plan, the document design, revision and editing decisions, and supporting multimedia.

2. The **selection of a lead person** who can motivate, coordinate, and deliberate among the members. This person may be chosen by the group but may be named by management. This leader need not necessarily be the senior employee, but the one who manages other people well—a "people person." He or she should review for the group the advantages and disadvantages of team writing so that each member is aware of the strengths and weaknesses of collaborative writing. The leader should recognize the skills of each team member; for instance, one may be stronger on the actual writing, one on layout and graphics, one on revising and editing, and another on multimedia support materials, but ideally all members will participate in each activity. Generally, the leader makes the final decisions on the activity plan, the document design, the revision and editing changes, and the necessary multimedia support materials.

3. The **development of an activity plan** with specific assignments and deadlines. It must include a number of informal consultations, scheduled group meetings, and ongoing interactive computer hookups in order to stay on track. A calendar of assignments and deadlines should be provided in writing to all team members. Progress reports, preferably written, should be given by all partici-

pants at each group meeting or circulated via e-mail on a regular basis. All decisions and transactions should be written down, possibly in memo form, following each group meeting and distributed to all participants as definite reminders of the group's progress.

4. A **sender/receiver awareness** based on the theories discussed in this chapter. All team members must understand the issues of the sender and the perceptions of the audience.

5. The **development of a document design plan** that includes paper quality and size, page orientation, bindings, graphics, art, print size, attachments, and so forth. In addition, the plan should indicate whether there will be a title page, a table of contents, a bibliography, an index, pockets, and other presentation matters. These particulars are discussed at length in Chapters 3 and 4.

6. A **revision and editing plan** indicating who will read each draft, who will suggest changes, who will be in charge of final editing, who will check all visuals, and other details. Ideally, all members will participate in all revision and editing decisions.

7. A **production plan** determining whether the document will be an in-house publication or one that is jobbed out. An assigned production manager should oversee the final production.

8. An **evaluation system** that feeds back the lead person's and management's assessments of the effort and the final document. An evaluation form could touch on the members' regularity of attending meetings and actual participation, the exact contributions of each member, the members' diplomacy, tact, and teamwork ability, and final appraisal of the document.

These considerations put to work should result in an effective document that all members endorse and one that will effectively explain, persuade, solve a problem, or fulfill another predetermined purpose.

Capabilities of the Sender or Team

The individual writer or the collaborative team members must conduct self-analyses in terms of the communication project at hand.

- How much do I (we) know about the subject?
- Do I (we) need to do research before proceeding?
- How credible am I (are we) in terms of my (our) background?
- What do I (we) need to do to be absolutely authoritative in the presentation?

Recognizing and rectifying your deficiencies will make your communication work.

Knowledge?   Credibility?

Research?   Authority?

Purpose

By keeping one clear-cut purpose in mind, you will write a better document. Purpose is your intention, aim, or plan. Is your purpose to query? To explain? To analyze? To define or instruct? To persuade? Not only should your purpose be clear to you, but it should also be established as quickly as possible in your writing. Begin your document with such statements as:

> I am seeking additional information on your Corel Clipart software.
>
> This pamphlet has been prepared to instruct you in the operations of your BellSouth 25-Channel Autoscan Cordless Telephone, Model 33112.
>
> The feasibility of purchasing ten Mustek computer scanners for use in our ten advertising agencies is presented in this study.
>
> The purpose of this proposal, regarding a salary index scale, is to provide incentive to our employees to remain with the company and to eliminate the charge of "favoritism" that has contributed to low employee morale.

Knowing and establishing your purpose will keep your message "on track" and will give your reader a clear sense of how to handle your information.

AUDIENCE/RECEIVER

Anticipating your audience's needs, biases, prior knowledge, multicultural differences, and other particulars is a major key to your communication project.

Just as you would make natural adjustments in vocabulary, sentence structure, and overall message in explaining the operation of a toaster to a child or to an adult, so will you make subtle changes in written instructions of any sort to a potential customer or to a group of factory assemblers. Consider the differences in the following two excerpts—one written for an advanced audience and the other for a novice audience:

> The buccal cavity is abounded laterally by the cheeks, anteriorly by the lips, superiorly by the hard and soft palates, inferiorly by the tongue and associated muscles, and posteriorly by the pharynx and uvula. It contains the teeth, which masticate food; the tongue, which aids in mastication and deglutition; and the salivary glands, which manufacture and excrete salivary amylase, an enzyme that attacks starch and hydrolyzes it into maltose and mucin, a very sticky substance that mixes with the food to form the bolus.

The vocabulary and sentence structure are, perhaps, too sophisticated for a novice reading audience. The following rewrite is more appropriate for an uninformed audience.

> The mouth cavity is bounded on the sides by the cheeks, in front by the lips, on top by the roof of the mouth, on the bottom by the tongue, and in the back by the throat. It contains teeth to chew food, and a tongue which helps to chew food and to swallow. There are also salivary glands, which produce a chemical that attacks starch and breaks it down into a sugar, forming a sticky substance that holds food in a ball.

You must ask yourself still more questions to determine the characteristics of your audience:

- Is the audience informed or uninformed on the subject?
- Does the audience have a grasp of the specialized vocabulary of the field?
- Is the audience singular or a group?
- Is the audience a subordinate, a peer, a superior, or a layperson?
- Does the audience have different international or intercultural communication practices?

Audience's Knowledge

Carefully consider the degree of knowledge your readers already hold. Sometimes a supervisor or colleague is less informed about the writing project than you assume. This lack of knowledge may be why you were asked to prepare a communication in the first place. Perhaps the audience is of a different nationality or different culture, which may require careful investigation of language and multicultural perceptions. These differences may be crucial for understanding even a simple document. Ask yourself these questions:

- How would preliminary knowledge differ for the technician, the marketing member, research and development personnel, an interested layperson, or a potential customer?
- For the uninformed, what detailed statements are necessary to clarify purpose, background data, technical information, and other special information before proceeding?
- What tone, style, and sentence structure will best enable the receiver to understand the message upon completion?
- What, if any, multicultural and multinational differences should be considered both inside and outside the United States?

The Audience's Technical Level

In this technological age you must consider the **specialized training** and **vocabulary** of your reading audience and make the necessary adjustments. Your audience may have familiarity with your content or be completely new to the subject. These considerations call for more analysis:

- Is the audience's professional background similar or dissimilar to mine?
- What training in the field can be accurately expected?
- What technical terms, words, abbreviations, and graphics are appropriate?
- What definitions, synonyms, descriptions, explanations, and other helpful details are needed?

Irrespective of the technical ability of your audience, plain English is always necessary, but in international communications other considerations, such as more flowery openings, literary allusions, and complimentary honorifics (titles and terms of respect) may be appropriate.

Whether your audience consists of **subordinates, peers, superiors,** or **laypeople** (everyone outside your field) should have a bearing on your document or presentation. More questions are in order:

- What are the common denominators among the readers? What are their disparities?
- In the organization hierarchy, are my readers considerably or only slightly above me or my team, about the same, or slightly or considerably below?
- If the audience is within the organization, should the tone be deferential, gracious, familiar, or supportive?
- If the audience is outside my field and/or personal organization, such as readers of software manuals, instructions, and the like, what is the *lowest* level of understanding within that group to which I must address the message?

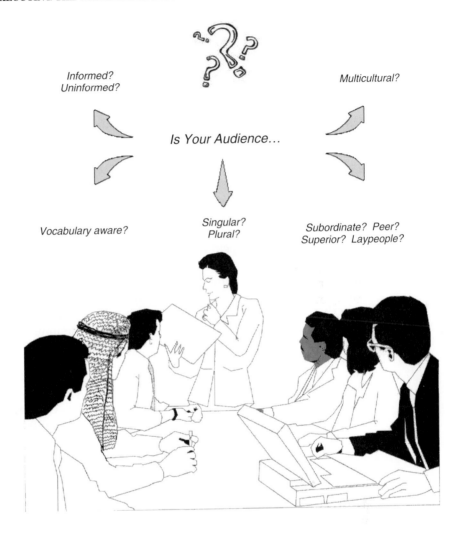

Informed?
Uninformed?

Multicultural?

Is Your Audience...

Vocabulary aware?

Singular?
Plural?

Subordinate? Peer?
Superior? Laypeople?

International and Multicultural Aspects

Other special considerations arise if your audience is comprised of members of **different nations** or **dissimilar cultures,** even within your country. Today no organization can overlook these differences. Tens of thousands of U.S. citizens are employed by overseas companies. In addition, we export many thousands of products, with heavy implications for the operation manuals, installation guides, and maintenance procedures. There is a wide range of other types of intercommunication with foreign consumers. Over 30,000 U.S. companies export products abroad. Our export sales account for billions, perhaps one-third, of the U.S. gross national product. The European Economic Community is an enormous market, as is Japan and other parts of Asia.

According to the U.S. Department of Labor, about 30 percent of our workforce is composed of rather recent immigrants to our country. Among our established citizenry are vast differences in educational systems, ethnic groups, and regional practices. In 1996 the school board of Oakland, California, passed a resolution to teach Ebonics, an amalgam of *ebony* and *phonics,* the prevalent black dialect spoken in America, which was proclaimed by the school board as the "primary language" of African Americans. Further research reveals, despite some definable differences, that the dialect may or may not have West African roots and may or may not derive from old Southern white speech. The point is that some African Americans have difficulty speaking and writing in Standard English, a fact communicators must recognize. Add to this that more than 150 languages are spoken in U.S. schools; the figure in Los Angeles alone is 108.

Even English-speaking people in different parts of the world attach **variant meanings to words, vary word order, vary spellings, and associate different values to colors**. For instance, *toilet* usually suggests a waste-flushing apparatus to an urban American, but may suggest anything from an outhouse to a hole in the ground or a ditch to a rural American or foreigner. To an American, *jumper* denotes a one-piece outfit, and *torch* means a burning flame. To an Australian or a Brit, a *jumper* is a sweater, and a *torch* is a flashlight. A *fortnight* to a British citizen means two weeks, but means nothing to an American. Americans write *color* and *curb* while Brits spell the same words *colour* and *kerb*. The word *American* itself must be used cautiously because any member of a South American country considers himself or herself to be an American. Blue indicates first prize in the United States, but red indicates first prize in Great Britain. In the United States the color red indicates dangers or warnings, but such usage may confuse people of other nations.

Even foreigners who have studied English are confused by our homonyms (words that are spelled the same but mean different things), such as "The farm was used to produce produce," "The dove dove into the water," "There was a row among the oarsmen on how to row," and "too close to the door to close it." Awareness of these homonyms will help you to communicate clearly.

Numbers and dates are handled differently in different countries. In the United States we write dates in the order of month, date, year (February 17, 2000 or 2/17/00), but most European countries write dates in the order of date, month, year (17 February 2000 or 17/2/00). Misuse of dates to international companies can result in late deliveries or worse. Also, in the United States we use commas and decimal points in numbers as follows: 3,927.50, but some Europeans reverse the punctuation (3.927,50). Unawareness of these differences can lead to great misunderstandings.

Translations are difficult for many reasons, even for the person who has good command of more than one language. Companies may assign native-speaking translators to produce products for international markets, yet assessment of translation quality must be "purpose-based" rather than word for word. The translation must meet the original meaning and intent of the document.

Americans seldom know another language well enough to critique other-language documents. You may possess a vocabulary of another language but not be well versed in the sentence structure conventions, word placement rules, idioms, and other practices. In Antwerp, Belgium, students in technical communications are required to be certified to communicate in at least one, or sometimes two, other languages (French, German, English, or Spanish). Learning other languages is the norm in other countries.

Translations into other languages get mangled. In Taiwan, the Pepsi slogan, "Come alive with the Pepsi Generation," translates as "Pepsi will bring your ancestors back from the dead." In Chinese, the Kentucky Fried Chicken slogan, "finger-lickin' good," translates as "eat your fingers off." When General Motors introduced the Chevrolet Nova to South America, the company was unaware that *no-va* mean "it won't go." After GM figured out why it wasn't selling any cars, it renamed the car for its Spanish-speaking markets. The Dairy Association's huge success with the campaign "Got Milk?" prompted them to expand advertising to Mexico. It was soon brought to their attention that the Spanish translation read, "Are you lactating?" One last example of many amusing mangled translations concerns Clairol's introduction of the Mist Stick, a curling iron, into Germany only to find out that *mist* is slang for manure. Not too many Germans had use for the Manure Stick.

Rhetorical strategies differ from country to country also. Even though the international language of technology is English, the strategies of communication among nations may vary. The U.S. technical writer asserts a main idea and supports it with facts and evidence. The writer most often uses a "you" perspective, which does not mean the use of the second person pronoun but that the writer puts himself or herself in the reader's place and writes naturally, concisely, and objectively. This is not so in all other nations. For instance, Arabs may not include multiple facts or evidence but will use repetition to provide emphasis and persuasion. Some Hispanics do not include straight expository evidence but rely heavily on appeals to emotions to get across an idea. The French include a great deal of literary language and literary allusion in even a straightforward explanation. Chinese may begin letters and other material with a great deal of attention to honorifics and flowery generalizations about friendship, the joy of doing business together, and so on. In Japan, documents are written in ways that maintain distance and status rather than exhibit the U.S. emphasis on personal warmth and democratic approaches to all hierarchies.

International ethics vary from ours, which accounts for the differences in rhetorical strategies just discussed. The need to familiarize ourselves with the ethics of other countries if we are to promote products, persons, and ideas is critical. For instance, the Chinese usually adhere to Confucian ethics, which value, above all, Goodness (benevolence, love, or humanity). Goodness offered to others is determined according to the benefits one has received from others. The hierarchy of relationships has clear levels: family, colleagues, friends, and then neighbors. Individuals with whom a Chinese person has an existing relationship will usually be preferred to those who are strangers. Thus, a foreigner might interpret Chinese behavior and preferences as showing favoritism or being strongly prejudicial, whereas Chinese people are actually simply exercising the Goodness Ethic by choosing the product, person, or idea favored by one who fits highest in the hierarchy.

Even **graphics layout** bears consideration. We read left to right, but other language groups read right to left. Our culture scans multiple graphics clockwise. Other cultures may scan graphics counterclockwise. Multiple graphics in a document may prove disconcerting to others. For instance, we would scan the following multiple graphics in the direction of the arrows:

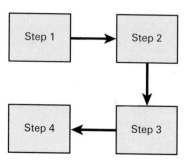

but an Arab-speaking or Chinese-speaking reader would possibly scan the graphics like this:

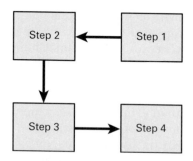

The typical icons used in computers do not always "travel" well. In Asia it is rude to point; therefore, the typical finger that "grabs" material from the toolbars or pointing finger icons indicating direction or continuation may be offensive. The mailbox and trash can icons for sending and receiving mail and deleting a document do not necessarily send universal messages. After all, the mailbox is a largely rural picture, whereas trash cans do not look the same the world over; in Indian and African cultures, the trash can is probably a woven basket of some sort. The technical writer who writes for international and multicultural audiences must develop a sharp awareness of such differences.

MESSAGE

Your message is the explanation or analysis, response, request, set of instructions, descriptions, recommendations, questions, proposal—the "meat" of your communication—that will accomplish your transaction. All the word processing, graphic inclusions, and layout skills in the world have little or no value if your message is not effective in itself. Developing the message involves the writing processes of prewriting (preliminary notes, fact lists, organization, outlines, voice, tone, and style selections), writing the draft, and postwriting (editing and revision, grammar and mechanics check, and so forth). Figure 2.3 is a flowchart of the steps and substeps of the writing process:

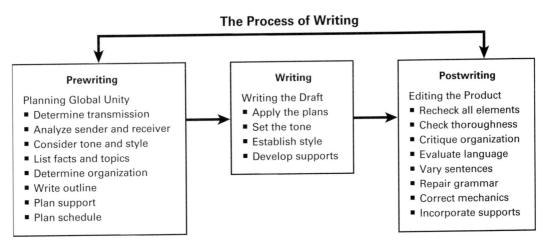

The Process of Writing

Prewriting

Planning Global Unity
- Determine transmission
- Analyze sender and receiver
- Consider tone and style
- List facts and topics
- Determine organization
- Write outline
- Plan support
- Plan schedule

Writing

Writing the Draft
- Apply the plans
- Set the tone
- Establish style
- Develop supports

Postwriting

Editing the Product
- Recheck all elements
- Check thoroughness
- Critique organization
- Evaluate language
- Vary sentences
- Repair grammar
- Correct mechanics
- Incorporate supports

FIGURE 2.3   *The writing process—a flow chart*

The experienced writer or team will include document design elements, graphics and other visuals, and presentation elements throughout each of the steps, but these considerations will be dealt with in the discussion on channel, noise, and encoding/decoding in this chapter and in Chapters 3 and 4, which cover graphics and visuals and document design.

PREWRITING

An effective writer seldom composes a finished product by merely typing a message off the top of the head. Besides the analysis of the sender and the audience, you must concentrate on the overall purpose of your message, the organizational factors, the appropriate tone and style, as well as the mechanics of the finished product. You must aim for a *global* whole, an overall unity. Picture your reader as a busy person, and you will soon develop the habit of coming to the point quickly.

TOPIC LIST

With your audience and overall purpose in mind, think over the material and list the topics to be covered. Research these topics for the facts and data to be included. Brief topic entries will get you started. This chapter was conceived with the following initial topic list:

introduction	Bell Telephone Model	parallelism
tone	feedback	redundancies
prewriting	interpretation	signal words
style	mechanics	agreements
organization	collaborative writing	copyediting
Rhetorical Method	gender	outlines
repairs	multiculturalism	roundabout phrases
jargon	writing theories	pronouns
sender	flow	document design
receiver	induction/deduction	draft writing
channel	wordiness	writing theories
encoder/decoder	process steps	columns
noise	grammar	postwriting
spelling	usage	noise
infinitives	sexist language	voice

As you review your list and research the topics, you will likely discover additional topics to complete your list.

OUTLINES

The next step is to organize the material into a tight presentation. Arrange your final topic list entries into an outline that conveys logical order and completeness. Word processing systems have an outline tool that will help you to edit, move, and subordinate the outline entries. A partial, organized outline with necessary subordination follows:

    I.   Introduction
        A.   Theories
        B.   Process
   II.   Basic Communication Theories
        A.   Rhetorical Method
        B.   Bell Telephone Laboratories Model
             1. Sender
             2. Receiver/audience
             3. Message
             4. Channel
             5. Noise
             6. Encoding/decoding
             7. Interpretation
             8. Feedback
        C.   Application of model
  III.   Sender: Individual or Collaborative Team
        A.   Advantages and disadvantages
        B.   Collaborative writing strategies
             1. Selection of team
             2. Selection of lead person

Once you have organized the items from your initial list and are satisfied with its logical flow, you will find your outline extremely helpful in writing your first draft.

WRITING THE DRAFT

Writing the rough draft is simply a matter of expanding your outlined topics into sentences and paragraphs. Beyond global considerations of format, tone, and style, do not worry too much at this time about the refinements of language or the nitty gritty mechanical aspects in your draft. Refinements are the task of revision and repair. Try to write with a speedy flow, using simple, conversational words and sentences.

Induction/Deduction Formats

Do decide whether to present the information in a particular paragraph inductively or deductively. *Inductive* presentation cites particular facts or individual cases followed by a general conclusion. *Deductive* presentation cites a general conclusion followed by the supporting facts or individual cases. Modern professional writing leans more and more to the deductive method without, however, entirely abandoning the other.

Draft 1 (Inductive)

Gross margins declined from 13.8% in the first quarter of 199X to 13.1% in the corresponding 199X quarter. There was a drop of 26% in sales volume between these two periods. At the branch level new unit margins dropped from 15.8% to 12.2%, but 50% of all units sold in the first quarter of 199X were from inventory made prior to 199X. Profits have been adversely affected by a decline in sales volume and a drop in gross margin.

Draft 2 (Deductive)

Company profits have been adversely affected by a decline in sales volume accompanied by a drop in gross margin. Sales in the first quarter of 199X were 26% below the volume in first quarter 199X. In the same period gross margin decreased from 13.8% to 13.1%. The principal reason for the lower gross margin in 199X was that 50% of the 199X sales were from 199X inventory rather than new production.

The first draft loses the reader in a mass of figures while the second draft presents the key thought in the first sentence. It presents a global purpose first so that the reader is able to understand the statistics quickly.

The general conclusion placed at the beginning or ending of concrete supporting facts is also called the **topic sentence.** It organizes all of the other sentences. The topic sentence employs abstract generalized words expressing an opinion or conclusion. The supporting sentences are concrete, provable facts that support the opinion or conclusion stated in the topic sentence.

Tone

Almost all professional writing demands a factual and objective tone. Tone is the word choice and phrasing which expresses your attitude toward the subject. Professional writing is marked by its lack of emotionalism, editorializing, sarcasm, or even overt enthusiasm. Good writers avoid humor, satire, anger, irony, and bitterness. Consider this partial text of an incident report illustrating inappropriate tone:

On Wednesday, 26 May 199X, at 10:42 A.M., I had the misfortune of witnessing Keypunch Operator Polly Black take a nasty fall in the 5A West office area.

While not looking where she was going, Polly clumsily caught her heel on a CRT tri-stand and crashed to the floor. The paramedic on call at security rendered first aid and transported her to Mercy Hospital. The accident was due to sheer carelessness.

The tone in this report is sarcastic *(had the misfortune)*, judgmental *(while not looking where she was going, clumsily)*, emotional *(a nasty fall, crashed)*, and pompous *(CRT tri-stand, rendered, transported)*. The following, a revised version of the same incident, illustrates an objective tone:

> On Wednesday, 26 May 199X, at 10:42 A.M., I witnessed Keypunch Operator Polly Black fall in the 5A West Office area.
> While approaching the door, Ms. Black caught her heel on an equipment stand and fell on her left side. The paramedic on call at security treated her bruised knee and took her to Mercy Hospital.

The tone of professional communication should be factual and impartial.

Style

Technical writing uses an impersonal, simple, and nonsexist style. Style is the manner or mode of expression in language, your way of putting thoughts into words. Unless your audience is technically sophisticated, you will want to use simple, concrete words, uncomplicated sentence constructions, and short paragraphs. Avoid overtechnical terms, all jargon, shoptalk, gobbledygook, and overblown language. You must also avoid wordiness, redundancy, ambiguity, and unnecessary passive voice constructions. Inappropriate style presents more problems in technical and professional writing than does any other consideration. Some computer software programs offer grammar checks to help you examine and analyze your document, and some offer suggestions on style and tone. Too much reliance on these functions may prevent you from analyzing your own writing sufficiently. If your audience cannot understand your language, your message is lost.

SEXIST LANGUAGE

Sexist language favors one sex at the expense of the other. Our language tends to emphasize the role and importance of men over women. Today, sensitive writers consider carefully the usage of gender-linked nouns, titles, expressions, and pronouns.

Sexist Nouns	*Nonsexist Nouns*
manpower	human resources, work force
mankind	humankind, people
modern man	modern society, modern civilization

Sexist Titles	*Nonsexist Titles*
chairman	chairperson, chair, presiding officer
congressmen	members of Congress, representatives
fireman	firefighter
stewardess	flight attendant
policeman	police officer
salesman	sales agent

Sexist Expressions	*Nonsexist Expressions*
founding fathers	pioneers, founders
gentleman's agreement	informal agreement, oral contract

Sexist Pronouns	*Nonsexist Pronouns*
Each nurse treats *her* patients with care.	Each nurse treats patients with care.
A president sets *his* own agenda.	Presidents set *their* own agendas.
Every employee should sign *his* own clock card.	All employees should sign *their* own clock cards.

Sometimes plurals (*presidents, employees,* etc.) can be unclear.

> **Unclear** Drivers should study *their* manuals. (Does each driver have more than one manual?)
>
> **Clear** Each driver should study *his or her* manual.
>
> **Smoother** Each driver should study the appropriate manual.

Sometimes the use of *his or her, he or she, him and her* can be awkward, but they are, at least, clear and nonsexist. Usually, a little thought can produce a nonsexist, smooth expression of the idea.

Drop the female suffix *-ess* in *authoress* and similar words. In professional letters avoid *Miss* and *Mrs.* unless you are sure; use *Ms* to indicate a female just as you use *Mr.* to indicate both single and married men. Notice that there is no period after *Ms* because it isn't an abbreviation, but a new title of respect. You will, however, often see the period used optionally, as in *Ms. D. C. Grady.*

Sexism in language is not a trivial matter. Responsible writers are careful not to be offensive.

CONCRETENESS

Avoid general, abstract words that represent broad categories, unlimited number, and immeasurable concepts. Eliminate vagueness by using concrete and definite words to prevent misinterpretation and to increase readers' interest. Here are some abstract words and vague phrases along with suggested concrete terms and restructured sentences for clarification:

Vague	buildings
Concrete	bank, supermarket
Vague	devices
Concrete	laptop computer, IBM laser printer
Vague	expensive
Concrete	$70.00 $8 million
Vague	many
Concrete	5,000 to 20,000
Vague	Some of our competitors have very good businesses.
Concrete	Both Sunbelt Instruments, Inc. and Ohio Testing Laboratories grossed over $6.2 million during the fourth quarter of last year.
Vague	As we discussed recently, I have the figures on the project.
Concrete	I have the comparative costs of three word processing computers, which you requested in our telephone conversation last Friday.
Vague	The policy change will affect us adversely.
Concrete	New Policy 1204.05 (Leaves) will decrease our allowable sick days from 10 to 8 per year.
Vague	We will fill your order within the next few weeks.
Concrete	We will ship C.O.D. your order for three, 4-drawer, 36 in. high by 45 in. deep by 14 in. wide, beige filing cabinets 2 September 199X. You should receive them no later than 30 September.

JARGON, SHOPTALK, AND GOBBLEDYGOOK

Jargon is the specialized vocabulary and idiom of those in the same work. The jargon of one field often spreads to the professional world at large. Other words and phrases are intelligible only to those in the same line of work and may be classified as *shoptalk*. Another type of jargon, which may be called *gobbledygook,* is characterized by unintelligible, pompous, or stiff language. General jargon and shoptalk may be acceptable in very personal oral communication, but a competent writer eliminates all jargon from written communication.

GENERAL JARGON

The following words and phrases are used rather widely in informal professional communications:

ballpark figure	mother of . . .
bottom line	optimization
finalized	output
game plan	parameters
handlers	politically correct
impacted	played
input	time frame
interface	viable

Moderate use of such terms is common in verbal transactions but should be revised for written messages. The following sentences demonstrate typical jargon exchanges and suggested written revisions:

Jargon Give me a ballpark figure on the new office furniture.

Revision Give me a price estimate on the new office furniture.

Jargon We'll use the input of each department to finalize our game plan.

Revision We will consider the suggestions of each department to complete our programming.

Jargon The bottom line is that the recession has impacted on our hiring time frame.

Revision The key point is that the recession has affected our hiring schedule.

Jargon The parameters for departmental interfacing must be viable.

Revision The guidelines for departmental boundaries must be realistic.

Jargon His running mate is a heavyweight debater, and if he doesn't dribble around in circles or suffer a late-inning letdown, he is sure to deliver the knockout punch.

Revision His vice-presidential candidate is a fine debater, and if he doesn't include too many details or get discouraged, he will win the debate.

Jargon The last vote was the mother of all elections.

Revision Over 100 million people voted last Tuesday in the largest turnout in history.

Jargon	It just isn't politically correct to suggest a purchase from a company that is played.
Revision	It just is not smart to suggest a purchase from a company whose sales are falling.

SHOPTALK

The use of shoptalk, the more technical slang of those in the same profession, becomes second nature to the users but should never be used in writing. Every occupation has its own shoptalk.

Television shoptalk	He shot the bridge with a minicam and then bumped up the tape.
Translation	He filmed the connecting segment between the news items with a small camera and then machine-processed the videotape to a larger size.
Aviation shoptalk	He checked the pax list, activated the SATCOM, and prepared the PIREP.
Translation	He checked the passenger list, turned on the satellite communication system, and prepared the pilot's report on meteorological conditions.
Internet shoptalk	He went online to chastize the newbie who flamed a chat group last night.
Translation	He connected to the Internet to chastize a new user who used abusive and rude language in a group conversation last night.

GOBBLEDYGOOK

Besides avoiding jargon and shoptalk, the skillful writer should avoid gobbledygook—unintelligible, pompous, and stiff language. Gobbledygook may sound more official or important but rarely states the message clearly.

Gobbledygook	At this juncture, the aforementioned procedure should be utilized.
Plain English	The plan that we discussed should be used now.
Gobbledygook	It would be prudent to consider expeditiously the provision of instrumentation that would provide an unambiguous indication of the level of fluid in the reactor vessel.
Plain English	We need a more accurate device to measure radioactivity.

Gobbledygook	The pilot deployed a vertical antipersonnel device and then activated his aerodynamic personal decelerator before the B29 underwent an involuntary conversion.
Plain English	The pilot dropped a bomb and ejected with a parachute before the bomber crashed.

WORDINESS

Effective professional writing is characterized by its brevity. The concise writer avoids roundabout phrases, redundancies, and sluggish passive-voice constructions.

Following is a checklist of shorter words and phrases to replace wordy, roundabout phrases:

Roundabout Phrases	*Concise Expressions*
a downward adjustment	cut, decrease
a great deal of	much
a majority of	most
accounted for the fact that	because
affix a signature to	sign
after the conclusion of	after
as a result	so, therefore
as a result of	because
as per your request	as you requested
as soon as	when
at which time	when
at all times	always
at an early date	soon
at a much greater rate than	faster
at the present time at this time	now
at the time of	during
avail yourself of	use
based on the fact that	because
be acquainted with	know
be of assistance to	assist, help
brief in duration	short, quick
by way of	by, to
came to an end	ended
consensus of opinion	opinion, everyone thinks
despite the fact that	although, though

due to the fact that in view of the fact that	because
enclosed please find	here is
for the purpose of	for, to
for this reason	so
for the reason that	since, because
give encouragement to	encourage
he was instrumental in	he helped
higher degree of	higher, more
in a manner similar to	like
in a position to	can
in accordance with	by, under
in favor of	for, to
in lieu of	instead
in reference to in relation to	on, about
in the amount of	of
in the nature of	like
in the vicinity	near, around
is dependent upon	depends on
is situated in	is in
it is necessary that	you must
it is recommended	we recommend
miss out on	miss
not infrequently	often
on account of	because, due to
on the part of	from, of
preparatory to prior to	before
provided that	if
pursuant to	under, with, following
referred to as	called
so as to	to
through the use of	by, with
to the extent that	as far as
until such time as	until
with reference to with regard to	on, about
with the exception of	except
with the result that	so that

Avoid "Due to the fact that your reader has a great amount of other work to account for, it is necessary that you write so as to eliminate wordiness in your writing, through the use of concise words." Write "Because your reader is busy, write concisely."

REDUNDANCY

A *redundancy* is a phrase that says the same thing twice *(repeat again)*, contains obvious expansion *(square in shape),* or doubles the idea *(each and every).* Such repetition is pointless, wordy, and distracting. Consider the redundancies in the following sentences:

Redundant	It is *absolutely essential* in *this day and age* to *completely eliminate bigotry and prejudice.*
Revision	Bigotry must be eliminated now.
Redundant	The sheriff's department and the city police *cooperated together* in *the month of May to devise and develop a totally unique drug and narcotics* control program.
Revision	The sheriff's department and the city police cooperated in May to develop a unique narcotics control program.
Redundant	There are *many in number* who consider the *total understanding of basic fundamentals* a *good asset.*
Revision	Many consider the understanding of basics an asset.

VOICE

Verbs have two voices: active and passive. In an active-voice expression the subject of the sentence performs the action stated by the verb.

Mr. Jones *conducts* the plant tours.

The president *presented* the budget.

The firm is *spending* $50 million this year to promote the new beer.

The new price *will increase* our profit.

In passive-voice expressions the performer appears in a postponed sentence element or not at all.

The plant tours *are conducted* by Mr. Jones.

The budget *was presented* by the president.

Fifty million dollars *will be spent* this year to promote the new beer.

Our profit *will be increased* by the new prices.

Generally, the active voice suggests immediacy and emphasizes the subject. Because the passive-voice verb is always at least two words (the verb plus a form of *to be*), passive-voice expressions tend to be wordy. The passive voice may bury your main ideas in sluggish sentences.

Edit your writing to determine which voice permits your desired emphasis in the fewest number of words.

> The secretary typed the report. (Emphasis on secretary)
>
> The report was typed by the secretary. (Emphasis on report)

PARALLEL CONSTRUCTIONS

Parallel constructions are those that place related ideas in a sentence into the same grammatical forms. Adjectives should be parallel with other adjectives, verbs with verbs, phrases with phrases, and clauses with clauses.

Nonparallel	Tungsten steel alloys are tough, ductile and have strength.
Parallel	Tungsten steel alloys are tough, ductile, and strong.
Nonparallel	He must use parallel constructions and look to avoid ambiguities.
Parallel	He must use parallel constructions and avoid ambiguities.
Nonparallel	The process of writing demands four activities: prewriting, message composition, the use of additions, and the ability to postwrite.
Parallel	The process of writing demands four activities: prewriting, writing the message, adding graphics and other art, and postwriting.

All of these style considerations and more are discussed in Appendix A. This would be a good time to review the appendix.

POSTWRITING

Efficient revision and editing will repair your rough draft and produce a polished product. The first tip is to allow time between your draft writing and your revising and editing. Do something else for a few minutes, hours, or even days for a cooling-off period. You will find yourself more objective and critical when you return to the writing task. You are not alone in your urge to just get the paper off your computer, off your desk, and off your mind. But take the time to rethink, rewrite, and revise. The goals of postwriting are

- Rechecking all elements for audience appropriateness
- Establishing that all of the facts and topics are covered thoroughly

- Critiquing the organization and paragraphing
- Evaluating language and word choice
- Scrutinizing sentence construction and variation
- Repairing grammatical errors
- Correcting punctuation, spelling, and other mechanics

The idea here is to consider the relationship of the whole to the parts and the parts to each other. Add what is needed and purge unmercifully what is not. If you need to review conventions of sentence construction, grammar, and usage, consult Appendix A in the backmatter of this book. Appendix B covers punctuation and mechanical conventions, including common technical abbreviations and symbols, number handling, frequently misspelled words, and more. The checklist at the end of this chapter may aid you in your revisions and editing.

Review the copyediting symbols in Table 2.1. As you read your entire draft, use the appropriate symbols in your text. Alternatively, put a check in the margin when you find a questionable word, an awkward or unclear sentence, an uncertain grammatical construction, a mechanical complexity, or a design discrepancy. Then begin a systematic revision of your work.

Overall Revisions

You need the basics before the shine. Study your document to assure that the first three goals (audience appropriateness, thoroughness of content, and logical organization) are met. Once again question if everything in the document is appropriate to the audience. Determine that the tone and style are consistent. Examine the facts for completeness and accuracy. Make absolutely certain that the overall purpose and emphasis are immediately clear. Establish that each paragraph or section has a clear focus and the necessary supporting information. Eliminate any digressions, needless repetitions, or padding. If you are team writing, have other members examine these first three areas and make all of the necessary revisions.

Language and Word Choice Revisions

The analysis of language and word choice is the next task of revising and editing. Review your language to determine if it is too technical for the audience. Consider multinational and multicultural differences that may distort your message. Review the material on sexist language, concrete wording, jargon, wordiness, redundancy, and voice in this chapter in order to apply these considerations to your revision. Revise your document accordingly. Next, examine the need for signal words, precise word choice, and usage corrections.

TABLE 2.1   *Copyediting Symbols*

Symbol	Meaning
(35)	Spell out
(two hundred)	Use numerals
Scientific	Use lowercase
french	Capitalize
Techn*i*cal	Insert letter
misspelled~~mispellede~~	Mark out grossly misspelled words and print the correct spelling above
saample	Delete and close up
clean ~~clear~~ the ribbon	Delete and join to remaining text
He typed the report⊙	Insert period
to carefully read	Transpose words
thier	Transpose letters
. . . noise. When taping	Start new paragraph
the company's car	Insert apostrophe
run-onwords	Insert space between words
to provide aide (stet)	Retain the original
know the symbols⊙ ⌐They will help . . .	Close up or join
What did he mean?	Insert question mark
feet, inches and pounds	Insert comma
yourss	Delete letter

SIGNAL WORD REVISIONS

Add signal words—those that indicate time, additions, results, summary, comparisons and contrasts, and other indications of what is to follow—for paragraph to paragraph and sentence to sentence coherency. Consider the following paragraph:

> The writer should follow a number of steps to revise and edit a paper. The writer must recheck for audience appropriateness. The writer must establish that the facts and topics are complete. The writer should assiduously critique the organization and paragraphing. The writer must evaluate the language and word choice. The writer must scrutinize sentence construction and variation. He must make grammatical corrections. He must diagnose and correct punctuation, spelling, and other mechanics. He must reexamine and enhance the document design and additions. He must prepare the final document.

In the revision, signal words have been added for unity and coherence and some other language changes have been made.

You should follow a number of steps to revise the paper. **First,** you must recheck the document for audience appropriateness. **Next,** you must establish that the facts and the topics are complete, **and** you should carefully recheck the organization and paragraphing. **Additionally,** you must evaluate the language and word choice. **Equally important,** you should make grammatical corrections, **and then** you must diagnose and correct punctuation, spelling, and other mechanics. **Then,** you should reexamine and enhance the document design and additions. **Finally,** you should prepare the final document.

Besides the addition of the signal words, *you* replaces *the writer* and *he* to personalize the message and to eliminate the sexist *he* or *he and she*. Furthermore, the word *carefully* replaces *assiduously* for a more consistent style.

A list of signal words to make connections and transitions follows:

Time Signals	*Addition Signals*	*Results Signals*
soon	again	hence
then	and	therefore
finally	besides	consequently
previously	furthermore	accordingly
first, second, etc.	also	thus
next	additionally	as a result
last	moreover	finally

Summary Signals	*Contrast Signals*	*Comparison Signals*
in brief	however	similarly
finally	nevertheless	likewise
in conclusion	yet, and yet	correspondingly
to conclude	but	equally
summing up	still	equally important
on the whole	on the other hand	in the same manner
lastly	though, although	in the same way

Other Signals

simultaneously	conversely
meanwhile	unfortunately
for example	to begin with
for instance	so
to demonstrate	in the past
indeed	in the future
in other words	eventually

WORD CHOICE CHECK

If you are composing and editing your document using word processing software, you probably have a **thesaurus** tool to seek out more precise meanings to words and to find synonyms for weak or overused words. You may even have a **dictionary** providing exact meanings, synonyms, and antonyms. If you do not have these tools on your computer, a desk dictionary and word thesaurus are absolutely necessary for the effective writer. Pay particular attention to your verbs; select the word with just the right nuance. For example, here are two lists of verbs requesting action. Notice how each column increases in intensity from top to bottom:

(would) appreciate	hinted at
(would) suggest, propose	implied, intimated
prefer, would like	suggested
request, ask	indicated
recommend, urge	signified
remind	demonstrated
advise	confirmed
require	substantiated
insist	proved
demand	established

Likewise, noun synonyms can suggest minute shades of meaning.

aid	sketch
assistance	plan
help	program
full support	design
championship	project

USAGE CHECK

English is a complicated language containing homonyms (words that sound alike: *there, their, they're*) and other confusing similar words *(insure, assure, ensure)*. The astute reader is disturbed over common usage errors. Check carefully such words as *its/it's, accept/except, advice/advise,* and the like. The adverb *hopefully* is frequently misused; the writer usually means "I hope." The best advice is to avoid *hopefully* altogether. Your audience may miss the impact of your message if you fail to edit these problems carefully. Familiarize yourself with the **Usage**

Glossary in Appendix A, which covers 41 sets of confusing words, with explanations and an exercise to test yourself. Keep this glossary on hand every time you write or revise a document in order to repair your mistakes.

Sentence Repair

Consider the parts carefully for necessary sentence repairs and variation. It is all too easy to write sentence fragments, faulty parallel constructions, and run-on or comma-spliced sentences in your draft. Read through slowly for sentence sense. Find the main sentence elements (subject, verb, and object) to ensure that your sentences convey the ideas you intend. Again, Appendix A discusses basic sentence constructions and errors and provides exercises to test your recognition of each error. Review that material attentively. Some reminders follow:

Fragment	Lithium has many uses in bioengineering. *Especially* in the development of pacemakers.
Repair	Lithium has many uses in bioengineering. *It is especially* useful in the development of pacemakers.
Parallel problem	We need adjuncts *to handle* peak-hour activity, *to free* full-time employees from routine duties, *to relieve* assembly workers for lunch breaks, and *for the replacement of* vacationing employees.
Repair	We need adjuncts *to handle* peak-hour activity, *to free* full-time employees from routine duties, *to relieve* assembly workers for lunch breaks, and *to replace* vacationing employees.
Run-on	If the Roman government at the height of its power, and at a time when means of communication had been greatly improved, showed anxiety for the food supply of that Italy which was dominant in the Mediterranean world, it may be imagined that in the period preceding the great economic organization introduced by the Roman Principate the peoples of the Mediterranean region peoples no one of which at the height of its power had controlled the visible food supply of the world so widely or so absolutely, had far graver cause for anxiety on the same subject, and this was an anxiety such as would be, under ordinary circumstances, the main factor, or, even under the most favorable circumstances possible in those ages, a main factor, in molding the life of the individual and the policy of the state.
Repair	If we understand that the Roman government at the height of its power showed anxiety for the food supply, we can imagine that earlier Mediterranean peo-

ples had far graver cause for anxiety. The Roman Principate had greatly improved communications and economic organization to handle the food supply of Italy, the dominant country in the Mediterranean world. No earlier people in the region had controlled the visible food supply so widely or so absolutely as had the Romans. Under ordinary circumstances the food supply was a main factor of concern. Under even the most favorable circumstances, anxiety over the food supply would be a main factor in molding the life of the individual and the policy of the state of those earlier peoples.

Comma-splice Cellular telephones are available in a wide variety of styles, each manufacturer offers other special features.

Repair Cellular telephones are available in a wide variety of styles; each manufacturer offers other special features.

Sentence variation is another area for examination and revision. Too many simple or compound sentences may not only be dull but also suggest that you do not have a good command of the alternative compound and compound/complex constructions. Review the sections on basic sentence elements and secondary elements in Appendix A. Then revise your material to lend a variety of constructions with logical participial phrases and dependent clauses. Consider the following paragraph, which employs five simple sentences:

> The writer must carefully consider the concerns of prewriting. He must consider the audience. He must then devise a topic list of the areas to be covered in the document. Next, he must organize the list into a logical outline. He may then move to the next step, writing the draft.

Consider how the use of participial phrases and dependent clauses adds variety of sentence construction and sophistication to the paragraph:

> To write well you must carefully consider the concerns of prewriting (Simple). Considering the needs of the audience, you will have direction for planning the document (Complex). Next, you must devise a topic list of the areas to be covered in the document, and then you can devise a logical outline, which will guide the preparation of the draft (Compound/Complex). Finally, you are ready to write the draft (Simple).

The word *you* replaces *the writer* to offer a more personal tone and to avoid the sexist *he*.

Too many very long sentences or too many very short sentences will bore or burden the reader. An easy-to-read sentence is usually 12 to 25 words long, but variety adds smoothness and emphasis. Consider the following:

> I prepared for the Olympics by exercising my body with weights and aerobics. I repeated my routine at least 15 times a day. I asked my coach to point out even the slightest movement that could be improved. I designed and redesigned my costume for the event continually. Finally, it was my turn for my ice skating routine, and I won.

In this paragraph each sentence contains 10 to 13 words, which is acceptable, but they are boring in their construction and rhythm. Look at the following revision:

> I prepared for the Olympics by exercising my body with weights and aerobics and repeated my routine at least 15 times a day. I asked my coach to point out even the slightest movement that could be improved. Furthermore, I designed and redesigned my costume continually until finally it was my turn for my ice skating routine. I won.

In this example sentences have been combined so that the variation is from 15 words to 23 words, followed by the dramatic final two words.

Overuse of the indefinite phrases *there is* and its related forms (*there are, there will be, there may be, there might have been,* and so on) weaken sentences by delaying the appearance of the subject. Eliminating most of these constructions will make your sentences more effective.

Consider the following:

Ineffective	There are specific steps to take to make your sentences more effective.
Effective	Specific steps will make your sentences more effective.
Ineffective	There will be a fire alarm scheduled for this afternoon.
Effective	A fire alarm is scheduled for this afternoon.

Sometimes, however, the construction adds emphasis or avoids the verb *exists*.

Weak	A very good reason to rest between exercise sets exists.
Stronger and eliminate *exist*	There is a very good reason to rest between exercise sets.

Grammar Repair

Once again, your computer may have a **grammar check** tool that ferrets out grammatical errors, unnecessary repetitions of words, as well as some mechanical errors. This tool is valuable for a first scan of grammar problems, but it cannot replace your obligation to know correct grammatical conventions. The most common grammar errors include subject/verb disagreements, pronoun/antecedent disagreements, general pronoun misuse, modifier misplacement, and split infinitives. Appendix A addresses these problems at length and provides some exercises to test your knowledge, but here are some tips.

SUBJECT/VERB CHECK

A verb agrees with its subject in number.

> **Incorrect** *Each* of the insurance companies *are* reviewing the application.
>
> **Correct** *Each* of the insurance companies *is* reviewing the application.
>
> **Incorrect** The *manager* as well as the supervisor *have* been promoted.
>
> **Correct** The *manager* as well as the supervisor *has* been promoted.

PRONOUN/ANTECEDENT CHECK

Similarly, pronouns must agree with their antecedents in number.

> **Incorrect** *Each* of the women bought *their* own computer software.
>
> **Correct** *Each* of the women bought *her* own computer software.
>
> **Incorrect** *Ace Company* is sending 20 of *their* employees to the seminar.
>
> **Correct** *Ace Company* is sending 20 of *its* employees to the seminar.

PRONOUN USAGE CHECK

Pronouns are grouped into cases. Case 1, the nominative, is used for sentence subjects and predicate nouns, nouns that follow a linking verb and identify the subject. Appendix A covers pronoun case fully. Some common errors follow:

> It is *I.* (not *me*)
>
> It will be *he* who goes. (not *him*)
>
> *He and she* were appointed to the committee. (not *him and her* or *he and her*)
>
> Ms Johnson and *I* attended the meeting. (not Ms Johnson and *me*)

Case 2, the objective, is used for direct objects, indirect objects, and objects of prepositions.

Direct object	Catherine's secretary took her boss and *me* to lunch. (not *I*)
Indirect object	Ken sent Charles and *me* the document. (not *I*)
Object of preposition	The plan divides the work between Mr. Howard and *me*. (not *I*)

Avoid using reflexive pronouns *(myself, himself, herself, themselves, ourselves)* as subjects or objects.

Incorrect	Jane and *myself* wrote the manual.
Correct	Jane and *I* wrote the manual.

Incorrect	The work was completed by the secretary and *herself*.
Correct	The work was completed by the secretary and *her*.

MODIFIER PLACEMENT CHECK

Check the placement of modifying words, phrases, and clauses.

Incorrect	The *little brick doctor's* house.
Correct	The *doctor's little, brick* house.

Incorrect	The accountant *only* comes in on Tuesdays.
Correct	The accountant comes in on Tuesdays *only*.

Incorrect	The computer was broken *in the closet*.
Correct	The computer *in the closet* was broken.

Incorrect	*Speaking before a crowd of people,* my knees shook.
Correct	*Speaking before a crowd of people,* I had shaky knees.

SPLIT INFINITIVE CHECK

An infinitive, you remember, is a verb form using *to (to work, to write, to go)*. Traditionally, it has been considered an error to split an infinitive with an adverb *(i.e., to further investigate),* but modern practice teaches that some splitting of infinitives is not only acceptable but also often offers a clearer and more natural expression. As a general rule, avoid splitting infinitives with adverbs or adverb clauses. However, split infinitives are sometimes necessary to avoid awkward construction, but they should be used sparingly and with careful thought.

Incorrect	For her *to* never *complain* seems unreal.
Correct	For her never *to complain* seems unreal.

Incorrect	The writer needs *to,* no matter how busy she is, *proofread* the report thoroughly.
Correct	No matter how busy she is, the writer needs *to proofread* the report thoroughly.
Awkward	We need *to investigate* safety measures further that will prevent electric shocks. *(Further* separates *safety measures* from its modifier *that will prevent electric shocks.)*
Awkward	We need further *to investigate* safety measures that will prevent electric shocks. *(Further* now suggests that we also need to investigate safety measures.)
Clear	We need *to* further *investigate* safety measures that will prevent electric shock. (The meaning is now clear that safety measures must be further investigated.)

Mechanics Repair

Mechanics are the spelling, punctuation, abbreviations, capitalization, hyphenation, italics, and number and symbols conventions. There are some differences in acceptable styles among specific organizations and professional associations. For instance, in **standard style,** commas are used as follows:

> designers, technicians, and manufacturers

In other styles, such as newspaper writing, commas are used as follows:

> designers, technicians and manufacturers.

Most technical writers adhere to standard rather than open conventions. To establish consistency within an organization or a journal, professional organizations present their preferred styles in **style manuals.** Appendix B in this textbook reviews standard mechanics that you should review carefully. In addition, many dictionaries contain valuable mechanical style information. Following are some published mechanical style sources:

- *Webster's Standard American Style Manual,* 1995. Springfield, Mass.: Merriam Webster.
- The *fifth edition* of the *MLA Handbook for Writers of Research Papers,* 1999. New York: The Modern Language Association of America.
- The revised edition of *The Associated Press Stylebook and Libel Manual: The Journalist's Bible,* 1996. New York: Addison-Wesley Publishing Company, Inc.

- The fourteenth edition of *The Chicago Manual of Style,* 1993. Chicago: University of Chicago Press.
- *Government Printing Office Style Manual,* 1993. *Supplement on Word Division,* 1994. Washington, D.C.: U.S. Government Printing Office.

The Council of Biology Editors, the American Chemical Society, and the American Psychological Association, to name a few specialist organizations, also publish their own style manuals. When writing for publication, the publisher will usually advise you of the preferred style manual. All of these manuals are updated periodically.

PUNCTUATION

As previously mentioned, Appendix B reviews the standard style in detail. Be particularly alert to the use of an **apostrophe** to indicate the plurals of letters, numbers, designated words, and symbols as in *r's, 7's, and's,* and *#'s;* the use of **brackets** within a quotation to add clarifying words as in, "The author stated, 'They [writers] need to use standard mechanics'"; the use of a **colon** to introduce a phrase or clause that explains or reinforces a preceding clause as in, "The writing process consists of four main steps: prewriting, writing, adding visuals, and postwriting." **Comma usage** demands careful attention. It is best to use **exclamation points** guardedly; make your emphatic message through the use of strong words.

Very few periods are used in technical abbreviations. Appendix B includes a list of **Common Technical Abbreviations.** Technical usage differs from conventions in other types of writing as in *fl oz* instead of *fl. oz.* And *rpm* instead of *r.p.m.* Further, in technical writing the plural abbreviations are written in the same form as the singular as in *5 hr, 10 cc, 100 lb.* Review in the appendix **capitalization, hyphenation, italics,** and **numbers.** Table B.3 lists some common technical symbols, such as || for parallel, = for derived from, and x for snow.

SPELLING

Most word processing software includes a spelling tool that automatically underlines misspelled words. It can scan through your finished document and suggest corrections for replacement. This is a valuable tool, but it will not indicate words that are correctly spelled but incorrect in particular usage. Consider this little poem:

> Eye halve a spelling checquer,
> It came with my pea sea;
> It plainly marques four my revue
> Miss steaks eye kin knot sea.

Your spell checker will not pick up these mistakes. In addition to your software spell check, use a dictionary to check all doubtful spellings and refer to Table B.4 "Frequently Misspelled Words" in Appendix B. This list includes those demons *amateur, benefited, calendar, contemptible, exaggerate, foreign, medieval, occasionally, rhythm, separate, sophomore,* and *weird.*

CHANNEL

It is important to determine the best means of transmitting your message. Print alone may not be sufficient. The use of space as a visual indication of content can enhance your document. Besides space variations, text design elements may include column layout, margins, justifications, indentations, headings, font and type sizes, emphatic feathers, graphics, and other art. Chapters 3 and 4 discuss graphics, document design, and other art. Further, the communication project may suggest oral presentations, computer-enhanced presentations, a slide program, videotapes, and so on. The strategies of verbal and visual communications are covered in Chapters 17 and 18.

NOISE

Noise includes **mechanical** interference, such as illegible print copy due to poor typing, an outdated printer, faded print, or poorly reproduced duplicates. In addition, missing sentences or paragraphs, inappropriate graphics or other art, confusion of fonts and type sizes, and the like can be considered noise. Mechanical noise in verbal messages includes static or background noise in messages transmitted by phone and missing sentences or paragraphs.

Noise also includes **semantic** problems, such as the use of ill-defined terms, incorrect or inappropriate language, poor grammar, questionable usage, incorrect spelling, or the omission of essential background information. Other noise or barriers to communication can arise from divergent backgrounds of the participants (e.g., differences in education, sex, race, nationality, intelligence) or the mental and physical stress of the sender or receiver at the time of the communication. Consider all of the possible barriers that might hinder effective communication.

ENCODING/DECODING

Encoding/decoding calls for you to determine the symbols (letters, numbers, words, graphics) and layout (space, headings, fonts, print size, and other design elements) that are most potent for conveying your message.

In other words, you encode the message. Your receiver (the decoder) probably has piles of paperwork to handle—letters requiring responses, memos giving instructions for action, bulletins and reports imparting information to retain or react to, and more. That audience must be able to decode your intent exactly.

Bear in mind, too, that the last thing a beleaguered receiver wants is to read a "gray" message, a page or pages of uninterrupted type. To avoid the "gray" look, use the following encoding devices:

- Be as brief as possible.
- Divide your message into an introduction, body, and closing.
- Write in short paragraphs (four to five typed lines).
- Consider spacing on the page.
- Use headings (Completed Work, Work in Progress, Work to Be Completed).
- Vary the font and type size for divisions and headings with discretion.
- Use **boldface,** *italics,* or <u>underlining</u> for emphasis.
- Use numbers or bullets (• _____) for lists of items.
- Insert graphics (tables, bar charts, line graphics, circle graphs, organization and flow charts), clip art, icons, and photographs for emphatic understanding.

All of the encoders that pertain to text design and layout are discussed in Chapter 4. Although a message may be as brief as two sentences, it may also be many pages. Careful attention to visuals will make your message more readable by dividing the text into logical units, adding emphasis to essential points, and helping the reader to relocate quickly those passages requiring action.

INTERPRETATION

Language Clarity.   Review all of your work thus far with an eye to interpretation. Eliminate all multiple meanings, possibilities of double interpretations, idioms, or other barriers to understanding. Make clear if directions are being given from left to right or top to bottom. Indicate time zones if necessary. Clarify your authority (years of experience and/or research documentation). Consider if translations of parts or the whole should be offered in other languages. Remember that your clarity depends on a collaboration between the sender and the receiver.

FEEDBACK

Feedback from your audience constitutes an entirely new message. To determine if all of your communication is interpreted correctly, be alert to letters and phone calls that seek additional information. Be sure to include addresses, phone and fax numbers, and e-mail addresses in your message. Encourage feedback by asking for questions, being available for comment, or enclosing a questionnaire concerning your communication for return to you, the sender. The writers of operations manuals for electronic equipment, such as computers and their software, microwave ovens, VCR players/recorders, and the like often include a questionnaire/comment sheet. Figure 2.4 shows a typical evaluation and comments sheet designed to elicit feedback.

Web pages on products and manufacturers also include questions and answers, ongoing discussion, and other means for interaction between user and manufacturer.

EVALUATION AND COMMENTS SHEET

FROM:

NAME: _____   TITLE: _____

BUSINESS

ADDRESS: _____

_____

PHONE: _____ _____

TITLE, PUBLICATION NO., AND DATE _____

1. BASED ON YOUR EXPERIENCE, RATE THIS MANUAL

	Good	Average	Poor
Manual Organization	____	____	____
Manual Readability	____	____	____
Supporting Illustrations / Diagrams	____	____	____
Installation Procedures	____	____	____
Operator's Procedures	____	____	____
Maintenance / Troubleshooting Procedures	____	____	____

2. PLEASE WRITE COMMENTS IN THE SPACE PROVIDED BELOW INDICATING ERRORS YOU HAVE FOUND. INCLUDE PARAGRAPH OR FIGURE NUMBER AND THE PAGE NUMBER.

_____

_____

_____

_____

_____

3. WHAT WOULD YOU SUGGEST WE ADD TO MAKE THE MANUAL MORE INFORMATIVE AND USEFUL?

_____

_____

_____

_____

_____

FIGURE 2.4   *Typical feedback solicitation form*

CHECKLIST

Applying Theory, Prewriting, Draft Writing, and Postwriting Activities

THEORY

☐ **1.** Do I recognize the parts of the Bell Telephone Laboratory model?

    ☐ Sender?

    ☐ Receiver/audience?

    ☐ Message?

    ☐ Channel?

    ☐ Noise?

    ☐ Encoding/decoding?

    ☐ Interpretation?

    ☐ Feedback?

PREWRITING

☐ **1.** As an individual or collaborative writer, have I assessed my (our) capabilities?

    ☐ Knowledge?

    ☐ Credibility?

    ☐ Research adequacy?

    ☐ Authority?

☐ **2.** If I am participating in a writing team, do I understand the strengths of collaboration?

    ☐ Wider knowledge?

    ☐ Wider skills?

    ☐ Divergent opinions?

    ☐ Wider responsibility?

    ☐ Increased productivity of participants?

❑ **3.** If I am participating in a writing team, do I understand the weakness of collaboration?

    ❑ Workload disparities?

    ❑ Interpersonal conflicts?

    ❑ Longer production?

    ❑ Disunified document?

    ❑ Groupthink?

    ❑ Gender and multicultural misunderstandings?

    ❑ Lack of appreciation?

❑ **4.** Have I (we) assessed the audience and their needs?

    ❑ Former knowledge of subject?

    ❑ Vocabulary level?

    ❑ Singular or multiple readers?

    ❑ Subordinate, peer, superior, or laypeople?

    ❑ International/intercultural differences?

❑ **5.** Am I clear on the exact purpose of my message?

    ❑ To instruct?

    ❑ To persuade?

    ❑ To inform?

    ❑ To describe?

    ❑ To question?

❑ **6.** Have I devised a topic list?

❑ **7.** Have I devised a complete and logical outline?

DRAFT WRITING

❑ **1.** Is my purpose clearly evident?

❑ **2.** Have I written with a flow, knowing I will edit the work later?

❑ **3.** Is the inductive or deductive format appropriate?

❑ **4.** Have I maintained a factual, objective, and unemotional tone?

❏ **5.** Is the style effective?

  ❏ No sexist slips?

  ❏ Concrete expression?

  ❏ Lack of jargon, shoptalk, gobbledygook?

  ❏ Precise words?

  ❏ Correct word usage?

  ❏ Lack of wordiness and redundancies?

  ❏ Signal words?

❏ **6.** Is the sentence pattern varied?

POSTWRITING

❏ **1.** Have I considered overall revisions?

  ❏ Have I rechecked everything for audience appropriateness?

  ❏ Are all of the topics and facts covered accurately and thoroughly?

❏ **2.** Are there any errors in sentence construction?

  ❏ Fragments?

  ❏ Faulty parallelism?

  ❏ Run-ons?

  ❏ Comma splices?

  ❏ Sufficient variation?

  ❏ Appropriate length?

  ❏ Avoidance of *there is* and *there are*?

❏ **3.** Are word revisions necessary (sexist terms, jargon and the like, wordiness, redundancy, word usage, voice)?

❏ **4.** Have I added appropriate signal words?

❏ **5.** Is the grammar correct?

  ❏ Subject/verb agreements?

  ❏ Pronoun/antecedent agreements?

❏ Pronoun usage?

❏ Modifier placement?

❏ Split infinitive elimination?

❏ **6.** Are there any mechanical errors?

❏ Punctuation?

❏ Capitalization?

❏ Hyphenation?

❏ Necessary italics?

❏ Number usage?

❏ Spelling?

SENDING THE DOCUMENT

❏ **1.** Have I selected the appropriate channel?

❏ **2.** Have I eliminated mechanical and semantic noise?

❏ **3.** Have I logically encoded the document?

❏ **4.** Have I eliminated all other barriers to interpretation?

❏ **5.** Have I elicited feedback?

EXERCISES

1. **Inductive/Deductive Organization.** Rewrite the following inductive paragraph to deductive organization:

   Bankall reports net income of $310 million, up from $159 million a year ago. Securities gains account for $204 million of the showing. Earnings per share were $1.29, up from 70 cents in the first quarter of last year. Noninterest income, such as bank-card income and deposit fees, rose 12 percent. Results of the merger indicate successful operation figures.

2. **Tone.** Rewrite the following paragraph in an impersonal, objective tone:

   Last Tuesday, Ivan Brown, the arrogant, young director of personnel, brazenly fired three clerks and a computer programmer. The inexperienced computer programmer had established with the clerks an insidious plan to pilfer inkjet computer paper, Emory address labels, pens, scissors, and other supplies from the dimly lit supply room. Brown was selling these supplies to unwary students in front of the Student Union every Monday for a huge profit.

3. **Sexist Language.** Rewrite the following paragraph to eliminate the sexist language:

> Neither the draftsman nor the designer could complete his project until he had manpower to assist him. John Ackerman became the self-appointed spokesman on how to reapportion the man hour duties. Everybody involved gave his opinion on how his work could be made easier.

4. **Concreteness.**   Rewrite the following paragraph to eliminate the vague, abstract words. Make up your own details:

> The Director of Technical Writing vetoed the plan to update the software in the department. He felt that the proposal asked for too many things because the department is adequately supplied with enough software. He stated that the high expense would curtail certain other expensive plans in the works.

5. **Jargon/Shoptalk/Gobbledygook.**   Rewrite the following paragraph to eliminate jargon, shoptalk, and gobbledygook:

> He accessed the Emergency Room and was asked for input on his accident. "In a nutshell," he said, "I was cruising down the Interstate when this dude tailgated me so closely that I decelerated. He deployed his pick-up into the aft of my conveyance resulting in my broken schnoz. My car was totaled."

6. **Wordiness/Redundancies.**   Rewrite these sentences to eliminate wordiness, roundabout phrases, and redundancies:

    a. Pursuant to our talk and discussion, we are inclined to make the suggestion that 150 auto assemblers be furloughed until such time as the recession shows a strong indication of recovery and upward climb.

    b. Due to the fact that the seat belt broke, the passenger sustained a high degree of injury.

    c. Either one or the other of the word processing programs is absolutely essential for the secretaries and typists to work quickly and efficiently.

WRITING PROJECTS

1. **Audience Analysis.** Locate an upper-level technical writing sample in one of your most technical textbooks. Rewrite this sample for a junior high school student. Consider that student's knowledge and vocabulary. What technical terms must be simplified? How must the message be reworded? Do you need more or fewer graphics,

numbers, bullets, headings, font and type size changes, and so forth? Attach a photocopy of the original to your paper.

2. **Foreign Product Document Analysis.** Find an English-language manual, brochure, or advertisement for an item manufactured by a foreign company. Critique it for ambiguity, passive voice, conversions of one part of speech into another, and other language problems. Attach a photocopy of the document to your paper.

3. **International Correspondence Differences.** Write to a foreign embassy to ask about cultural business letter practices: the level of formality needed, the appropriate tone, organization, word choice, level of details, and so forth. Write a few paragraphs about your findings or attach a copy of a business letter written by a foreigner and evaluate it for differences from a typical business letter written by a North American. (See Chapter 12 "Composing Correspondence.")

4. **Collaborative Project.** Form a group of four. Each of you are to watch the same Sunday morning news show. Meet with your group to analyze the effectiveness of the show. What was the central message? Was it clear and factual? Assess the style and tone. Did gender or cultural differences affect the participants or your team members' understanding? Collaboratively write your findings in a four- to five-paragraph paper.

5. **Collaborative Multicultural Project.** Form a team of four. Interview two or three foreign students enrolled at your school. These may be classmates, dorm mates, or students in your school's English as a second language program. Preplan your interview questions to determine cultural differences. Determine variances from Standard English. Do some English words have variant meanings to members of other nations or cultures? How does a foreign language word order vary from English word order? Do spellings vary? How does the interviewee handle numbers? Color values? What major language or interpersonal problems do foreign students face? Can you determine any rhetorical strategy differences? How do ethics vary in family, personal, and business relationships in contrast to those of the United States? Present your findings in a collaborative document.

NOTES

CHAPTER *3*

Utilizing Graphics and Other Visuals

DILBERT   by Scott Adams

DILBERT reprinted by permission of United Feature Syndicate, Inc.

S K I L L S

After studying this chapter, you should be able to

1. Recognize the purpose and function of tables, bar charts, line graphs, pie charts, flow charts, organization charts, drawings, maps, photographs, text art, clip art, and icons.
2. Name and practice the conventions of incorporating tables and figures.
3. Prepare random and continuation informal tables.
4. Prepare formal tables.
5. Prepare a variety of bar charts.
6. Prepare several types of line graphs.
7. Prepare several types of pie charts.
8. Prepare a flow chart.
9. Recognize the positive and negative aspects of three-dimensional charts.
10. Prepare an organization chart.
11. Prepare a simple, exploded, or cutaway drawing.
12. Incorporate a map or photograph into a document.
13. Recognize text art and construct it by hand or by computer.
14. Recognize clip art and incorporate a piece into a report if you use a computer.
15. Recognize a variety of icons and determine where they may be used in a report.
16. Critique the construction and incorporation of visuals in technical reports.

INTRODUCTION

Graphs (*tables, bar charts, line graphs, pie charts, flow charts,* and *organization charts*) plus other visuals (*drawings, maps, photographs, text art, clip art,* and *icons*) are characteristic of professional and technical documents. Large companies may employ graphic artists to assist writers in graphic illustrations, while computers enable thousands of writers to create and incorporate their own graphs and other visuals into their documents to enhance the text.

PURPOSES

Illustrations are never merely random nor decorative. They must serve to

- Speed up a reader's comprehension
- Emphasize a point
- Assist as a method of quick reference
- Reveal differences at a glance
- Allow for easy comparisons
- Provide more detail than is actually discussed for full examination
- Serve as a multicultural/international language
- Aid the reader's retention
- Break the monotony of "gray" page data
- Add to the attractiveness of the report

Each type of graph and visual serves a distinct purpose. For instance, a table displays data in vertical columns that would otherwise involve lengthy prose sentences, that, in turn, might be difficult to comprehend or to interpret. A bar chart illustrates comparisons of parts, while a pie charts shows not only comparisons of parts but also the relationship of each part to the whole. Drawings, maps, and photographs can show details that words cannot adequately describe. Clip art may serve to enhance a title or chapter page or to emphasize a concept. Icons serve to indicate warnings and directions, particularly in instructions. You must consider not only when to use a graph or other visual but also which type will best serve your purpose. Clearly, then, visuals are not merely decorative but essential considerations of the technical writer's responsibility.

FREEHAND COMPUTER GRAPHICS AND VISUALS

Even if you do not have access to an expert nor have mastered your own computer graphic capabilities, it is your task to decide, at least, on the appropriate graphic and visuals to convey your data, and a rough idea of the layout (see Chapter 18 for some visual design principles). In fact, many simple graphics and visuals can be handled by the novice without assistance. Graph paper, a ruler, and a compass are the tools you should have on hand; however, most illustrations can be done right on your computer. A variety of software programs allow you to create tables, charts, pictures, and maps, and to add clip art, icons, and other visuals to your documents. With a scanner, you can copy visuals from magazines, brochures, newsletters, manuals, and other documents. The scanner imports the image to your computer, where you can crop, resize, or otherwise alter it, then place it into your document.

The Internet has thousands of Web sites with clip art; photographs; icons; maps; science and math symbols and pictures; electrical and mechanical engineering symbols, drawings, and schematics; and more. Many of these images are free and may be used without copyright infringement. Once you have downloaded or clicked and dragged an image to your screen, you may modify and customize it with software such as Paint, PaintShop, PhotoShop, CorelDRAW, Adobe Illustrator, or Freehand LV Pro. With these tools, you can change the color of an image; add text, objects, and shadows; add hatch marks, stripes, and other background designs, and use overlays, multiangle views, three-dimensional transforms, and countless other effects to enhance the image. You can also crop, skew, resize, and perform numerous other modifications to fit your needs.You may then store the image on a disk or on your hard drive.

Other graphics software offers tools for

- Creating visual effects of a mathematical model
- Writing equations to explain the sequences of a tornado or earthquake and the like
- Constructing models and testing for outcomes using a number of variables
- Integrating computer-assisted design (CAD) with computer-assisted manufacturing (CAM) to design architectural blueprints and to direct machinery operations
- Hypothesizing about concepts without doing the calculations
- Creating animations for presentations

This chapter addresses the uses of a wide variety of graphics and the overall and specific conventions for their construction and incorporation into your text.

GENERAL CONVENTIONS

Conventions of design, placement, titling, numbering, and referencing govern all graphics and visuals and include the following:

1. **Design.**   Graphics should be planned ahead, well thought out, and "print ready." Do *not* try to put too much information into any one graphic. Place all explanatory notes, keys, and legends within the graphic or beneath it in the left-hand position but above the number and title. There are many graphic software programs that allow you to construct graphics and visuals with a few clicks. You may also construct graphic figures on separate paper and pho-

tocopy or scan them into a space that you have allowed in your text. For hand-constructed figures, consider making your working graphics 1½ times as large as you intend for the final graphic or visual; photo reduce the work by 60 to 70 percent to obtain the proper proportion for your text. Use rulers, bow compasses, protractors, templates of geometric figures, and possibly transfer or stencil letters. Use Times Roman, Helvetica, or Univers type styles because they are easiest to read.

2. **Incorporation.**   Print all of your tables and figures (all other graphics and visuals) in the final draft of your document. Prepare your graphics with the top on the vertical plane of your paper, reduce it with a copy machine or scanner, and incorporate it so that no labels, legends, or other parts extend beyond your margins. If your computer and printer are not capable of incorporating graphics, or if you do them by hand, tape or glue the properly sized graphic to the page and then photocopy or scan each page for inclusion in your document.

3. **Placement.**   Graphics and visuals that are included only as a supplement to a report or those that do not appear near to their textual reference tend to lose their impact, so all graphics and visuals should immediately follow their initial reference. That is, once attention is drawn to them, they should be placed directly within the text at the point of reference or no later than at the end of the section in which they are mentioned. In a document printed with facing pages (manuals and books, for instance), they may appear on a facing page. They may also be referenced again later in the text without the art work being repeated. It is unconventional to place graphics sidewards (landscaped) on a page, but exceptions are sometimes neccessary.

4. **Titles.**   Usually include a precise noun phrase with a numbered designation for each graphic. A graphic may be taken out of a report for photocopying and distribution and will obviously require a title line. Some examples follow:

     TABLE 1.   Cost Comparison of Transportation Modes

     TABLE 2.   Smoke Detector Ratings

     Figure 1.   Cross section of a typical speed bump before and after modification.

     Figure 2.   Proposed Transit System Routes

5. **Numbering.**   In addition to the title, number your formal graphics. Always number and title a formal table *above* the data. The word *table* is capitalized. If only one table occurs in the entire document, it will not require a number nor usually even a title. Refer to all other graphics as figures and number and title them *beneath* the graphic. Clip art and icons are neither necessarily numbered nor titled. When you use a number and title for a table, *center the data* or

place it flush left to the margin of the table. Be consistent within any one document.

TABLE 4

MEAN SALT CONCENTRATION

IN VARIOUS SOURCES OF WATER

or

TABLE 4 Mean Salt Concentration in Various Sources of Water

6. **Continuations.**   In the case of tables that require more than one full page, begin the second page with the table number and the word *Continued.*

7. **Number Sequencing.**   If more than one table or more than one figure is used in a document, number each in order of the appearance throughout the material, but number tables separately from graphics. Use Arabic numbers:

   Table 1

   Table 2

   Figure 1

   Table 3

   Figure 2

   If the report contains numbered chapters, use a decimal numbering system to indicate both the chapter and sequential number of the graphic:

   Figure 7.1

   Table 7.1

   Figure 7.2

   Table 7.2

   Figure 7.3

8. **Periods and Capital Letters.**   Notice that the numbers and titles in Conventions 4, 5, and 7 indicate a variety of acceptable uses of periods and capital letters. It is essential to be consistent within a document, however. If you decide to use a period after the graphic or table number, do so for all graphic designations throughout your report. You may capitalize an entire title, capitalize initial letters of each work, or capitalize only the initial letter of the first word.

9. **Spacing of Lines.**   Single space titles that require more than one line. Align second and consecutive lines under the first word of the title, not under the word *table* or *figure:*

   FIGURE 5 Operation Manual for a Motorola Dimension 1000 Binary
             GSC pager (Courtesy of Motorola, Inc., Fort Lauderdale,
             Florida)

10. **Referencing.**   Use an introductory sentence or reference notation by graphic number to explain the purpose of each graphic before you include it. If the graphic immediately follows its sentence reference, use a colon at the end of the sentence as illustrated; if the graphic may be inserted after more written material or on another page, use a period.

    Figure 2 shows a line drawing made by tracing a photocopy:

    The available equipment, features of each, prices, and warranties are shown in Table 3:

    As you examine Figure 4, notice the high unemployment rate for even those with some college:

    Sometimes you may reference a graphic in parenthetical notation:

    Set the control dial (Fig. 1) to one of the six speeds.

    A document may be printed in two columns (see Figure 8).

    Figure 10.3 (see page 510) shows a title page for a proposal.

    Try to place your graphics as close to their references as possible.

11. **Commentary Line.**   Usually, follow the graphic with a sentence or two of comment or interpretation:

    It is evident that the introduction of the scanner product put Austin Technology in the lead for 2000 sales.

    The dotted line between the Comptroller and the Vice-president for Finance indicates that the Comptroller may report directly to the Vice-president for Finance without reporting first to the Treasurer.

12. **Acknowledgments.**   Identify the source of borrowed graphics in parentheses after the title:

    Figure 1   Sample formal proposal (Courtesy of Charles E. Smith Jr., Robert Heller Associates)

    Table 7   Average Yearly Salaries by Sex and Race [Source: Catherine Brown, *Discrimination in the Work Place,* New York: Silver Press, 2001 (8)]

    Figure 2   Fire Ground Injuries by Cause [From "Fire Ground Injuries in the United States during 2001" (12). Reprinted by permission.]

TABLES

Tables are visual displays of numerical or nonnumerical data arranged in vertical columns so that the data may be emphasized, compared, or contrasted. Tables may be **informal** (random and continuation) or **formal.** Certain specific conventions govern each.

Informal Random Tables

Informal random tables display brief lists of figures, dates, personnel, important points, and the like in vertical columns for visual clarity and quick reference. Figure 3.1 illustrates two informal random tables. The first shows an informal table of dates, places, and purposes, and the second shows a numbered list.

Informal Random Table Conventions

1. Use random tables only for brief data.
2. Introduce each by an explanatory sentence.
3. Indent the data 5 to 10 spaces from the left- and righthand margins of the page.
4. Consider column headings, numbered data, or bullets.
5. Do not include a table designation number or title.

Most software packages include a table feature that enables you to create columns and rows; set height, width, and margins; and format column heads and table body text. You can even convert text into a table and vice versa.

Informal Continuation Tables

A continuation table, another informal table, contains prose data in a displayed manner. It reads as a continuation of the text and includes the same punctuation marks that would be required if the data were presented in paragraph form. Figure 3.2 shows a sample informal continuation table.

Informal Continuation Table Conventions

1. Use a continuation table to present an alignment of figures, dates, or other data.
2. Introduce each by a sentence followed by a colon if the last introductory word is *not* a verb.
3. Indent the tabular data 5 to 10 spaces from the left and right margins.
4. Punctuate the data by standard commas, semicolons, and periods as if the material were presented in paragraph form. Note the word *and*.

> The Training Center announces the beginning of a mini-course, "Write It Right—Write It Well," for senior executive secretaries and administrative assistants. Dates, locations, and purposes follow:
>
> | Date | Room | Purpose |
> |------|------|---------|
> | May 11 | 102 | to review grammar/usage |
> | May 13 | 102 | to review brief report forms |
> | May 18 | 101 | to review graphics |
> | May 20 | 102 | to review manual components |
> | May 25 | 103 | to review formal reports |
> | May 27 | 101 | to critique individual writing |
>
> To register, fill out the attached form and forward it to Julie Wood, Training Specialist, Room 608.

> Regardless of what kind of accident is being reported, certain information must be reported objectively and specifically:
>
> 1. What the accident is
> 2. When and where the accident occurred
> 3. Who was involved
> 4. What caused the accident
> 5. What were the results of the accident (damage, injury, and costs)
> 6. What has been done to correct the trouble or to treat the insured
> 7. What recommendations or suggestions are given to prevent a recurrence
>
> Information required for the accident report has become so standardized that many companies have designed accident report forms.

FIGURE 3.1   *Two sample informal random tables*

Our insurance policy Allstate #17B-445-9100K will cover the cost of the fire damage. Repairs and replacements total $844.00. This price includes

$144.00	for carpet replacement (9 sq ft @ $16.00 per ft; Carl's Carpets),
75.00	for labor for removing burned carpet and replacing (5 hr @ $15.00 per hr),
115.00	for cleaning solution for wall (115 sq ft @ $1.00 per sq ft),
180.00	for labor for cleaning and repainting wall (10 hr @$18.00 per hr),
105.00	for labor for reupholstering armchair (7 hr @ $15.00 per hr),
160.00	for fabric for brown leather armchair (8 yd @ $20.00 per yd),
20.00	for new magazine stand from Pier One, and
45.00	for fire extinguisher replacement.

FIGURE 3.2   *Sample informal continuation table with sentence punctuation*

Formal Tables

Formal tables are used to present statistical information or to categorize and tabulate other written information. The conventions for formal tables are listed below and demonstrated in Tables 3.1 and 3.2.

Formal Table Conventions

1. Place titles flush left or centered *above* the table.
2. Include horizontal lines from margin to margin.

TABLE 3.1   *Time/Cost for Aerial Photograph Searches*

Time Frame	Searches #	Time @ 30 min ea (hr)	Cost @ $100.00 per hr
Daily	3	1.5	150.00
Weekly	15	7.5	750.00
Monthly	60	30.0	3,000.00
Quarterly	180	90.0	9,000.00
Yearly	720	360.0	36,000.00

TABLE 3.2   *Troubleshooting Chart for Heath Kits*

Difficulty	*Possible Cause*
Receiver section dead	Check V1, V3, V4, V7, and V8
	Wiring error
	Faulty speaker
	Faulty receiver crystal
	Crystal oscillator coil mistuned
Receiver section weak	Check V1, V2, and V3
	Antenna, RF, or IF coils mistuned
	Faulty antenna or connecting cable
Transmitter appears dead	Check V5 and V6
	Wiring error
	Recheck oscillator, driver, and final tank coil tuning
	Dummy load shorted on open

3. *Do not close* the sides of formal tables.

4. Always use vertical columns: the first body column is called the **stub.**

5. Use a **box head** of vertical column headings and symbols in parentheses, i.e., ($), (rpm), (hr), (ft). Notice the lack of periods in the standard abbreviations; refer to Table B.2 in Appendix B to study more usage of abbreviations in technical writing that differ from standard abbreviations.

6. Following modern practice, do not use **leaders** (spaced periods to aid the eye in following data) from column to column.

FIGURES

As previously mentioned, all graphics other than tables are referred to as figures both in textual reference and in titling. If you use a computer, construct your graphic figures on your graphics software, import them into your document, and place and size them with your cursor.

Bar Charts

Use bar charts to show differences in quantity and quality visually and instantaneously. The bars show quantities of the same item at different times, quantities and qualities of different items for the same time period, or quantities of the different parts of an item that make up the whole. They are compiled from statistical data. Conventions in addition to those that govern all graphics govern their construction.

You may plot your bars vertically or horizontally. Normally, use vertical graphs for bars that represent monetary units. If you cannot do a bar chart on a computer, use graph paper to plot your chart. The scale you select is critical to the success of your chart. Do not use grids that will not accommodate some portion of your bar. Scale all grids to equal increments, such as 0, 1, 2 or 0, 5, 10, 15, but not 0, 5, 7, 12.

Types of bar charts include **horizontal, vertical, multiple, stacked, deviation, creative**, and **histogram.** Overlapping bars or 3-D charts may also serve the more technical and sophisticated audience, but avoid overuse of 3-D graphics unless they actually contribute to the import of the graphic. The following conventions guide the proper construction of bar charts, and Figures 3.3 through 3.10 show eight variations of one-dimensional bar charts.

Bar Chart Conventions

1. If possible, *box* in all the bars, headings, legends, and so on.
2. Use bars of *equal width and design* within a chart.
3. If the order of the bar placement is not controlled by a sequential factor, *place the longer bars to the bottom of a horizontal chart* or *to the right in a vertical chart*; this placement avoids a top-heavy or one-sided appearance.
4. Use *vertical grid lines* or *tick marks* for horizontal bar charts and *horizontal grids* or *tick marks* for vertical bar charts; never use both in a single chart.
5. Use *partial cutoff lines* to separate headings from grid or tick notations.

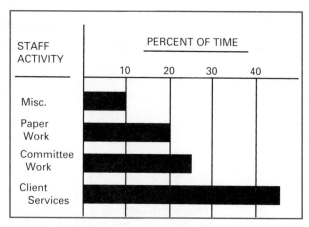

FIGURE 3.3   *Typical horizontal bar chart—
hand constructed*

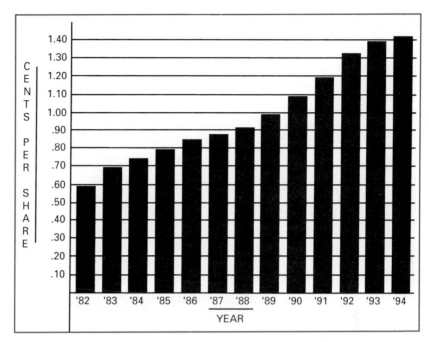

FIGURE 3.4   *Typical vertical bar chart—hand constructed*

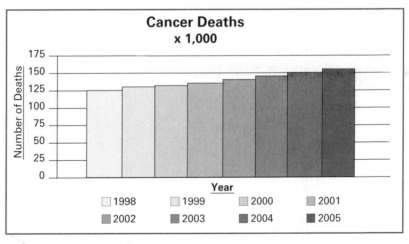

FIGURE 3.5   *Another typical vertical bar chart with legend at bottom—computer generated*

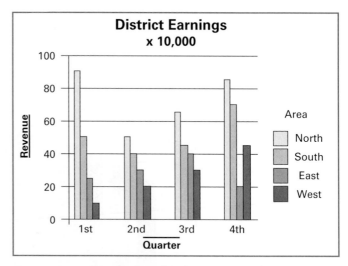

FIGURE 3.6   *Typical multiple bar chart with logical vertical bars to indicate increasing amounts of money—computer generated*

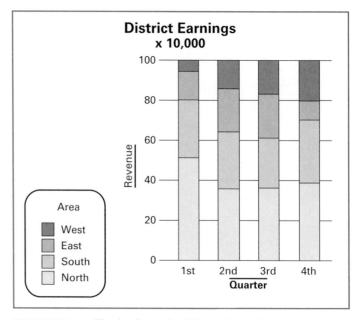

FIGURE 3.7   *Typical stacked bar chart depicting comparisons by area at a glance— computer generated*

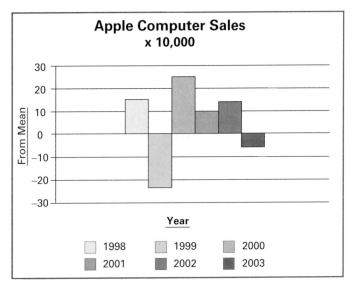

FIGURE 3.8 *Typical deviation bar chart depicting yearly sales deviation from normal—computer generated*

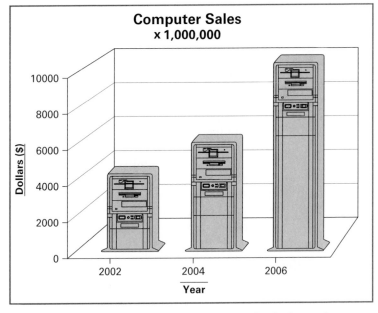

FIGURE 3.9 *Typical creative bar chart depicting sales increases in dollars x 1,000,000—computer generated*

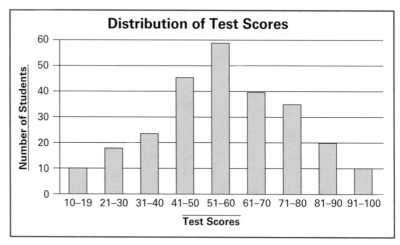

FIGURE 3.10   *Typical histogram indicating distribution of test performance scores of 259 students—computer generated*

6. Include a *heading* to indicate what the grids or tick marks show: hours, months, years, number of sales, amounts, activities, and so forth.

7. *Center grid notations* on the grid lines, not just above or just below.

8. *Center bar notations* on the bars.

9. When displaying multiple bars with various colors or textures, use *legend boxes* to distinguish the differences.

3-D Bar Charts

It is possible to construct overlapping and 3-D bar charts, which add visual emphasis and are engaging to view; however, they can confuse the layperson, making interpretation and emphasis difficult. Consider your audience's technical ability to "read" such graphics intelligently. Figures 3.11 through 3.13 show three sophisticated overlapping and 3-D bar charts (an overlapping, a vertical 3-D, and a stacked 3-D) to demonstrate the visual impact of 3-D graphics, but you are cautioned to use such examples sparingly.

Line Graphs

Line graphs, or curves, are used to show changes in two values. Most commonly, they show a change or trend over a given period or performance against a variable factor. They are standard *Cartesian* graphs named after the French mathematician and philosopher who invented

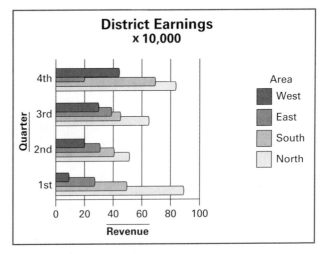

FIGURE 3.11   *Typical overlapping bar chart depicting quarterly earnings by area*

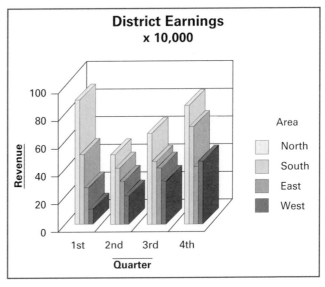

FIGURE 3.12   *Typical 3-D vertical bar chart that accents earnings blocks*

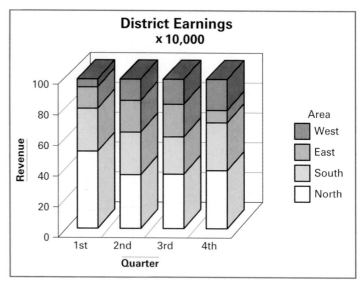

FIGURE 3.13   *Typical 3-D stacked bar chart accentuates quarter stacks by quarter*

them, Rene Descartes. Points are plotted algebraically on the vertical (y) axis and the horizontal (x) axis. The conventions for line graphs and Figures 3.14 and 3.15, illustrating a typical line graph and a typical multiple line graph, follow:

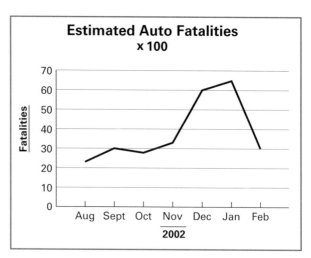

FIGURE 3.14   *Typical line graph depicting chronological rise and fall of auto fatalities over seven months*

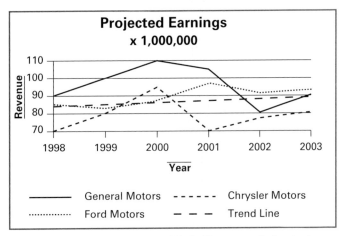

FIGURE 3.15   *Typical multiple line graph depicting rise and fall of four companies' revenues over six years*

Line Graph Conventions

Line graphs adhere to the same conventions as bar charts with some additions.

1. Always plot your curves *from left to right.*
2. Indicate the grids with tick marks; do *not* include grid lines because they become confused with the curves themselves.
3. *Capitalize* major headings; *capitalize* only the initial letters of subheadings and tick mark notations.
4. Use tick marks (not grid lines) on each line to indicate how many points have been used to plot the graph.

3-D Line Graphs

Once again, line graphs may be three-dimensional for visual appeal and interest, but should be used sparingly because they may confuse some viewers (i.e., in *band graphs* the area beneath each plotted line is filled in with variable shadings or patterns, but it is the *top* of each line which indicates the total). Figures 3.16 and 3.17 show, respectively, a 3-D multiple line graph and a 3-D multiple band graph.

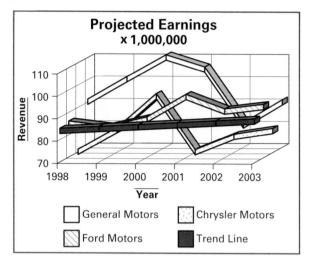

FIGURE 3.16   *Typical 3-D multiple line graph, which enhances the lines for easier comprehension*

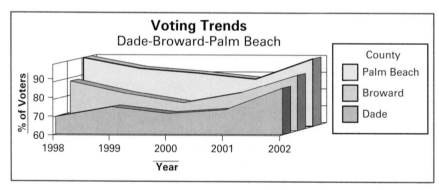

FIGURE 3.17   *Typical 3-D multiple band graph, which emphasizes the block of voters per county*

Pie Charts

Pie charts, also known as circle charts, circle graphs, pie diagrams, or sector charts, are used to compare the relative proportions of various factors to each other and to the whole. The circle represents 100 percent while the segments indicate the proportionate percentage of each factor. They are useful for illustrating financial information, survey results, and other results indicating parts as related to the whole.

If you use a computer, you can simply enter the data, and the software will construct the pie chart. If you need to figure out and construct the chart, consider the following: Since a circle contains 360 degrees, you must first convert your real data to percentages of the whole and then calculate the number of degrees needed to represent each segment or wedge. Use the following format to calculate the number of degrees for each segment:

Item	Raw Data	Frac- tion	4-place decimal	3-place decimal	Percent	Percent rounded	Degrees ($\% \times 360$)	Degrees rounded
Rent	$400	400	.1739	.174	17.4%	17%	61.2°	61°
		2300						
TOTAL: $2,300					**TOTAL**: 100%			**TOTAL**: 360°

There is a quick formula for these calculations: Multiply the 4-place decimal by 360 degrees. However this short method is not totally accurate, and you will probably have to make arbitrary adjustments to make sure your final calculations total 360 degrees.

If you are constructing your pie chart by hand, you will need a compass or circle template and protractor to draw the circle and to divide it into segments. As a rule of thumb, use a 3″-diameter circle on standard 8½″ × 11″ paper. This will make your circle large enough for emphasis yet small enough to allow space for labeling the wedges outside of the pie. Think of your circle as a clock, and plot your largest wedge in the upper right-hand quadrant from the 12 o'clock position. Then place the wedges from the next largest proportionately to the smallest clockwise.

A pie chart is effective without wedge differentiations, but color, shadings, and hatches can add effectiveness. Pie charts may indicate cutaway portions to emphasize that wedge in a context, or they may be three-dimensional to suggest bulk. Creatively, in specific contexts, it may be appropriate to use a round fruit, pizza, clock, or other naturally round object instead of a plain circle. The conventions for pie charts follow, and Figures 3.18 through 3.20 show a typical pie chart, a chart with a separated wedge, and a 3-D pie chart with a separated wedge.

Pie Chart Conventions

1. Normally, do not present a pie larger than 3 inches in diameter on an 8½″ × 11″ page.
2. Place the largest segment in the upper right-hand quadrant, with the segments decreasing in size clockwise.

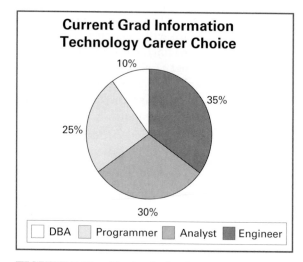

FIGURE 3.18   *Typical pie chart*

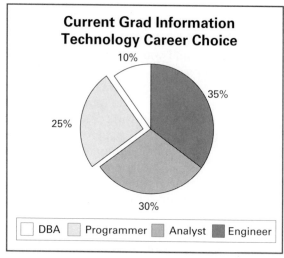

FIGURE 3.19   *Typical separated pie chart, which emphasizes program-mer group*

3. Write headings along with the percentages *outside* each wedge to avoid crowding.

4. *Center* each label on the radius of each wedge or use a *tag line* to aid the eye.

5. Type labels on a horizontal plane.

6. Contain all labels within the left- and right-hand margins.

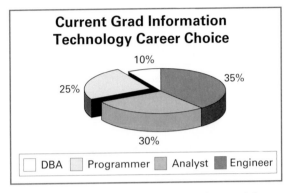

FIGURE 3.20   *Typical 3-D pie chart with separated wedge emphasizing quantity of graduate groups*

Flow Charts

A flow chart is used to show pictorially how a series of activities, procedures, operations, events, or other factors are related to each other. A flow chart shows the sequence, cycle, or flow of the factors and how they are connected in a series of steps from beginning to end. Such a chart condenses long and detailed procedures into a visual chart for easy comprehension and reference. The information is qualitative rather than quantitative as in bar, line, and pie charts.

The components of a flow chart may be diagramed in horizontal, vertical, or circular directions. Computer software uses templates that contain a variety of shapes symbolizing various activities, such as

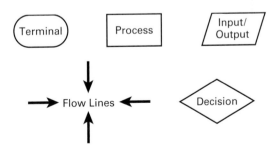

Usually, boxed steps are arranged in sequence and connected by arrows to show the flow. Flow charts may be simple, complex, or pictorial. Following are flow chart conventions and Figures 3.21 through 3.23, which illustrate simple, complex, and pictorial flow charts.

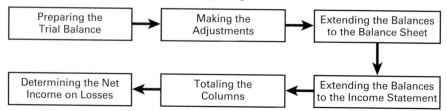

FIGURE 3.21 *Simple flow chart*

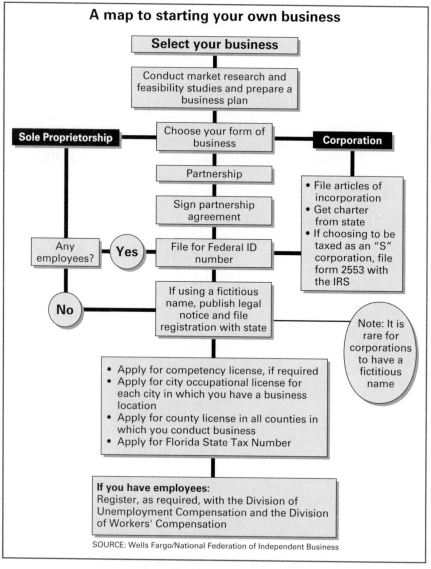

FIGURE 3.22 *Typical complex flow chart*

Computer future

The Broward County School Board's five-year plan for technology envisions a computer system that will allow students to talk to other students, parents to communicate with administrators and teachers, and educators to trade ideas with peers worldwide.

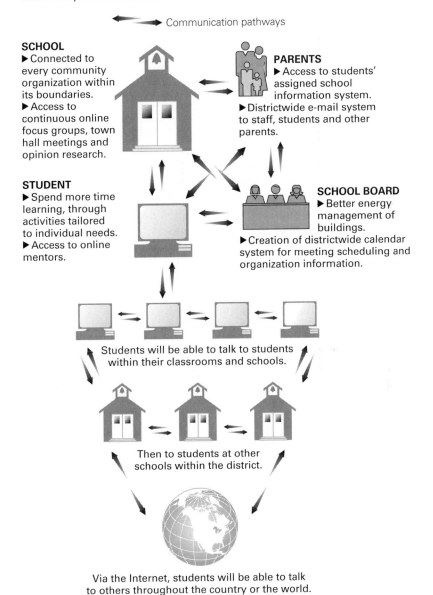

Communication pathways

SCHOOL
▶ Connected to every community organization within its boundaries.
▶ Access to continuous online focus groups, town hall meetings and opinion research.

PARENTS
▶ Access to students' assigned school information system.
▶ Districtwide e-mail system to staff, students and other parents.

STUDENT
▶ Spend more time learning, through activities tailored to individual needs.
▶ Access to online mentors.

SCHOOL BOARD
▶ Better energy management of buildings.
▶ Creation of districtwide calendar system for meeting scheduling and organization information.

Students will be able to talk to students within their classrooms and schools.

Then to students at other schools within the district.

Via the Internet, students will be able to talk to others throughout the country or the world.

FIGURE 3.23 *Complex pictorial flow chart* (Reprinted with permission from the *Sun-Sentinel*, Fort Lauderdale, Florida)

Flow Chart Conventions

1. Employ squares, boxes, triangles, circles, diamonds, and other shapes to enclose each step.
2. Lay out your flow chart in a horizontal, vertical, circular, or combination of directions.
3. Name major activities within the shapes.
4. Use lines or arrows of various dimension to connect the shapes and to indicate the flow.

Organization Charts

Like flow charts, organization charts show qualitative, rather than quantitative, material. An organization chart is used to show the relationship of the organization's staff positions, units, or functions to each other.

A **staff organization chart** shows the chain of command of the staff positions, such as president, vice-presidents, directors, comptroller, and salespersons. A **unit organization chart** depicts the relationships among such units as public relations department, research division, financial office, and personnel section. A **function chart** shows the span of control of such functions as planning, engineering, technical writing, marketing, and production. These three aspects—staff, unit, functions—should not be mixed together in the same chart.

Either a horizontal or a vertical emphasis can be imparted to an organization chart by the layout as shown:

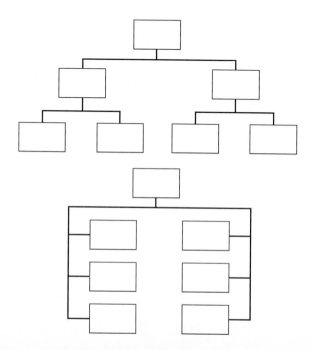

Figure 3.24 shows a staff organization chart for a manufacturing company. Notice within the chart that the Accounting Director is subordinate to both the Treasurer and the Comptroller. The solid line indicates that the normal chain of command is from the Comptroller to the Accounting Director; the dotted line indicates that the Treasurer may direct inquiries or report important information directly to the Accounting Director. The lines also show the normal and possible chains for the Accounting Director to direct questions or to issue information upward.

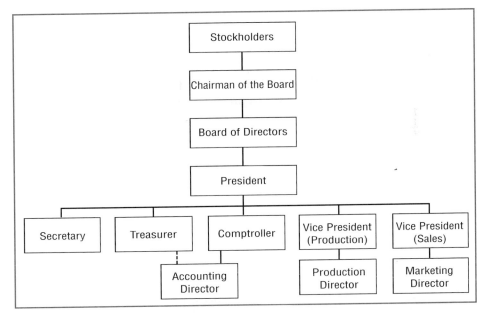

FIGURE 3.24 *Typical organization chart*

OTHER VISUALS

Besides tables and graphics, technical writing is characterized by drawings and illustrations, maps, photographs, text art, clip art, and icons. Each of these is used to emphasize and clarify points within your text. Computer software provides drawing capabilities, a variety of map types, text art, clip art, and icons. In addition, a scanner can copy photographs or other materials directly into your documents. Computer programs allow you to draw any shape imaginable and add shading, hatches, color, and labels to your work. Computer Assisted Drafting (CAD) requires special training; CAD is used primarily by drafters, engineers, and architects, but you may well find a use for such software in your writing.

Drawings/Illustrations

A variety of simple drawings may be executed by the novice. A simple line drawing is often clearer than a photograph because it can give emphasis to important features. Diagrams of procedures can clarify instructions. Exploded-view illustrations show the proper sequence in which parts fit together, and cutaway drawings show the internal parts of a mechanism or a piece of equipment. Electricians and electronic technicians use schematics and wiring diagrams to illustrate concepts.

Complicated or specialized drawings should be attempted by hand only if you are skilled in commercial art or drafting, but anyone can draw the types of art shown in this section. Poorly constructed drawings confuse the reader, exactly the opposite of well-produced drawings.

Figures 3.25 through 3.29 show typical procedural, labeled line, exploded view, cutaway, and schematic drawings that can clarify descriptions, definitions, and process analyses.

Drawing Conventions

1. If you do not use a computer drawing program, use grid paper and a ruler for careful drawings.
2. Keep your drawings uncluttered, properly ruled, and carefully labeled.
3. *Type,* do not hand letter, all labels and symbols.

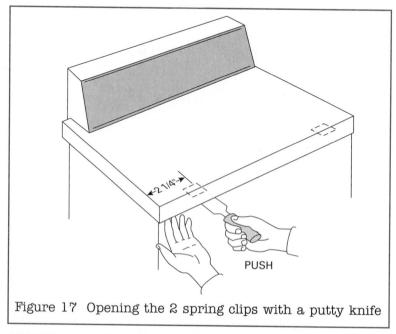

Figure 17 Opening the 2 spring clips with a putty knife

FIGURE 3.25 *Typical procedural drawing*

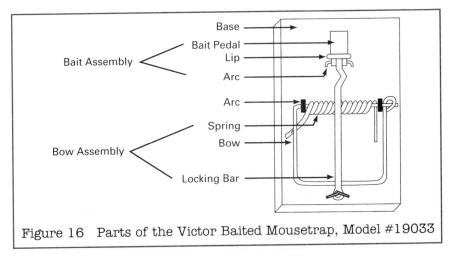

Base
Bait Pedal
Bait Assembly
Lip
Arc

Arc
Spring
Bow
Bow Assembly

Locking Bar

Figure 16   Parts of the Victor Baited Mousetrap, Model #19033

FIGURE 3.26   *Typical line drawing with labels*

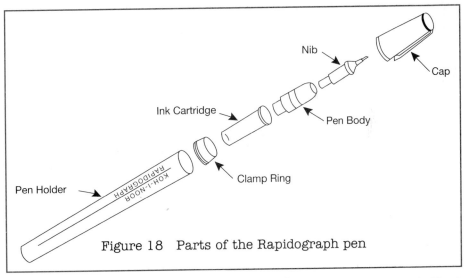

Nib
Cap

Ink Cartridge
Pen Body

Pen Holder
KOH-I-NOOR RAPIDOGRAPH
Clamp Ring

Figure 18   Parts of the Rapidograph pen

FIGURE 3.27   *Typical exploded-view illustration* (Courtesy of student Jennifer Woper)

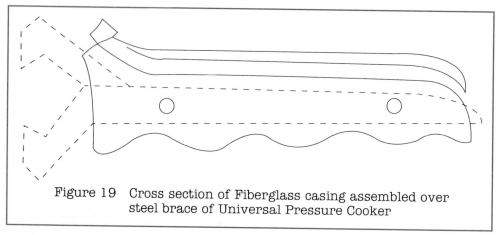

Figure 19   Cross section of Fiberglass casing assembled over
steel brace of Universal Pressure Cooker

FIGURE 3.28   *Typical cutaway drawing*

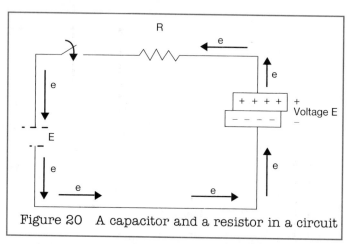

Figure 20   A capacitor and a resistor in a circuit

FIGURE 3.29   *Typical schematic drawing*

Maps

Maps can show sites, routes, and comparisons by geographical location. Maps may be **large scale** (highlighting an area) or **small scale** (showing a large area with position notations or comparisons). Computers offer all kinds of maps, but you may hand draw, photocopy, or scan others into your documents. As with other visuals you may add color, shadings, hatches, or patterns to distinguish regions from each other. Figures 3.30 through 3.32 show two small-scale maps and one large-scale map.

CENSUS REGIONS AND DIVISIONS OF THE UNITED STATES

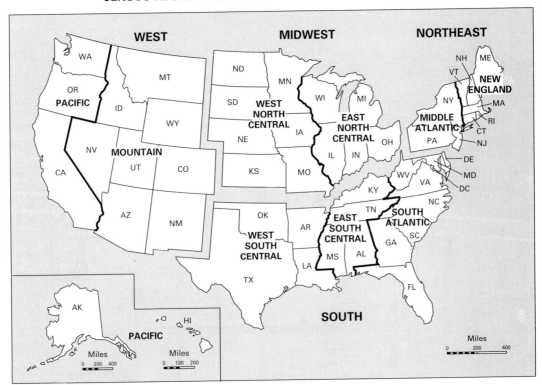

FIGURE 3.30 *Typical small-scale map* (Downloaded from maps—United States Department of Commerce—Economics and Statistics Administration. Bureau of Census)

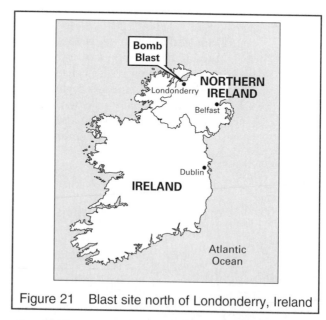

Figure 21   Blast site north of Londonderry, Ireland

FIGURE 3.31 *Typical small-scale map* (Fort Lauderdale *News/Sun Sentinel* map by Keith Robinson)

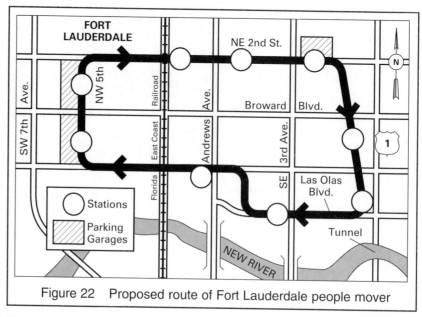

Figure 22   Proposed route of Fort Lauderdale people mover

FIGURE 3.32 *Typical large-scale map*   (Fort Lauderdale *News/Sun Sentinel* map by Tom Alson)

Photographs/Line Art

Photographs allow you to provide overall views, and they may be cropped to focus on detail. You must be careful that the photo is not cluttered with unnecessary and distracting elements. You must choose well-focused photos for reproduction. Large companies with sophisticated technical writers and computer capabilities use scanners to produce from actual photographs digitized line art that can be incorporated directly into a document. Line art is as clear as a photograph. Figures 3.33 and 3.34 show a photograph and a digital line art production from a photograph.

It may help to provide a scale to your photo by including a person, a hand, or even a ruler. Consider ethics. For example, an insurance photograph can selectively make a $300 dent appear to be a $1,000 dent with selective light and shadow. Furthermore, digital imaging technol-

FIGURE 3.33 *Photograph of a 900 Melz digital cordless telephone* (Courtesy of Lucent Technologies, Parsippany, New Jersey)

FIGURE 3.34   *Line art produced on a computer from an actual photograph*
(Courtesy of Motorola, Boynton Beach, Florida)

ogy allows storage of photographs on a disk, which can then be unethically altered. The photographs you use must be clear and accurate and without unethical alterations.

Text Art

Text art may be used for titles or within manuals for special emphasis. Computers allow you to write text, contour it to a variety of shapes, use various fonts and type sizes, insert lines in a variety of formats, and add color. In technical writing, you must guard against the purely decorative, but if such art adds meaning to your message, it may be effective. Figure 3.35 shows a few words produced by the text art feature of a word processing program. Notice the different fonts, type sizes, and contours:

NOTICE>

DANGER!

Caution

FIGURE 3.35   *Typical text art in various fonts, type sizes, and contours*

Clip Art

Clip art is also known as quick art on some computers. There are thousands of clip art possibilities. This writer's computer contains a library of more than 40,000 items. Such a library allows you to add professional images into your documents. These images include arrows, professional supplies, business representations, equipment including computer items and science objects, faces and people, transportation and space-age pictures, symbols, borders and banners, buildings, maps, plants, animals, birds, and a multitude of others. You can even edit these pictures by adding, for example, text or facial features. Again, these images should be used sparingly, but they can enhance your documents in many ways. Images can be sized, dragged to any place on your document page, and enhanced with more detail. Figure 3.36 shows a brief sampling of simple clip art images.

FIGURE 3.36 *Typical clip art images* (Corel WordPerfect 7)

Icons

An icon is a picture or image with a conventional meaning, such as a hand with a thumb up for approval, two clasped hands for agreement, or a round sign with a diagonal line through it meaning "not allowed."

These may be carefully drawn or abstracted from your computer clip art or quick art programs, as well as from fonts such as Wingdings, Milesto, Holiday Caps, and IC. These are not to be used decoratively but to convey essential meaning in documents such as operation manuals. Colors, textures, shadows, and other features may be added to icons. Figure 3.37 shows a number of easily recognizable icons.

FIGURE 3.37   *Typical recognizable icons*

FINAL WORDS ON EFFECTIVENESS

The level of detail, clarity, texture and color, size, and orientation on the page must all be considered when producing effective graphics and other visuals. Be careful not to try to include too much detail in any one piece of work. Figure 3.38 shows a graphic that is both overdesigned and lacking in vital information.

Use an inkjet or laser printer for **high resolution.** A dot matrix printer is generally not effective because of its low resolution and broken lettering. Do not try to reproduce computer graphics or art on regular copier paper because the details may be fuzzy, muddy edged, and dull. Use a premium paper, available from your office supply store.

The **size** of any graphic should not overwhelm or underwhelm the text. It takes some practice to determine a compatible size; give size your most careful attention. If the graphic or visual is too small or too large in relationship to the type size and other elements of the document design, it will lose effectiveness.

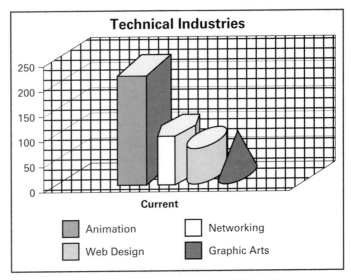

FIGURE 3.38 *Overembellished graphic with inadequate titles and headings, too many grid lines, and unnecessary multiple shapes*

Further, the **orientation** on the page is important. A graphic (with the possible exception of a table) should always be contained on one page. Changing the size slightly may enable you to do this. If the size or configuration will not fit well into its place of reference, consider placing it on a facing page. Decide whether your visual should be centered, flush left, or in some other position on the page. Do not exceed the margins of your document; leave some white space around the graphic or visual.

Textures and colors must also be sharp and logical. Textures and colors that "bleed" from one to another are not acceptable. Use both sparingly; remember, less is more. Use texture and color, perhaps, for headings or to draw attention to important or additional information, such as a checklist. Study a color wheel to determine how colors are related and to discern which colors interact. Those that lie directly opposite each other on the wheel are dynamic and exert a push/pull effect. Monochromatic color schemes (one color with variable tints and shades) tend to unify a document. Colors that lie closely together on the color wheel (yellow, orange, rust) create harmony. Use bright colors for accent and pale colors as background. Unnecessarily brash colors are disconcerting and distracting. Stick to the primary colors for most graphics, and do not overdo the number of textures in any one graphic or visual.

Textures may include background shadings (gray tone or color); patterns of horizontal, vertical, or diagonal straight lines placed close together or farther apart; a variety of wavy or curvy lines; large or small dots; checkered, brick, or beehive patterns; or other designs.

CHECKLIST

Graphics and Visuals

OVERALL CONSIDERATIONS

☐ 1. Have I decided on a graphic that speeds the reader's comprehension, adds credibility to the document, reveals content and differences at a glance, and adds to the attractiveness of the document without being merely decorative?

☐ 2. Have I chosen the most appropriate graphic to convey my data?

☐ 3. Is the size proportionate to the text?

☐ 4. Are all of the numbers correct?

☐ 5. Are all elements (line thickness, notes, labels, legends, titles) positioned within the margins?

☐ 6. Is my graphic incorporated into the correct place in the document after the initial textual reference?

☐ 7. Is there a sentence of reference for the graphic?

☐ 8. Is it properly numbered and titled?

☐ 9. Is the sequencing of graphic numbers correct?

☐ 10. Am I consistent in my use of periods and capital letters in my numbering and titles?

☐ 11. Is there a commentary line following the graphic?

☐ 12. Are my sources acknowledged correctly?

DESIGN

☐ 1. Are the elements of my graphic correct (table columns, bars, lines, pie wedges, flow chart arrows, organization boxes and arrows, drawings, photographs, maps, etc.)?

❏ 2. Is the graphic large enough to convey the data but not overwhelm the text?

❏ 3. Is the material indented and spaced well within the document?

❏ 4. Are table sides open but other graphics boxed?

❏ 5. In a bar chart, are the bars of equal width and design?

❏ 6. In a line graph, are the curves plotted from left to right?

❏ 7. In a pie chart, do the wedge percentages total 100 percent?

❏ 8. In a pie chart, are the wedges placed clockwise from the largest in the upper right-hand quadrant to the smallest?

❏ 9. In a pie chart, are the labels outside of the wedges and placed in the center of each radius?

❏ 10. In a flow chart, are the shapes appropriate, and are the directions of flow clear?

❏ 11. In an organization chart, are the boxes evenly spaced and connected correctly?

❏ 12. Are my drawings and maps accurate, sharp, and labeled in size appropriate to the document?

❏ 13. Are my photographs and line art appropriately cropped and rid of extraneous material?

❏ 14. Is my text art correct and integrated in design to the purpose of the document?

❏ 15. Is my clip art well-placed and helpful in conveying the message?

❏ 16. Do my icons instantaneously convey the correct meaning?

EXERCISES

1. **Graphic Choices.**   Which graphic would specifically illustrate the following (a random, continuation, or formal table; a simple, multiple, stacked, or creative bar chart; a simple, exploded, or cutaway drawing, etc.)?

    a. The changing cost of your telephone bill over a six-month period.

    b. The number of students enrolled in each department of a community college division (i.e., communications division: expository writing, technical writing, literature, speech).

    c. An indicator to continue to the next page.

    d. A comparison of three types of sneakers by cost, design, material, and durability.

    e. The installation directions for a VCR.

    f. A specific model of Sony stereo receiver.

    g. Tools and materials needed for installation of a Westinghouse clothes dryer.

    h. Assembly of a halogen lightbulb into a lamp.

    i. A list of times, room numbers, and places for class registration.

    j. The percentages of time spent in eight job-related duties.

    k. A storm-damaged roof.

    l. A breakdown of presidential vote choices by county in your state.

    m. The chain of command where you work.

    n. The steps in the process of preparing a technical instruction manual.

    o. A progress report that indicates work completed, work in progress, and work to be done.

2. **Table.**   Determine the percentages and devise a table to present the following data obtained from a corporate study on sex discrimination. Include the job titles, numbers, and percentages of the total.

    Of 215 employees, 103 are women

    Of 36 executives, 12 are women

    Of 9 managers, 3 are women

    Of 7 assistant managers, 5 are women

    Of 18 administrative assistants, 3 are women

    Of 6 technical writers, 0 are women

3. **Bar Chart.**   Determine how many hours you spend on a typical Monday on each of the following activities. Present this information in a standard bar chart. Number and title your graphic.

Travel	Grooming
Study	Class attendance
Leisure	Work
Meals	Sleep

4. **Pie Chart.**   Using an imaginary monthly income of $2,180, devise a budget for the following expenses and present your data in a pie chart. Number and title your graphic.

Rent/mortgage	Insurance
Utilities	Leisure
Auto Expenses	Credit payments
Food/meals	Clothing
Education	Miscellaneous

5. **Line Graph.**   Devise a line graph showing your grade point average on a term-to-term basis since your freshman year, your salary increases (or decreases) at work since your employment, the number of hours worked per day in a week or two-week period, the high and low temperature ranges for one week (a multiple line graph), the distances run or daily times spent in other exercise in a month, or other subject of your choice.

6. **Flow Chart.**   Devise a flow chart depicting the line of legislative command from the president of the United States to the members of the House of Representatives. Do not confuse this with an organizational chart. Number and title your graphic.

7. **Drawing.**   Select a small mechanism, hand tool, or kitchen implement and construct a simple or exploded drawing of it. Label all parts, including nuts, bolts, rivets, tabs, and handles. Number and title your graphic.

8. **Map.**   Obtain a bus route map and write a brief paragraph into which to insert the map. Introduce, number, title, and comment on the graphic.

9. **Text Art.**   Design a title page for a longer report you may be submitting based on the information in Exercise 3 or 4. Design some text art by hand or on your computer.

10. **Clip Art.**   If you have clip art on a computer, print some images that could be used in a technical document. Write a few sentences to explain your choice and intention.

11. **Icons.**   Show three icons that may be commonly used in technical writing. Draw these by hand or locate these icons on a computer. Explain what each icon indicates.

WRITING PROJECTS

1. **Graphic Evaluation/Positive.**   Select a graphic that you find effective in a magazine, newspaper, textbook, manual, newsletter, or brochure. Write a few paragraphs about its effectiveness. How does it adhere to the general conventions of graphics and visuals? Is it the best type of graphic for the situation? What, if anything, could be improved?

2. **Graphic Evaluation/Negative.**   Select a graphic that you find ineffective in any of the sources listed in the previous project. Write a few paragraphs about its weaknesses. How does it deviate from the general conventions of graphics and visuals? Why doesn't it work well? What needs improvement? Would another type of graphic work better? Explain.

3. **Collaborative Project—Sales Promotion Materials.**   Form a group of four students. Imagine that you are a team of marketing specialists at National Motor Corporation. Select a lead person, decide on your responsibilities, and set deadlines. Write a memorandum report (see Chapter 12) to the vice president of marketing, concerning data on the new-model Econo automobile. Your purpose is to suggest facts that should be stressed in the new-model Econo sales promotion materials. Use at minimum an informal table, a formal table, a pie chart, a bar chart, and a line chart in the text of your memorandum. If other visuals (a drawing, photograph, text art, clip art, or icons) are appropriate, you may add them, but not as decoration. Introduce your graphics, number and title them correctly, and comment on each.

> The Econo was the most popular car compared to four other competitive mid-size cars on the market last year. Of total sales, 30 percent of buyers chose Econo, 23 percent chose Car A, 20 percent chose car B, 17 percent chose car C, and 10 percent chose car D.
>
> The Econo has dramatically increased its fuel efficiency in five years. Four years ago the Econo was rated at 15 miles per gallon (mpg); 3 years ago the Econo averaged 18 mpg; 2 years ago the fuel efficiency increased to 19 mpg; a year ago it increased to 20.5 mpg; and this year it has a 23 mpg rating.
>
> Despite base price increases, the new Econo is only $10.49 more per month to own and to operate as it was five years ago due to increased fuel economy. Five years ago the base sticker price was $20,000. A 15 percent down payment of $3,000 resulted in a $17,000 balance to finance. The monthly cost of financing for 60 months was $310.25 at 9.5 percent. The fuel expense per month for 1,250 miles of driving (15,000 miles annually) at $1.30 per gallon was $108.33 at 15 mpg. The cost per month over 60 months (finance charge plus fuel cost) was $418.58. The new Econo has a base sticker price of $22,000. A 15 percent down payment of $3,300 leaves $18,700 to finance. The monthly cost of financing over 60 months is $358.42 at 9.5 percent. But fuel expense is reduced to $70.65 at 23 mpg. Therefore, the new monthly cost of owning a new Econo over 60 months (finance charge plus fuel cost) is $429.07, only $10.49 more per month than five years ago.

NOTES

CHAPTER 4

Designing and Producing Documents and Presentations

DILBERT   by SCOTT ADAMS

DILBERT reprinted by permission of United Feature Syndicate, Inc.

S K I L L S

After studying this chapter, you should be able to

1.   Name at least ten elements of document design.
2.   Recognize and employ a number of layout designs with spatial variations.
3.   Recognize and employ tabs and columns to enhance a document.
4.   Use variable line lengths, leading, and kerning where necessary.
5.   Understand the value and use of headings in various type sizes and fonts.
6.   Recognize and use a number of fonts, such as Arial, Helvetica, Times New Roman, and Goudy.
7.   Know when to use *serif* and *sans serif* fonts.
8.   Recognize and employ variant margins and indentation options.
9.   Distinguish between and recognize the values of a ragged-edge and justified page.
10. Recognize and employ numbered or bulleted lists.
11. Recognize the value of and employ boldface, italics, underlining, all capitals, small capitals, and drop caps.
12. Know how to use a window for emphasis.
13. Employ symbols, icons, and reversed type for emphasis and eye appeal.
14. Recognize the value of color and employ it when possible and appropriate.
15. Know when to use borders, fills, and watermarks.
16. Recognize and employ headers and footers.
17. Be able to devise and number correctly front presentation matter: a title page, a table of contents, a list of illustrations, and separate chapter title pages.
18. Be able to devise back presentation matter: a glossary, an appendix, and an index.
19. Recognize a summary, abstract, and bibliography or works cited listing.
20. Recognize factors that determine the size, quality, and orientation of paper.
21. Evaluate the use of folds and/or bindings, flaps, pockets, perforations, and windows in documents.

INTRODUCTION

Document design, final presentations, and production options are increasingly more the task of a writer or collaborative team than the job of a professional printer or publisher. Awareness of the capabilities of computer enhancement of design is leading even those writers without computer literacy to pay more attention to the layout of the document, its special presentation features, and its production options.

Ideally, you have considered and taken advantage of design elements as you wrote your rough draft, edited and revised, and added illustrative graphics and visuals. Besides decisions about the page layout, spacing, headings, emphatic features, and other design features, you will need to make decisions on whether to include a title page, table of contents, a glossary, an index, or other presentation features. Finally, you need to make production decisions regarding the size and quality of the paper, the page orientation, bindings, flaps, pockets, and so on.

DOCUMENT DESIGN OPTIONS

Word processors, desktop publishing software, and other page design programs are focusing more attention on text design. Heretofore, just typing a double-spaced document in a Times New Roman 12-point font, using one-inch margins all around, and indenting new paragraphs five spaces sufficed. Your document probably looked like the document on p. 116.

The layout does not invite nor aid the reader; such layout is useful for scholarly journals and academic papers, which are read by choice, not necessity. For letters, reports, proposals, and manuals that must be read by busy people in ongoing business transactions, writers must design documents that attract and aid the readers.

Most word processors have templates or style sheets for memos, business letters, proposals, resumes, and reports, including term papers. Companies can devise other templates for newsletters, flyers, expense reports, order forms, legal pleadings, calendars, brochures, booklets, and the like. By devising a variety of templates complete with logos, icons, and other identifying elements, a company can assure an overall similarity of appearance to provide a corporate identity "look."

Following are some simple possibilities of document design and layout that illustrate the points to be discussed in this section (see pages 116–120). These are just a few of the many document design possibilities that may serve as models for you.

Look carefully at the page designs of this textbook, including the separate parts division pages, the chapter title pages, the skills listings and checklists, page headers, text placement, division headings and their subordination, sizes of type, different fonts, bold face and italic type,

Uninteresting "Gray" Document Designs

Text, text. Text, text, text, text, text, text, text, text, text, text, text, text, text, text, text, text, text, text. Text, text. Text, text, text, text, text, text, text, text.

Text, text, text, text, text, text, text, text, text, text, text, text, text, text, text, text, text, text. Text, text. Text, text, text, text, text, text, text, text, text, text, text, text, text, text, text, text, text, text. Text, text, text, text, text, text, text, text.

Text, text, text, text, text, text, text, text, text, text, text, text, text, text, text, text. Text, text, text, text, text, text, text, text, text, text, text, text, text, text, text, text, text. Text, text.

Text, text, text, text, text, text, text, text, text, text, text, text, text, text, text, text, text. Text, text. Text, text, text, text, text, text, text, text, text, text, text. Text, text.

Text, text, text, text, text, text, text, text, text, text, text, text, text, text, text, text, text, text, text. Text, text, text, text, text, text, text, text, text.

Dense blocks of text without headings, separating white spaces, and other design elements present a dull, too dense page appearance. Except for paragraph indentations, white space is not used to define sections.

Document Page Designs

HEADING

Text, text. Text, text, text, text, text, text, text, text, text, text, text, text, text, text, text, text, text, text, text, text.

Text, text, text, text, text, text, text, text, text, text, text, text, text, text. Text, text, text, text, text, text, text, text, text, text, text, text, text. Text, text, text. Text, text, text, text, text, text.

HEADING

Text, text, text, text, text, text, text, text, text, text. Text, text, text, text, text, text, text, text, text, text, text, text, text, text, text, text, text, text, text, text. Text, text, text, text, text, text, text, text, text, text, text, text, text. Text, text, text, text, text, text, text, text, text, text, text, text, text, text, text, text, text, text.

Text, text, text, text, text, text, text, text. Text, text, text, text, text, text, text, text, text, text, text, text, text, text, text, text, text, text, text. Text, text, text, text, text, text, text, text, text, text, text, text, text, text, text, text, text.

Text, text, text, text, text, text, text, text, text, text, text, text, text, text, text, text, text, text, text. Text, text. Text, text, text. Text, text, text, text, text, text, text.

Text, text, text, text, text, text, text, text, text, text, text,text, text. Text, text, text, text, text, text, text, text, text, text, text, text, text, text, text, text. Text, text, text, text, text, text, text, text, text, text, text, text, text. Text, text, text, text, text, text, text, text, text, text, text, text, text.

HEADING

Text, text, text, text, text, text, text, text, text, text, text, text, text, text, text, text, text. Text, text, text, text, text, text, text, text, text, text, text, text, text, text, text, text, text.

Text, text, text, text, text, text, text, text, text, text, text, text. Text, text, text, text, text, text, text, text, text, text, text, text, text, text.

Text, text, text, text, text, text, text, text, text, text, text, text, text. Text, text.

HEADING

Text, text, text, text, text, text, text, text, text, text, text, text. Text, text, text, text, text.

Text, text, text, text, text. Text, text, text, text, text, text, text, text.

In this example a horizontal line separates the title from the text. The text is presented in two columns. Capitalized, boldface headings separate parts of the text. The paragraphs begin flush left to the margin, and a space is skipped between paragraphs. Right-hand margins are justified.

Document Page Layouts

Text, text, text, text, text, text, text, text, text, text, text, text. Text, text, text, text, text, text, text, text. Text, text, text, text, text. Text, text, text, text, text, text, text, text, text, text, text, text, text, text, text, text. Text, text, text, text, text, text, text, text, text. Text, text, text, text, text, text, text, text, text. Text, text, text, text, text, text, text, text. Text, text, text, text.

Text, text, text, text, text, text, text, text, text, text, text. Text, text, text, text, text, text, text. Text, text, text, text, text, text, text, text, text, text, text. Text, text, text, text. Text, text, text, text, text, text, text, text, text. Text, text, text, text, text.

Text, text, text, text, text, text, text, text, text, text, text. Text, text, text, text, text, text. Text, text, text, text, text, text, text, text. Text, text, text, text, text, text, text, text. Text, text, text, text, text, text, text, text, text, text, text. Text, text, text, text, text, text, text, text, text, text, text. Text, text, text, text, text, text, text, text, text, text. Text, text, text, text, text.

Text, text, text, text, text, text, text, text, text, text. Text, text, text, text, text, text, text, text. Text, text, text, text, text, text, text, text.

Text, text. Text, text, text,

text, text, text, text, text, text, text, text, text, text, text.

Text, text, text, text, text, text, text, text, text, text, text, text. Text, text, text, text, text, text, text, text, text, text, text. Text, text, text, text, text, text, text, text, text. Text, text.

Text, text, text, text, text, text, text, text, text, text, text, text, text. Text, text, text, text, text. Text, text, text, text, text, text, text, text, text, text, text, text.

Text, text, text, text, text, text, text, text, text, text, text, text, text, text. Text, text.

Text, text, text, text, text, text, text, text, text, text, text, text, text, text, text.

Text, text. Text, text, text, text, text, text, text, text, text, text, text, text. Text, text, text, text, text, text, text, text, text. Text, text.

Text, text, text, text, text, text. Text,

In this example the title is boxed and textured (shaded). The text is presented in two columns with a line separating the columns. Each paragraph is indented five spaces, and a space is skipped between each paragraph. The text is justified. One paragraph is textured to give it emphasis.

Document Page Design

Text, text, text, text, text, text. Text, text. Text, text, text, text, text, text, text, text, text, text, text, text, text, text, text, text, text, text, text, text. Text, text, text, text, text, text. Text, text, text. Text, text, text, text.

Text, text, text, text, text, text, text, text, text, text, text, text, text, text, text. Text, text, text, text, text, text, text, text, text, text, text, text, text, text, text. Text, text, text, text, text, text, text, text, text, text, text, text, text, text, text, text, text, text, text.

Text, text, text. Text, text, text, text, text, text, text, text, text.

Text, text.

Text, text, text, text, text, text, text, text, text, text, text, text. Text, text, text, text, text, text, text, text, text.

Text, text, text, text, text, text, text, text, text, text, text, text. Text, text. text, text, text.

Text, text, text, text, text, text, text, text,

text. Text, text, text, text, text, text, text, text, text, text, text, text, text, text, text. text, text, text, text.

Text, text, text, text, text, text, text, text, text, text. Text, text, text, text, text, text, text. Text, text, text, text, text, text, text, text. Text, text, text, text, text, text, text, text, text. text, text.

Text, text, text, text, text, text, text, text, text, text, text. Text, text, text, text, text, text, text, text, text, text, text, text, text, text, text.

Text, text, text, text, text, text, text, text, text, text, text, text, text, text, text. Text, text, text, text, text, text, text, text, text, text. Text, text, text, text, text.

Text, text, text, text, text, text, text, text, text, text, text, text, text, text, text, text, text, text, text. Text, text, text, text, text, text, text.

In this example the text is presented in two columns, and the title is bordered and placed in the first column. A drop capital T begins the first paragraph. The text is fully justified left and right. A visual is inserted into the text of the second column.

Sample Document Page Designs, Presentations, and Productions

Text, text, text, text, text, text, text, text, text, text, text, text, text, text, text, text. Text, text, text, text, text, text, text, text, text, text. Text. Text. Text, text, text, text, text, text, text, text, text, text, text, text, text, text, text.

Text, text, text, text, text, text, text, text, text, text, text, text, text, text, text, text, text, text. Text, text, text, text, text, text, text, text, text, text, text, text, text, text, text, text, text, text.

Text, text, text, text, text, text, text, text, text. Text, text, text, text, text, text, text, text, text, text, text, text, text, text, text, text, text, text. Text, text, text, text, text.

Text, text, text, text, text, text. Text, text, text, text, text, text, text, text, text, text, text, text, text, text, text, text, text, text.

Text, text, text, text, text, text, text, text, text, text, text. Text, text, text, text, text, text, text, text, text, text. Text, text, text, text, text, text, text, text, text, text, text, text, text, text, text.

Text, text, text, text, text, text, text, text, text, text. Text, text, text, text, text, text, text, text, text, text, text, text, text, text, text. Text, text, text, text, text, text, text, text, text, text, text, text, text, text, text, text, text, text, text. Text, text, text, text, text, text, text, text, text, text, text, text, text, text, text, text, text, text. Text, text, text, text, text, text, text, text, text, text, text, text, text, text, text, text, text, text. Text, text.

Text, text, text, text, text, text, text, text, text, text, text, text, text, text, text, text, text, text.

In this example the title is presented in two columns. It is separated from the text. The text is presented in a nearly three-quarter-page column with left and right justifications. Visuals are placed in the remaining column, which is approximately one-third of the page. The text is flush left without indented paragraphs.

Document Page Layout

COMPUTER VIRUSES

Text, text, text, text, text, text, text, text, text, text, text, text, text, text, text, text, text, text. Text, text, text, text, text, text, text, text, text, text, text, text, text, text, text, text, text, text. Text, text, text, text, text ,text, text, text, text, text, text.

    Quotation text, quotation text, quotation text. Quotation text, quotation text, quotation text. Quotation text, quotation text, quotation text. Quotation text, quotation text, quotation text. Quotation text, quotation text, quotation text. Quotation text, quotation text, quotation text.

VIRUS REMOVAL

Text, text, text, text, text, text, text, text. Text, text, text, text, text, text, text, text, text, text, text, text, text, text, text, text, text, text. Text, text, text, text, text, text, text, text, text, text, text, text, text, text, text. Text, text, text, text, text, text, text, text, text, text, text ,text, text, text, text, text, text, text, text, text, text, text, text, text.

    Text, text, text, text, text, text. Text, text, text, text, text, text, text, text, text, text, text, text. Text, text, text, text, text, text, text, text, text, text. Text, text, text, text, text, text, text.

    Quotation text, quotation text, quotation text. Quotation text, quotation text, quotation text. Quotation text, quotation text, quotation text. Quotation text, quotation text, quotation text. Quotation text, quotation text, quotation text.

PREVENTING INFECTION

Text, text, text, text, text, text, text, text, text, text, text, text, text, text. Text, text, text. Text, text, text, text, text, text, text, text, text, text, text. Text, text, text, text. Text, text, text, text, text, text, text, text, text, text ,text, text, text. Text, text, text, text, text, text, text, text, text, text, text, text, text, text, text, text, text, text, text.

In this example the title is landscaped (presented sideways) and underlined. Boldfaced, capital headings separate portions of the text. A line is skipped under each heading. Paragraphs are flush left against the margin but ragged in the right. Display quotes are indicated by the indented text. Because of their indentation, they do not require quotation marks.

Space Defines Divisions of Documents at a Glance

HEADING

Text, text, text, text, text, text, text, text, text, text, text, text, text, text, text. Text, text, text, text, text, text, text, text, text. Text, text. Text, text, text, text, text, text, text, text, text, text, text, text, text, text.

    Text, text, text, text, text, text, text, text, text, text, text, text, text, text, text, text, text, text, text. Text, text, text, text, text, text, text, text, text, text, text, text, text, text, text, text, text, text.

Subheading

Text, text, text, text, text, text, text, text, text, text. Text, text, text, text, text, text, text, text, text, text, text, text, text, text, text, text, text, text. Text, text, text, text, text, text, text, text, text.

    Text, text, text, text, text, text, text. Text, text, text, text, text, text, text, text, text, text, text, text, text, text, text.

Subheading

Text, text, text, text, text, text, text, text, text, text. Text, text, text, text, text, text, text, text, text, text. Text, text, text, text, text. Text, text, text, text. Text, text, text, text, text, text, text, text, text.

HEADING

Text, text, text, text, text, text, text, text, text, text. Text, text, text, text, text, text, text, text, text, text, text. Text, text, text, text, text, text, text, text, text, text, text, text, text. Text, text, text, text, text, text, text, text, text, text, text, text, text, text, text, text, text, text, text.

In this example, space clearly defines the divisions of the document. The title is boxed. Capitalized headings indicate major divisions of the text, while uppercase and lowercase headings indicate subdivisions. Headings align on the left-hand margin. First lines of text under headings are not indented, but second and successive paragraphs within a document division are indented.

Document Design with Emphatic Features

HEADING

Text, text, text, text, text, text, text, text, text, text, text, text, text, text, text. Text, text, text, text, text, text, text, text, text. Text, text. Text, text. Text, text, text, text, text, text, text, text. text, text, text, text, text, text, text, text, text, text.

  1.   Text, text, text, text, text, text, text, text, text, text, text, text, text, text, text, text, text, text, text.

  2.   Text, text, text, text, text, text.

  3.   Text, text, text, text, text, text, text, text, text.

Subheading

Text, text, text, text, text, text, text, text, text, text. Text, text, text, text, text, text, text, text, text, text, text, text, text, text, text, text.

  •   Text, text, text, text, text, text.

  •   Text, text, text, text, text.

  •   Text, text, text, text, text, text, text, text.

Subheading

Text, text, text, text, text, text, text, text, text, text, text. Text, text, text, text, text, text, text, text, text. Text, text, text, text, text.

CAUTION! Text text text text

Text, text, text, text, text, text, text, text, text. Text, text, text, text, text, text, text, text, text, text, text, text, text, text, text, text, text, text, text. Text, text, text, text, text, text, text, text, text, text, text, text, text, text, text, text. text, text, text, teet, text.

In this example, the title is separated from the text by a partial cut-off line. A capitalized heading indicates a major heading for a division of the text; the lowercase headings indicate subdivisions of the major division. Paragraphs are not indented under the headings, but numbered and bulleted items are indented. Notice that second lines of indented text return only to the first word of the entry, not to the number or bullet. A box is used to set off a caution. Boldface type is used for headings, numbers, bullets, and the caution note.

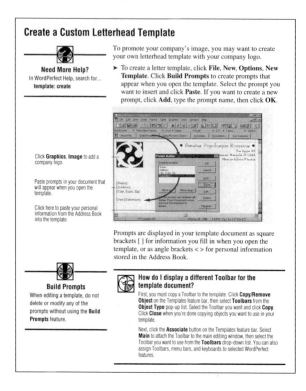

Create a Custom Letterhead Template

Need More Help?
In WordPerfect Help, search for...
template: create

To promote your company's image, you may want to create your own letterhead template with your company logo.

➤ To create a letter template, click **File**, **New**, **Options**, **New Template**. Click **Build Prompts** to create prompts that appear when you open the template. Select the prompt you want to insert and click **Paste**. If you want to create a new prompt, click **Add**, type the prompt name, then click **OK**.

Click **Graphics**, **Image** to add a company logo.

Paste prompts in your document that will appear when you open the template.

Click here to paste your personal information from the Address Book into the template.

Prompts are displayed in your template document as square brackets [ ] for information you fill in when you open the template, or as angle brackets < > for personal information stored in the Address Book.

Build Prompts
When editing a template, do not delete or modify any of the prompts without using the **Build Prompts** feature.

How do I display a different Toolbar for the template document?
First, you must copy a Toolbar to the template. Click **Copy/Remove Object** on the Templates feature bar, then select **Toolbars** from the **Object Type** pop-up list. Select the Toolbar you want and click **Copy**. Click **Close** when you're done copying objects you want to use in your template.

Next, click the **Associate** button on the Templates feature bar. Select **Main** to attach the Toolbar to the main editing window, then select the Toolbar you want to use from the **Toolbars** drop-down list. You can also assign Toolbars, menu bars, and keyboards to selected WordPerfect features.

In this sophisticated example from a word processing user manual, white space is used effectively to "chunk" sections of the text: the "tip" notation, the actual directions, the sample pages, and specific directions at the close. In addition, a page title, icons, arrows, partial cut-off lines, boldface type, and varying type sizes aid the reader. Line length and ragged right-hand margins enable easy reading.

numbered and bulleted items, color, and other design features. Your word processor can achieve most of the following document design options:

- Space
- Tabs and columns
- Line length, leading, and kerning
- Headings, type sizes, and fonts
- Margins, indentations, and justifications
- Lists, numbers, and bullets
- Emphatic features, icons, reversed text, and color
- Borders, fills, watermarks
- Headers and footers

Space

Look at your text as a series of chunks of information and organize the content into positive (ink-filled) and negative (white space) space to invite the reader's eye and to indicate parallel parts of a document (headings, paragraphs, numbered or bulleted items, boxed warnings, and the like). Use negative white space to separate sections of text. Double spacing of lines may waste the potential of white space to separate the ink-filled or positive spaces more dramatically.

Spacing provides many visual clues for the reader. You probably recognize the following as a memo:

XX: Xxxxxxx Xxxxxxxxx
XXXX: Xxxx Xxxxxx
XXXX: 00 Xxxx 0000
XXXX: Xxxxx Xx Xxxxxxx Xxxxxxx Xxxxx
_____

The To, From, Date, and Subject lines are usually followed by a partial cutoff line to separate the control data from the text, which is usually presented in block paragraphs. The memo design is just one possibility of space designed to organize materials.

Tabs and Columns

Subordinated text is easy to handle by presetting tabs so that just a click or two moves the carriage or cursor to the correct spot. Most word processors have preset one-inch tabs all across a line, saving the writer the time of establishing indentations. These may be changed with just a few clicks.

Columns are another way to display text. Usually, two columns suffice and can be set like newspaper columns or with parallel parts opposite each other or with block protects at the sides of text. Columns can be of equal or unequal widths, as can the spaces between the columns. Further, you can divide the columns by vertical lines or not. Charts, graphics, tables, photographs, and other artwork can be dragged by your cursor or moved around by tabs and placed into columns easily.

Line Length, Leading, Kerning

Line length, called *measure,* helps to determine text readability. Characters are the letters, punctuation marks, numbers, and other possible symbols. The fewer characters in a line, the easier it is to read. The rule of thumb for readability is 40 to 60 characters per line. About 65 characters—the usual number on a standard 8 ½" x 11" page with one inch margins—is the maximum for a serious document. This will average about 15 words per line. User manual instructions usually use far fewer characters per line so as not to confuse the reader/operator. A good actual line length for a specific text is dependent on how closely spaced the font is and how large the point size is. Larger fonts allow for longer lines.

Leading (pronounced "ledding" and named after the metal slugs that separated lines in hand-set type) provides interlinear space, the space between the bottom of the characters on one line to the top of the characters on the next line. Lines of type can be set tight or open. All word processors can adjust this space between lines with settings for single, space-and-a-half, double spaces, or other variables.

Consider these lines of type:

In this example the type is in 12 point New Times Roman and is single-spaced, very readable, and quite normal.

In this example the type is in 12 point NewsGoth BT with the second line set at a space-and-a-half below the first, a more open interlinear space that works well for ad copy, marketing information, or a sophisticated annual report.

IN THIS EXAMPLE THE TYPE IS 10 POINT GIOVANNI WITH THE SECOND LINE SET ONLY THREE-QUARTERS SPACE BELOW THE FIRST; IT IS MORE DIFFICULT TO READ BUT COULD BE USED WHEN SPACE IS AT A PREMIUM AND THE CONTENT IS NOT VITAL.

In this example the type is 8 point Arial, double-spaced, providing, perhaps, more lead-

ing than necessary, but it is very easy to read. Rule of thumb: the larger the font size,

the less leading is needed.

As a student of technical writing, you need only be aware of the possibilities. Experiment with your documents for the most readable leading, or line spacing.

Kerning is an advanced design tool that allows a typesetter to adjust the spacing between letters. That space can be open or tight. The letters AVW as typed on a word processor appear to vary the spacing, but this is due only to the oblique rather than rectangular shapes of A, V, and W. Due to their roundness, the letters O and Q often appear to be openly kerned, but they are not. A series of lines beginning with capital letters may not line up to the eye on the left margin because of their different shapes, but can be adjusted. Word processing options under your Format tool usually include Word/Letter Spacing and Manual Kerning capabilities. Reliance on a little kerning rather than trusting the eye can make a poster or flyer look more professional.

Line length, leading, and kerning considerations will help you to produce carefully fitted, highly readable copy into your documents.

Headings, Type Sizes, Fonts

Headings are not just decorative. They can

- Break up continuous text
- Indicate to the reader coherent sections of text
- Provide a clue to the upcoming content

Look at the headings in this and other chapters. The various sizes indicate major divisions, first-level subparts, and even third-level subparts. Notice, too, that the first- and second-level headings are printed in color, but third-level are all capitals in boldface black. Look at the headings in the sample document design pages where all of the major headings are capitalized boldface letters, and subordinate levels are indicated by uppercase and lowercase boldface letters. More than three levels of headings may tend to confuse the reader and so must be used cautiously.

You may also design your headings with white space clues. Generally, there are more blank lines above a heading than after. Judicious use of headings breaks up the "gray" text and makes the document more coherent and accessible.

The headings in this textbook are noun phrases, but you may use headings that name actions, ask or answer questions, or list steps. Consider the following headings:

WHAT ARE THE PARTS OF MY COMMUNICATION?
WHAT ARE MY STRENGTHS AS A SENDER?
WHAT ARE THE NEEDS OF MY RECEIVER?

PREWRITING THE DOCUMENT
Brainstorming the Topics
Writing the Outline

Another design concept concerns type sizes and typefaces, called fonts on your computer. The upper group of sample headings here are in 17 and 14 point Helvetica, and the bottom three are in 14 point Eurostyle bold. Your computer software may offer over 125 typeface fonts; however, you can buy software to add hundreds more fonts including scripts, stencils, symbols, foreign alphabets, such as Cyrillic, Greek, Arabic, and more. Most computers offer type size options from 4 point to 72 point or more. Familiarize yourself with your fonts and type sizes. Some are irresistible for newsletters and invitations, holiday cards, brochures, and the like, such as

BOXED IN

GALLIA

Poster Bodoni ATT

Shelly Volante BT

YEARBOOK OUTLINE

in various type sizes and colors. The technical writer, however, usually uses straight-edged fonts such as Eurostyle or Helvetica for headings and titles, and the more curly fonts, such as Times New Roman or Goudy for the body text. The straight-edged fonts are called *sans serif* (without a flourish at the end of the letters), and curly fonts are called *serif* (with a flourish at the end of the letters). Consider these elements in your documents. The professional technical writer will study typography extensively and use page design and desktop publishing software.

Margins, Indentations, Justifications

If a document is to be bound, a wide, empty inner margin will make the content easier to read. If annotations are expected, a wide, empty right-hand margin is beneficial. Text can be set flush left as is this

small portion,

centered on the page like
this brief example, or

flush right as is shown in
this brief sample of text.

Even paragraph indentation practices are changing. There is no need to indent the first line of text beneath a heading; the heading itself indicates that a new text division paragraph is to follow. Your word processor will probably automatically indent one inch for each tab, but one-quarter-inch indentations are more modern. If you learned to type on a typewriter, you learned the habit of double-spacing after closing punctuation, but modern word processors automatically adjust the space for you after a period, exclamation point, question mark, or colon. Your extra spacing is not necessary to indicate sentence endings and new beginnings.

Most typewriters and word processors automatically present a **ragged right margin,** meaning the text is not flush to the right hand margin as in

This sentence is set with a ragged right-
hand margin suitable for brief memorandums,
letters, short reports and similar documents.

Users of conventional typewriters compensated for excessively ragged right margins by splitting words with a hyphen, but sometimes too many words ended up hyphenated at the end of lines, creating an unaesthetic appearance. Word processors automatically wrap the overlong word to the next line, although you may use a setting that allows for end-word hyphenated divisions. Text may also be justified so that the right-hand margin is always flush as in

> This sentence is set for line justification to assure that the right-hand margin will be perfectly flush as in this book and in most multicolumned materials, such as news magazines and newspapers.

A **full justified block format,** with flush margins both left and right, is used throughout most of this textbook for the explanatory paragraphs, but a ragged right margin is used in the examples. Actually, the ragged right helps readers to track the text and to keep their place. Full justified text may also cause excessive white space when the extra spacing between words (kerning) forms noticeable white rivers running down the page. Full justification is used most often in textbooks, brochures, newspaper columns, and some flyers.

In many technical texts important stress and focus are signaled by structural levels in the document. Robert Kramer and Stephen A. Bernhardt explain

- Subordinated block indents from the left signal content structure and use white space effectively.
- Any indent pattern beyond two levels of subordination becomes merely decorative and most probably useless to the reader.
- Actually, the reader is likely to lose track of the level he/she is on, so the focus is lost.
- An indent cue that the reader doesn't understand only confuses that reader.

Lists, Numbers, Bullets

Often your material will suggest a textual list. By organizing material into lists, the content becomes emphasized and highly visual. Examine your text to extract lists and then write them in parallel structure, indent them from the rest of the text, and consider numbering or bulleting each item. In fact, what I have just written suggests a list:

> Organize the items
> Write each item in parallel structure
> Indent the list from the text margin
> Consider numbering or bulleting each item

Some lists reflect chronology or time sequence:

1. Lift the handset to turn on your phone.

2. Start your conversation.

3. Place the handset in the base unit to turn off your phone.

Numbers are visual clues for your reader to indicate unquestionable order. Lists that do not necessarily indicate a chronology may be bulleted, as in

- Definitions
- Mechanism Descriptions
- Instructions
- Process Analysis

Very often a list may become the headings of text sections. Software may include a variety of bullet options (✐,❑,☆,✔,●).

Emphatic Features, Icons, Reversed Text, Color

Emphatic features are those that style your text, such as boldface, italics, underlining, all capitals, small capitals, drop caps, and icons. Other emphatic features include windows, symbols, icons, reversed text, and color.

- **Boldface**, *italics,* underlining, and SMALL CAPITALS are used for action verbs and emphasis. Consider the following five sentences:

  Save your text before beginning a new document.

  Save your text before beginning a new document.

  Save your text before beginning a new document.

  Save your text *before* beginning a new document.

  Save your text BEFORE beginning a *new* document.

  The words *save* and *before* are the words requiring emphasis. The first example provides none and is not an effective instruction. The second example ignores *before*. The third and fourth are acceptable, but the fifth is overdone. Excessive use of these elements will make your document a meaningless hodgepodge.

- ALL CAPITALS are used for major headings and for emphasizing important words, such as WARNING, DANGER, CAUTION, or NOTE. Use them sparingly.

- SMALL CAPITALS are frequently used to indicate that a word in the text is included in a glossary. (Glossaries are discussed in this chapter.)

  The COLLATERAL is insufficient to secure the loan.

- **D**rop capitals are more decorative than emphatic but may begin a paragraph in a poster, flyer, or brochure.

- A window will set off a caution, warning, or note to make it emphatic as in:

> **Note:** Initial start up operation, may result in minimal smell and smoke (about 15 min.). This is normal. It is due to the protective substance on heating elements which protects them from salt effects during shipping from the factory.

- Symbols (*, ^, >, $,) may be used sparingly for emphasis.
- Icons, which are pictures, images, or other representations, (☎ ⏰ 🗓 ✉ ʊ), may provide emphasis and eye appeal but must be used sparingly. They are more appropriate in manuals, flyers, brochures, newsletters, and so forth rather than in letters, reports, and proposals.
- Inverse text, sometimes called reversed text, is the printing of white or light letters on a black, shaded, or colored background. It helps readers to access information.

Tip: Use Inverse Text Sparingly!

- Color is used for emphasis to make words or phrases stand out, such as red for *Danger* or *DO NOT*. Color is often also used for document titles and headings to make them more reader accessible. Bright, dark colors provide more contrast than do pastels to the surrounding text.

Borders, Fills, Watermarks

Full page borders, fills, and watermarks are other design elements. Borders are not usually used in technical writing except, perhaps, for title pages. They may be plain lines and may be thin, thick, double, or shadowed. Fancy borders, seldom used in technical writing, include ribbon effects in just the upper left-hand section of a page, a top and bottom fancy line, or decoration all around the page.

Fills are shadings (light, dark, or color), which may provide a background behind a title, emphasize a paragraph, or be used within a window or bordered page. Light gray fill is used in the document page layout sample on page 117 for both design (around the title) and for emphasis (over the paragraph). Fill options may be solid, waves, vertical, horizontal, or diagonally stripes, to name a few.

A watermark, a faint picture beneath the type, is typically a logo or a picture of a product. Technical writers may use them on title pages or elsewhere in the text. Figure 4.1 shows a title page with both a border and a watermark.

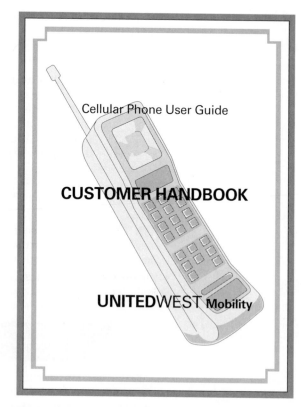

FIGURE 4.1   *Sample title page with watermark and border*

Headers and Footers

Headers and footers orient information such as page numbers, author names, publication information (volume and issue, chapter number and title). Notice that this textbook layout uses headers to indicate page, part number and title, and chapter number and title. It would be just as logical to move only the page number to the bottom and to retain the other infomation at the top. In this text the left-hand page places the page number flush left and the part number and title flush right. The right-hand page places the chapter number and chapter title flush left and the page number flush right.

Consider the logical headers and footers for your documents. Page numbers are usually essential, but other information serves as navigational tools to your reader as well. Documents can be numbered at the top or bottom and in either corner or the middle of the page. If you intend to bind pages, numbers can be set in alternating corners. Select an option (14, -14-, Page 14, Page 14 of 20) that suits your document. You may also include chapters (Chapter 4: Document Design, Ch.1 Pg. 14, or 1.14), to illustrate a few choices.

PRESENTATION FEATURES

Front Matter

Front matter may include a separate title page, a table of contents, and possibly a list of illustrations, a summary or abstract, and separate chapter title pages.

A long document, such as a proposal, a user manual, or a professional paper for publication, will require a separate **title page.** A title page usually includes at the least a centered, full title; the name of the author or corporate author; and usually a date. Other data may be included. The title is usually presented in a larger type size than the largest heading within the document. The font should usually be straight-edged *(sans serif)* and may be presented in all capital letters. If the title is the name of a product, you will include the model number, purpose, and other significant material to indicate fully the content to follow. Figures 4.2 and 4.3 show title pages of a proposal and a user manual. Figure 4.4 shows a number of sophisticated title pages using various fonts, type sizes, drop capitals, artwork, photographs, and watermarks.

**PROPOSAL TO IMPROVE
EMPLOYEE WORK SCHEDULES
IN THE
CUSTOMER INQUIRY DEPARTMENT**

**Prepared for
Andrea Brooks
Assistant Vice President
Operations Division
State Trust Charter Bank
Bigtown, New Hampshire**

**by
Monica Ferschke
Administrative Assistant
October 18, 2001**

FIGURE 4.2   *Sample title page*   (Courtesy of Monica Ferschke)

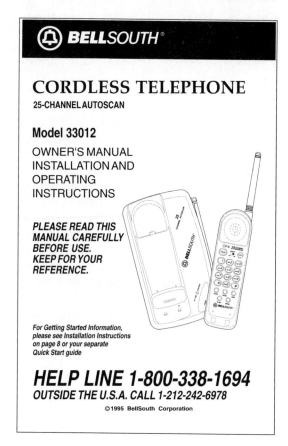

FIGURE 4.3   *Sample title page for a user manual*   (By permission of BellSouth Products, Inc.)

Although the title page is never numbered, all other front matter pages are numbered in lowercase Roman numerals (i, ii, iii, iv, v), which should be placed in the same position as the page numbers of the document body.

A **table of contents** in a long document (eight pages or more) will not only help your reader to turn rapidly to a particular section of the document, but also give the reader an initial indication of the organization, content, and emphasis of a document. Title the list *Table of Contents.* Some software will create this table for you by allowing you to type in the chapter titles and major headings plus the page numbers in a separate tool. The computer then generates the finished table. To create your own table of contents, take your headings and subheadings (or chapters and major headings if the document is divided into chapters) using indentation between the major headings and the subordinate

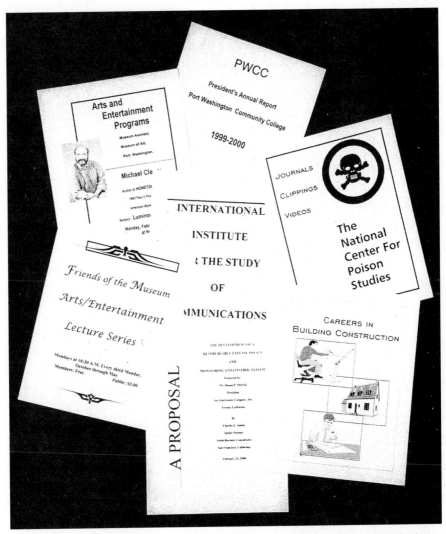

FIGURE 4.4 *Sample title pages utilizing a variety of document design options* (By permission of Mark DeJanon)

headings. Use spaced dots to connect the heading with the beginning page number of that section.

If your document contains **illustrations** (graphics, photos, maps, etc.), include a separate listing entitled *List of Illustrations* below the table of contents or on the next page. Include the illustration number and title and then connect that material with spaced dots to the exact page reference. Finally, list tables separate from figures (all other graphics, photos, maps, etc.). Figure 4.5 is an example of a table of contents for a proposal.

TABLE OF CONTENTS

LIST OF ILLUSTRATIONS

FIGURE 4.5 *Sample table of contents for a proposal*

You may wish to include an **abstract** or **summary,** a concise review of the parent document. Title this item *Abstract* or *Summary* and place it after the table of contents. Abstracts and summaries are discussed at length and samples are provided in Chapter 15.

If the material is divided into separate chapters, you will want to consider separate introductory **chapter title pages.** If the chapter title page is to include artwork or other significant prefatory material, use a separate chapter title page. If there is nothing other than the chapter number and title, place them on the first page of the chapter so as not to waste space. Notice the chapter title pages in this textbook include artwork.

Back Matter

Back matter, or supplements to the body of the document, may include a glossary, an appendix, a bibliography, and an index. They are placed at the back of the document in the order listed above and numbered consecutively, continuing from the document body.

A **glossary** is an alphabetized list of definitions of words used within your document that may not be understood by the general reading public. Title the first page *Glossary* and list the unnumbered words below. If there are only a few (five or so) specialized terms used in your document, include the definitions within the text or in a footnote at the bottom of the page where the word occurs. If you use a separate glossary, you may want to use small caps for the word or place a star after the term within the text when it is first used.

All terms may not be technical but may include laypersons' terms that you are using in a special sense within your document: "In this report a cell is a small group acting as a unit within a larger organization." Instructions for writing simple and expanded definitions appear in Chapter 8. Within the glossary, boldface or underline the term and add either a colon or extra white space before the definition. Include the glossary in the listing of the table of contents. Figure 4.6 shows a partial glossary from a college textbook on business practices.

An **appendix** contains material that expands information contained in the body of your document. Parts of the appendix material may be discussed within the document, but the entire block of information may clutter your message. Refer to the appendix within the text, where appropriate, if you want to alert your reader to its existence.

An appendix may include tabulated interview and survey results along with the original instruments (such as questionnaires), details of an experiment or investigation, financial projections, job descriptions and resumes of new personnel, lists of references, and so on. If the document includes artwork and sample forms, you may refer to them as **exhibits** (Exhibit 1. Map of Florida Voting Results). Exhibits require numbers, whereas appendixes use letters (Appendix A. Resume of Merrill C. Tritt). An appendix (its letter and title) or a list of exhibits (their numbers and titles) is included at the end of the table of contents. This book, you will notice, includes Appendix A, "Conventions of Construction, Grammar, and Usage," and Appendix B, "Punctuation and Mechanical Conventions." Some of the material is included in the text with references for full discussion in the appendixes.

A **bibliography** is an alphabetized list of all works consulted in the preparation of your document. The bibliography may be titled *Bibliography, References,* or *Works Cited,* depending on the documentation style appropriate to your paper. A full discussion of documentation styles along with examples of bibliographies and works cited pages are

GLOSSARY OF TERMS

Adequate protection   The protection that a debtor grants a creditor with court approval to prevent foreclosure on property. This can include a resumption of monthly payments.

Administrative creditor   All creditors holding claims for debts that were incurred after the bankruptcy was filed. This includes fees for legal and accounting professionals used by the debtor in possession.

Adversary proceeding   Matter of dispute brought before the court for determination. Many regard amount of debt, security interest, or use of cash as examples of issues to be settled by court action.

Automatic stay   Becomes effective on filing of a bankruptcy petition and prevents any creditors from taking further collection action against the debtor without court permission.

Bar date   The last date set by the court that a proof of claim can be filed to collect a debt owed by the bankrupt company.

Cash collateral   All proceeds paid or due to a company that have been pledged as collateral for a loan. Associated with the cash liquidation of inventory and accounts receivables.

Collateral   Assets that are used as security for a debt.

Confirmation   The action of the bankruptcy court approving the

FIGURE 4.6   *Sample Glossary*   (From *Saving Your Business* by Suzanne Caplan. Copyright ©1992. Reprinted with permission of Prentice Hall.)

included in Chapter 6. Figure 4.7 shows a sample works cited bibliography page.

An **index** is an alphabetized listing, along with the page references, of all of the key subject terms used in your document. It differs from a table of contents in that it will contain all of the key terms and their page references occurring throughout the text, not just the chapter headings. It enables the reader(s) to locate all of the references to a particular term quickly. Some software possesses tools to help you prepare an index. You type in the terms, set the directions for a manuscript search, and the computer generates a complete index. If you generate your own index, use $3'' \times 5''$ index cards, which can be alphabetized for quick reference. Review the entire manuscript, writing down each term to be indexed on a separate card along with the page number. Each time a term is used again throughout the text, include the page or pages (17, 19–24) on the appropriate card. You may even subordinate material under a particular key term. This list is then titled **Index** and included as the final back

Technical Communication Quarterly   325

Works Cited

Aristotle. *The Rhetoric and the Poetics of Aristotle.* Trans. W. Rhys
    Roberts and Ingram Bywater. New York: Modern Library, 1954.
Bazerman, Charles. "How Natural Philosophers Can Cooperate: The
    Literary Technology of Coordinated Investigation in Joseph
    Priestley's History and Present State of Electricity (1767)." *Textual
    Dynamics of the Professions: Historical and Contemporary Studies of
    Writing in Professional Communities.* Ed. Charles Bazerman and
    James Paradis. Madison: U of Wisconsin P, 1991. 13-44.
—. *Shaping Written Knowledge: The Genre and Activity of the Experi-
    mental Article in Science.* Madison: U of Wisconsin P, 1988.
Brasseur, Lee E. "Contesting the Objectivist Paradigm: Gender Issues
    in the Technical and Professional Curriculum." *IEEE Transactions
    on Professional Communication* 36.3 (1993): 114-23.
Brown, Stuart C. "Rhetoric, Ethical Codes, and the Revival of *Ethos*

FIGURE 4.7   *Sample works cited bibliography listing*   (From Gerald
Savage, "Redefining the Responsibilities of Teachers and the Social
Position of the Technical Communicator," *Technical Communication
Quarterly,* Vol. 5, No. 3 [Summer 1996], p. 325.)

matter in your document. Figure 4.8 shows a partial index from the last
edition of this textbook.

PRODUCTION DECISIONS

Production decisions include the size of the paper, the quality of the paper, the orientation on the page, watermarks, borders, folds or bindings,
flaps and pockets, and perforations and windows. Each of these entities
must be considered before the final publication of your document.

Factors that determine the **size of the paper** include the purpose
of the document, its intended distribution, and the size of packaging for
the product that the document is supporting. Most in-house documents
(those to be read by colleagues and management in your organization),
proposals, professional articles, and book manuscripts call for the usual
8½″ × 11″ paper. A book or journal publisher will determine the size of a
published page.

If the document is a user manual, a set of instructions, or a set of
specifications, the size of the paper may vary. If the document is to accompany a product, the size is determined by the size of the packaging.
You no doubt have a collection of user manuals in your possession. Note
how the size varies for the manuals that accompany your cellular phone,
your VCR, your fax machine, your computer and printer, and your

Index

A

FIGURE 4.8   *Sample partial index page*

kitchen appliances. Your cellular phone manual will probably fit into your pocket or even into the case that holds the phone.

The **quality of the paper** for in-house documents should be low-gloss, rag-bond, and white. Glossy paper produces glare and strains the eyes. Rag-bond paper has a high fiber content (25 percent minimum) and is heavier. Select a 20-pound or heavier weight paper but do not use anything as heavy as construction paper. Lower weight paper tends to be flimsy and crinkly. Out-of-house documents may require a slicker paper (like magazine pages) and a lower weight. The slicker paper allows for a sharper print, and the lower weight decreases the bulk of a document. White paper is the standard. Use color only for flyers, announcements, and the like.

Also consider the **orientation on the page.** Your document may be placed so that the narrower part of the page is at the top (called *portrait* in computer commands), or you may print out your document from left to right along the long side of your page (called *landscape* in computer commands). Landscaped lines may become so long that readers lose track as they scan back to the beginning of the next line. However, you may wish to landscape titles or other reference notations.

If your document is long, you may wish to consider **folds** or **bindings.** A brief, six-page manual may be folded into a three-panel brochure with printing on each of the six panels. A lengthier manual could be fanfolded. If some material is too extensive for a page but should be viewed all together, you may consider a foldout page.

Longer documents may require a binding. A document stapled in the corner is unprofessional. If additional pages are to be added over time, consider a **loose-leaf binder** with two or three holes punched into the margins of the document so that the pages may be inserted at the appropriate places. Inexpensive options are a **ring** or **spiral binding** or a **saddle binding,** which inserts internal staples to hold double pages. A professional printer can provide you with **perfect binding**, a glue bond as in a book. Spiral and saddle options are appropriate for flip bindings, pages joined along the short edge and flipped up page by page to read. Figure 4.9 shows five kinds of binding possibilities:

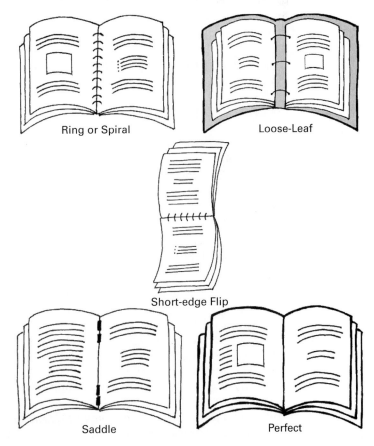

Ring or Spiral   Loose-Leaf

Short-edge Flip

Saddle   Perfect

FIGURE 4.9   *Bindings*   (Bindings by Heather Schaefer, ArtsyToo Creations. Reprinted by permission of the author.)

You may also consider **flaps** and **pockets.** Your separate binder may already contain front and back pockets. If you decide to include a cover harder than the content weight, you may wish to include harder internal dividers with flaps for identifying sections for easy access. You may add pockets if you foresee that additional material may be forthcoming. **Perforations** and **windows** may be decorative or functional. Consider how you might present a set of brief instructions on a cardboard format with a perforation to hang on a doorknob. Windows from one page to the next are a dramatic means of focusing attention on the framed material. You must consider all these options even if your document is being published by a professional.

CHECKLIST

Document Design, Presentation Features, and Final Production

DOCUMENT DESIGN FOR VISUAL MEANING

❏ **1.** Have I designed my document by adding visual meaning?

❏ **2.** Does my use of positive and negative space (columns, horizontal lines, borders, boxes, subordinations, and so on) make my document easier to read and access?

❏ **3.** Are my margins appropriate and do they leave enough room on the sides, top, and bottom?

❏ **4.** Have I centered any material?

❏ **5.** Should I use a ragged right edge or full text justification?

❏ **6.** Do my margins and indentations indicate major and subordinated material?

❏ **7.** Have I added headers and footers appropriately?

❏ **8.** Have I used headings to indicate major and subordinate divisions of the text?

❑ **9.** Are my headings parallel (noun phrases, questions, naming of actions)?

❑ **10.** Have I used varied type sizes to indicate major and subordinate headings?

❑ **11.** Have I used appropriate fonts for chapter headings, major and subordinate headings, and other material?

❑ **12.** Have I organized material into appropriate lists (plain, numbered, or bulleted)?

❑ **13.** Have I used emphatic features to advantage (boldface, italics, underlining, all capitals, small capitals, and drop capitals)?

❑ **14.** Have I used symbols, icons, reversed type, or color for emphasis and/or eye appeal?

❑ **15.** Have I considered borders, fills, and watermarks?

PRESENTATION FEATURES

❑ **1.** Does my document require a separate title page?

❑ **2.** Is my document long enough to require an abstract or summary before the text?

❑ **3.** Does my document require a table of contents? List of illustrations?

❑ **4.** Would separate chapter title pages be appropriate?

❑ **5.** Have I numbered my front matter with lowercase Roman numerals?

❑ **6.** Does my document require a glossary?

❑ **7.** Should I include an appendix or exhibits to add information that is helpful but not vital to the text?

❑ **8.** Does my document require a bibliography?

❑ **9.** Does my work warrant an index?

❑ **10.** Have I numbered my back matter correctly?

PRODUCTION DECISIONS

❑ **1.** Is 8 ½″ × 11″ paper appropriate or should I use a different paper size?

❑ **2.** Is the quality of the paper appropriate?

❑ **3.** Should I orient any text sidewards?

❑ **4.** Have I considered fanfolds or appropriate bindings?

❑ **5.** Does the document packaging require flaps, pockets, perforations, or windows?

EXERCISES

1. **Headings.**   Rewrite these headings to make them parallel:

   Extending Your Definition

   Write the Formal Definition

   Examples

   You Must Include Synonyms

   Providing Comparisons and Contrast

   Analogy

   Can You Provide the Origin of the Term?

   An Etymology Should Be Provided

   Are There Any Causes and Effects?

   Providing an Analysis of the Process

2. **Fonts.**   Which fonts would you consider appropriate for headings? Why?

   a. **WRITING THE DEFINITION**   e. **WRITING THE DEFINITION**

   b. WRITING THE DEFINITION   f. **WRITING THE DEFINITION**

   c. **WRITING THE DEFINITION**   g. *WRITING THE DEFINITION*

   d. **WRITING THE DEFINITION**   h. WRITING THE DEFINITION

3. **Justification.**   Name the major reason for using ragged right text printing.

4. **Emphatic Features.**   Which of the following uses of emphatic features do you think is most is the most effective? Why?

   a. *Press* the PAUSE key **to record** a cassette tape.

   b. **Press** the *pause* key TO RECORD a cassette tape.

   c. PRESS the **pause** key *to record* a cassette tape.

   d. **Press** the PAUSE key to record a cassette tape

5. **Presentation Features.**   Answer *true* or *false* to the following statements:

   a. All technical report documents require a separate title page.

   b. A table of contents is useful for a document that is eight or more pages.

   c. A list of illustrations should be used for graphics, exhibits, and other visuals.

   d. All technical documents require a glossary.

   e. The glossary is not listed in the table of contents.

   f. If there are only a few technical terms, they may be defined within the text.

   g. A bibliography is an alphabetized list of all of the works you consulted in your document preparation.

   h. An index is a list of page references, chapter by chapter.

6. **Size of Paper.**   Locate a user manual which is not printed on 8 ½″ × 11″ paper. State three reasons why you believe it is the size it is. Submit it or a photocopy of a few pages with your reasons.

7. **Quality of Paper.**   Locate a document that you think has been printed on the inappropriate quality of paper. Name three reasons. Submit the original with your reasons.

8. **Orientation.**   If you are using a computer for your assignments, print out a page in portrait and the same page in landscape. If you do not use a computer, find a partner who does. Name three advantages of one over the other. Submit the pages with your commentary.

9. **Flaps, Pockets, Perforations, Windows.**   Locate a printed publication that uses flaps, pockets, perforations, or windows. Name three ways these are advantages or only decorative. Submit the publication with your critique.

WRITING PROJECTS

1. **Document Design.**   Study the following material. Use your computer or typewriter to add at least six document design enhancements. Do not just be decorative. Make your design features impart

meaning to the document. You may reword sentences if you have a compelling need to do so.

> Prepare visuals to clarify and emphasize your oral report. Visual materials may include chalkboards, flip charts, posters of tables, charts, drawings, handout sheets, exhibits of models or equipment, filmstrips, or transparencies. The size of the room or auditorium, the kind of people in your audience, the available monies, the available equipment (chalkboards, projectors, tables, and so forth), and the nature of your speech are all factors for consideration in planning visual material. Effective visual materials are characterized by a number of factors, such as simplicity, unity, emphasis, balance, and legibility. Rehearse your speech a number of times before you actually give it. Practice before a mirror and use a tape recorder. If possible, give your speech to a small group of friends. Ask them to assess your poise, eye contact, voice, gestures, and rate. Your voice should be conversational, confident, and enthusiastic. Avoid at all costs a monotonous sound. Vary your pitch and intensity. Monitor the volume and rate of your speech. Judge the quality of your voice. Are you shrill, nasal, raspy, breathy, growly?

2. **Critique a Document Design.**   Photocopy two pages from another textbook or a user manual that you feel exemplify quality document design. Write a memo explaining why you feel the format helps to convey meaning. Be specific in your references to all of the discussed design options. Use your own document design elements in the memo.

3. **Table of Contents and List of Illustrations.**   Write a table of contents and list of illustrations for this chapter.

4. **Index.**   Select a partner among your classmates and prepare an index for just this chapter.

5. **Collaborative Project.**   Join the same students you worked with on collaborative Writing Projects 4 and 5 in Chapter 2. Select one of the papers and reproduce it with document design elements to enhance the meaning of the document. Submit both documents.

6. **Collaborative Project.**   Form a team of three or four classmates. Locate a user manual for a VCR, cell phone, or some other technical product. Pretend you are writing to the director of marketing of the firm, who desires your opinions on the manual. Critique the use of document design elements, the use of presentation features, and the production decisions. What would you change? Keep the same? Eliminate? Use document design elements to enhance *your* document. Do you need to include any presentation features? If one of your team is computer literate, make appropriate production decisions at least regarding the size and quality of the paper, the orientation on the page, and the binding.

7. **Collaborative Project.**   Use the same information in Collaborative Project 3 in Chapter 3, but instead of writing a memo, design a brochure with the same information. Try to include a team member who is computer literate. Use appropriate graphics, fonts, type size, emphatic features, orientation of the paper, fanfolds or bindings, and other document design options.

NOTES

PART *t w o*

The Research Strategies

CHAPTER *5*

Accessing Information

CATHY   by Cathy Guisewite

9:00 PM : CONNECTED TO IRS WEB SITE TO CHECK QUESTION ABOUT TAX FORM.

9:06 PM : EXPLORED JUPITER ON NASA WEB SITE.
9:20 PM : READ LETTERMAN'S "TOP 10" LISTS FOR JANUARY.
9:33 PM : CHECKED WEATHER IN FIJI.
9:46 PM : WANDERED AROUND LOUVRE.
10:15 PM : JOINED CHAT ROOM DISCUSSING BEET FARMING.
10:35 PM : PLAYED ONLINE SCRABBLE WITH COUPLE IN TOLEDO.
11:35 PM : VISITED LIBRARY OF CONGRESS.
12:19 AM : DOWNLOADED RECIPE FOR LOW-FAT GUACAMOLE.
12:30 AM : VISITED HOME PAGE OF GIRL SCOUT TROUP.
12:45 AM : ORDERED SIX CD'S AND AN EVENING GOWN.
1:10 AM : HAD TAROT CARDS READ.

1:30 AM : LOGGED OFF THE PROCRASTINATION SUPER HIGHWAY.

S K I L L S

After studying this chapter, you should be able to

1. Use a library computer retrieval system with its databases to access general books, reference works, periodicals, indexes and abstracts, full-text articles, and all other materials in the library.
2. Locate the book stacks, print indexes (if used), reference books, the microform files, the CD-ROMs, and the vertical files.
3. Understand and employ the Boolean operators to limit your subject searches.
4. Have a working knowledge of the Dewey Decimal and the Library of Congress systems for organizing and locating books on library shelves.
5. Name some special reference books (dictionaries, encyclopedias, handbooks, manuals, and almanacs) in your field of study.
6. Access indexes and abstracts for periodicals, patents, and technical reports by computer or in print.
7. Access and read microform materials.
8. Access information from CD-ROMs.
9. Access information using specialized subscription databases.
10. Access materials in vertical files.
11. Use a variety of search engines to explore the Internet on specific subjects.
12. Evaluate Internet Web site documents by considering authorship, publishing body, possible bias, scholarship of the literature, accuracy and verifiability of the details, and timeliness.
13. Access special reference materials and periodicals on your personal computer, including your personal CD-ROM library.
14. Understand and subscribe to electronic bulletin boards, such as USENET and Listserv.
15. Access specific authoritative Web sites directly.

INTRODUCTION

Technical writers and students rely on library computer retrieval systems and personal Internet searches to research information for informed reports and publications. Almost all of the papers you will write in this course will require some research. You need to know how to access print materials and electronic sources (secondary sources) as well as how to prepare questionnaires for surveys and polls, interview authorities, write letters of inquiry, conduct testing and field inspections, and use direct observation (primary sources). Essential to effective research is learning how to take notes, avoid plagiarism, and document your sources in the final paper.

Accessing and researching the source material involves four strategies:

1. Researching print and electronic sources
2. Preparing surveys, questionnaires, and other primary research techniques
3. Preparing the bibliography cards or computer bibliography file (see also Chapter 6)
4. Taking notes on cards or keyboard

Some instructors may require you to photocopy all materials and print out online sources to submit with a final paper.

RESEARCHING MATERIALS

In researching the source materials located through the library resources or by personal computers, you must accomplish five objectives:

1. **Skimming the material.**   Obtain the sources or their abstracts and skim read the material as you locate it to become familiar with the data on your subject. Look over the abstract plus the table of contents and the index. Read the preface and the introduction. Check the appendixes and glossaries for supplemental materials and definitions. Review the source's own bibliography listing, which may direct you to other related sources. Print out or save to a file or a disk all online text that you will use.

2. **Preparing preliminary bibliography cards or a computer file.**   The purpose of the preliminary bibliography cards or bibliography computer file is to compile a collection of possible sources of data for the research project after your skim reading has determined if they address the subject. The information to be recorded for each source is covered in this chapter. Alphabetize the cards or your computer file list for easy reference.

3. **Eliminating redundant and useless sources.**   Select those sources that best suit your specific, narrowed topic. Eliminate those that present redundant data, are outdated, or are of questionable value.

   4. **Preparing a working bibliography.**   Alphabetize the sources
      (cards or computer file) remaining in the preliminary bibliography
      set and prepare a working bibliography page (see later in this chap-
      ter and in Chapter 6).

LIBRARIES

Familiarize yourself with your school, company, or public library. Locate
the computer retrieval system workstations, the photocopy machines,
special indexes not on the computer, the book stacks, the microform files
and readers, the CD-ROMs, and vertical files if they are maintained.
Some libraries may still rely on the traditional card catalog for books or
for special holdings of rare books and the like. Some may rely still on
printed indexes for other materials (periodicals, essays within books,
special subject abstracts and indexes, and government publications).

Computer cataloging systems are still being installed around the
world and are constantly improving and updating access methods.
Therefore, it is impossible for any textbook to be up-to-date on all the ins
and outs of computer retrieval, but a general overview follows.

Library Online Computer Retrieval Systems

Most college, university, high school, large public libraries, and business
libraries now catalog their stock by computer retrieval systems that have
largely replaced card catalog drawers and periodical indexes. The systems list books, journals, magazines, newspapers, and special reference
books, as well as reports, government documents, manuscripts, type-
scripts, prints, photographs, maps, posters, animated illustrations,
CD-ROM listings, and more.

Library databases may give you access not only to the particular library's holdings but also to holdings of other nearby libraries (local, community college, and university), specialized research libraries, the
Library of Congress, state archives, and other locations. Further, libraries subscribe to any number of specialized databases on various subjects (sciences, technologies, business, health, government documents
and statistics, literature, education, law, psychology, and so on). These
databases have been developed around the world, and they are constantly updated. The information within them includes titles, authors,
abstracts, and full-text articles. A listing of the most important subscription databases follows in this chapter. Most of them may be accessed by
typing in your library card number.

Most library systems now allow free searches on the Internet using keywords, World Wide Web addresses, and hyperlinks to access and download research materials. Consult the librarians in any library for instruction.

At your computer workstation, begin your search for materials on your selected subject by seeking books, bibliography listings, abstracts, periodical abstracts, and full-text articles to skim read to determine their usefulness for your purpose. Check the items that seem pertinent and print out the selected lists and articles that you require. Use your printed lists and notes to prepare bibliography cards or a computer bibliography file (see also Chapter 6, "Documenting Research").

Typically, the computer will first display a screen that lists the categories of inclusion, such as the overall catalog of books, encyclopedias and other special reference books, magazine/journal indexes, newspaper indexes, and other accessible catalog classifications. This is usually called the **menu screen,** as shown in Figure 5.1.

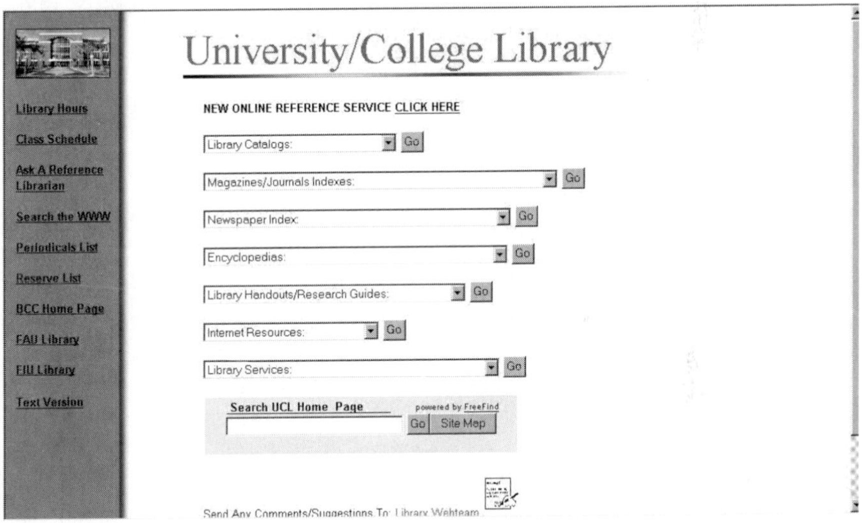

FIGURE 5.1   *A typical computer retrieval system menu screen*

When you click your mouse or trackball on a selected classification, your next screen will typically be a **dialog box,** as shown in Figure 5.2. In this instance, Magazines/Journals indexes on the menu page has been selected, and the dialog box allows you to enter a subject, author, title, or other categories for a search of holdings. Sometimes these dialog boxes will allow you to limit the search for full-text articles, limited date periods, specific journals, foreign language translations, and so forth.

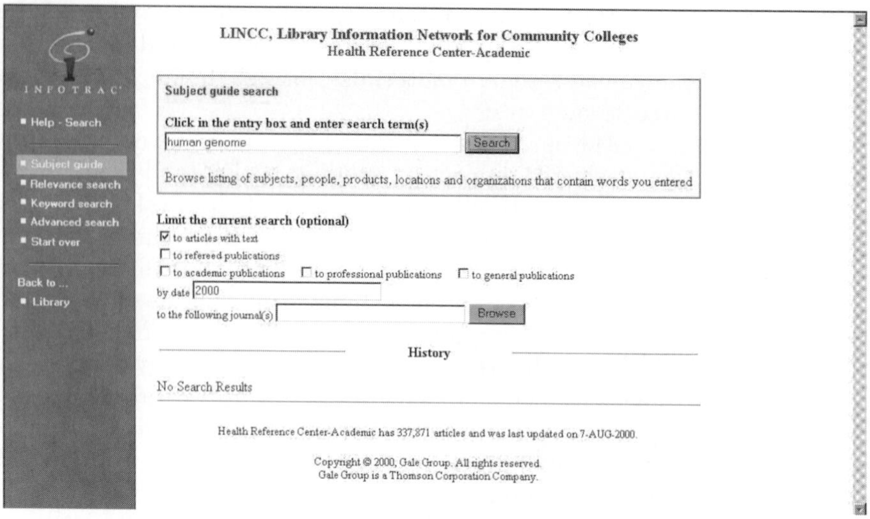

FIGURE 5.2 *A typical computer retrieval system dialog box with typed subject and checked limitations for full-text articles and date*

Boolean Operators

Most libraries use the Library of Congress Subject Headings (LCSH) for classifying books, videos, film, and other tangible holdings by subject, but you may, of course, use common sense to list keywords in your search. A single keyword may find just what you want (genome), or you may list several words (genome project). Your search engine system will probably allow you to make numerous combinations using the **Boolean operators:**

- AND. To search for two (or more) terms on the same page, type the word *and* between the terms or put a plus sign right before the second term (genome system *and* medicine) or (genome system+medicine) to narrow the search to holdings that discuss the impact of the mapping of our genes on medicine.
- OR. To search for two related subjects, type the word *or* between the terms (genome system *or* chromosomes), which may locate works on cloning, DNA, or other aspects of the genome system.
- NOT. To search for pages that include the first term and not the second, type the word *not* or put a minus sign right before the second term (genome system *not* discovery) or (genome system-discovery), which will eliminate works on how the system was mapped and concentrate on the findings.
- " ". To search for an exact phrase, enclose the phrase in quotation marks ("The most significant discovery of the century").

- ( ). To group parts of your search, enclose the terms in parentheses (genome system *and* (medicine or cancer)), which will return to you pages with both the words *genome system* and *medicine* or with both the words *genome system* and *cancer.*

When you click the *Search* or *Browse* icons on this particular dialog box, your next displayed page will be a title list page—in this case 30 citations on the human genome project, limited by full-text and the 2000 date selected on the dialog box. Figure 5.3 shows just one part of a title list page:

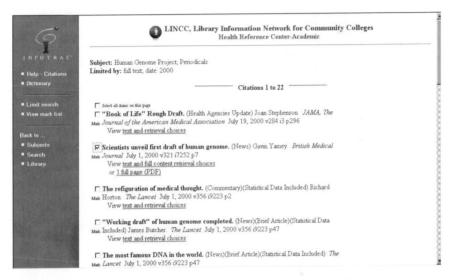

FIGURE 5.3   *A typical computer retrieval systems title list page limited to the Magazines/Journals indexes*

In order to access the full text, click on the box before the selected article(s), book(s), or other holding(s), depending on which type of holding you have selected. In this instance, the computer will next display the full text of the selected article, as shown in Figure 5.4.

By selecting other library databases for books, newspapers, or other classifications, you will locate lists of books (general or special reference), specific periodical indexes (i.e., *New York Times Index, Discover* magazine, and more), specialized abstracts and indexes (i.e., *General Science Index* or *Essay and General Literature Index,* and more), microform materials, CD-ROMs, and subscription databases on your subject. Many libraries will indicate if the books are available in other nearby libraries, whether or not they are checked out, whether or not online full text is available, and other information. Many libraries have interlibrary loan systems to obtain books that are not on site. Each system will differ, so do not hesitate to seek the help of a librarian.

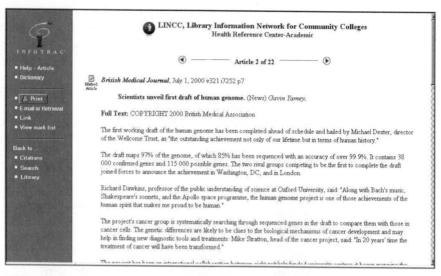

FIGURE 5.4   *A portion of a typical full-text article*

Books

Determine if your library catalogs its collection of books by the Dewey Decimal System or the Library of Congress System. In order to actually retrieve books on specific subjects from the shelves, it is helpful to have a good working knowledge of these two systems.

The **Dewey Decimal System** arranges all books and journals on the shelves by the following general subject categories:

000–099 General works (bibliographies, library sciences, general encyclopedias, publishing, manuscripts, and rare books)

100–199 Philosophy and psychology (metaphysics, paranormal phenomena, logic, ethics, ancient and Oriental philosophies)

200–299 Religion (theory of religion, Bible, Christian studies, comparative religion, and other religions)

300–399 Social sciences (political science, economics, law, public administration and military science, education, commerce, and customs)

400–499 Language (linguistics and all languages)

500–599 Natural sciences and mathematics (mathematics, astronomy, physics, chemistry, earth sciences, paleontology, biology, plants, and animals)

600–699 Technology/Applied sciences (medicine, engineering, agriculture, and home economics)

700–799 The Arts/fine and decorative arts (civic and landscape art, architecture, sculpture, drawing, painting, graphic arts, photography, and music)

800–899 Literature and rhetoric (American, English, Germanic,

Romance, Latin, Greek, and other culture literature)

900–999 Geography and history (travel, genealogy, geography, and Ancient World, European, Asian, African, and other continental general histories)

Each of these general categories is, in turn, divided into ten smaller categories. For example, the Technology/Applied sciences classification (600–699) is divided into the following:

600–609 Technology (Applied Sciences)

610–619 Medical sciences/Medicine

620–629 Engineering and allied operations

630–639 Agriculture and related technologies

640–649 Home economics and family living

650–659 Management and auxiliary services

660–669 Chemical engineering

670–679 Manufacturing

680–689 Manufacture for specific use

690–699 Buildings

Under each of these categories even more precise classifications apply. Familiarize yourself with the primary, secondary, and even third-level classifications within your field.

The **Library of Congress System** classifies books on the shelves by 21 letters to designate the following major subject categories:

A	General Works
B	Philosophy and Religion
C	History and Auxiliary Sciences
D	Universal History and Topography
E/F	American History
G	Geography, Anthropology, Folklore
H	Social Sciences
J	Political Science
K	Law
L	Education
M	Music
N	Fine Arts
P	Language and Literature
Q	Science
R	Medicine
S	Agriculture

T   Technology
U   Military Service
V   Naval Services
Z   Library Science and Bibliography

Each of these classes is subclassed by a combination of letters for subtopics. For example, T (Technology) contains 16 subclasses:

TA   General engineering, including general civil engineering
TC   Hydraulic engineering
TD   Sanitary and municipal engineering
TE   Highway engineering
TF   Railroad engineering
TG   Bridge engineering
TH   Building construction
TJ   Mechanical engineering
TK   Electrical engineering, Nuclear engineering
TL   Motor vehicles, Aeronautics, Astronautics
TN   Mining engineering, Mineral industries, Metallurgy
TP   Chemical technology
TR   Photography
TS   Manufactures
TT   Handicrafts, Arts and crafts
TX   Home economics

A number notation is also assigned to each book.

Reference Books

Reference books—general and specialized encyclopedias, almanacs, handbooks, dictionaries, histories, and bibliographies—may be available as books, system-specific electronic editions, online editions, or on CD-ROM. Increasing numbers of these materials are now appearing online on the Internet. You may access the online editions from your library workstation or your personal computer. There are reference books for every discipline. Following is a partial list of technical reference books, indicating which ones are currently online:

Online Dictionaries

American Heritage Dictionary
Biotech Life Science Dictionary
Computer and Internet Dictionary

Dictionary of Cell and Molecular Biology
Encarta World Dictionary
Harcourt's Dictionary of Science and Technology
Merriam Webster Collegiate Dictionary
Merriam Webster Medical Dictionary
New Oxford Dictionary of English
On-Line Medical Dictionary
Webster's New World Dictionary
Webster's Third International Dictionary of the English Language

More Dictionaries

Dictionary of Architecture and Construction
Dictionary of Engineering and Technology
Dictionary of Nutrition and Food Technology
Dictionary of Practical Law
Dictionary of Telecommunications
McGraw-Hill Dictionary of Scientific and Technical Terms
Paramedical Dictionary
Random House Dictionary of the English Language
Stedman's Medical Dictionary

Handbooks, Manuals, and Almanacs

Civil Engineering Handbook
Computer Almanac (online)
CRC Handbook of Marine Sciences
Fire Protection Handbook
Internet Almanac (online)
McGraw-Hill Computer Handbook
Nurse's Almanac
U.S. Government Manual
World Almanac and Book of Facts (online)

Online Encyclopedias

Compton's Encyclopedia
Encyclopedia Britannica
Encyclopedia of Analytical Instrumentation
Grolier Multimedia Encyclopedia
Lawcopedia
Science Hypermedia Encyclopedia
Smithsonian Encyclopedia

More Encyclopedias

Collier's Encyclopedia

Encyclopedia Americana

Encyclopedia of Careers and Vocational Guidance

Encyclopedia of Computer Science

Encyclopedia of Food Technology and Food Service Series

Encyclopedia of Management

Encyclopedia of Marine Resources

Encyclopedia of Materials Handling

Encyclopedia of Modern Architecture

Encyclopedia and Dictionary of Medicine, Nursing, and Allied Health

Encyclopedia of Urban Planning

McGraw-Hill Encyclopedia of Energy

McGraw-Hill Encyclopedia of Environmental Science

McGraw-Hill Encyclopedia of Food, Agriculture, and Nutrition

Special Indexes and Abstracts

Although periodical indexes for magazines, journals, and newspapers are now included in the computer catalog and may cover 15,000 periodicals, a library may still subscribe to selected periodicals. Whether the periodical indexes are on a computer database or in print form, they are constantly updated (daily for databases and every few weeks for print). The *New York Times Index* is the most comprehensive of the large periodical indexes and is most likely in your computer system. It indexes, by subject and author, stories and articles that have appeared in the *Times* since 1851. The index also summarizes articles and occasionally reprints photographs, maps, and other illustrations that accompanied the original article.

In addition, every newspaper indexes its own publication. Your major area newspapers will probably be included on your computer catalog. Companies also index their newsletters and will supply you with copies or printouts of articles you need.

If your periodical indexes are not available in the computer system, ask your librarian to show you the print indexes for magazines, journals, and newspapers, such as

Applied Science and Technology Index

Reader's Guide to Periodical Literature

Social Sciences Index

The newspaper indexes of the *Washington Post*, the *Chicago Tribune*, *New Orleans Times-Picayune*, and the *Los Angeles Times*.

Other sample listings for periodical indexes include

> *Agricultural Index*
> *Energy Index*
> *Environment Index*
> *Index to Legal Periodicals*

Technical report indexes from government and private-sector reports include

> *Scientific and Technical Aerospace Reports*
> *Government Reports Announcement and Index*
> *Monthly Catalog of United States Government Publications*

Patent indexes include

> *Index of Patents Issued from the United States Patent and Trademark Office*
> *NASA Patents Index*
> *World Patents Index*

Further, specialized indexes and abstracts for general and technical subjects may be included in the computer catalog or held by the library in print. Some of the most popular indexes are

> *Essay and General Literature Index*
> *Abstracts on Criminology and Penology*
> *Abstracts on Health Care*
> *Air Pollution Abstracts*
> *Criminal Justice Abstracts*
> *Engineering Abstracts*
> *General Science Index*
> *International Aerospace Abstracts*

Don't overlook these special indexes and abstracts as possible sources of information vital to your research.

Microforms, CD-ROMs, and Specialized Databases

If the periodical or specific articles you seek are not on the computer or in print form, the material may be on a microform, a CD-ROM, or a specialized online database. Microform technology reproduces printed information and stores it on microform reels or flat packets of microfiche. These materials are read on special machines that magnify the reduced image. They include newspapers, government documents, technical reports, business directories, and worldwide documents in translation. If you are not familiar with microforms, ask your librarian for assistance.

Index CD-ROMs are electronic bibliographies of thousands of article titles, abstract services, and reference texts that can be accessed by subject words or keywords. There are two general types of CD-ROM indexes: index only and full-text. The computer catalog will list the CD-ROMs that your library holds. Following are a few of the thousands of indexes on CD-ROM:

Applied Science and Technology

Government Publications Index

Infotrac (monthly listings of articles from more than 900 business, technical, and general magazines and journals)

Investext (index of company, industry, and market analysis with more than one million research reports from more than 500 investment firms, brokerage houses, market research organizations, trade associations, and worldwide companies)

LegalTrac (monthly listing of entries from 750 legal publications)

Population Statistics

Ulrich's International Guide to Periodicals

In addition to the catalog databases, most libraries subscribe to specialized independent online databases, which can access indexes, journals, books, dissertations, and reports. Subscription databases are usually more specialized than CD-ROMs, and are updated daily, as opposed to CD-ROMs, which are updated weekly, monthly, or quarterly. These subscription databases may be accessed free or require your library card number for access; others will require individual paid subscriptions. Databases may include bibliographies for specific field publications, abstracts, or full-text articles. Still others provide specialized facts on up-to-the-minute stock quotations, weather, new patents, major company credit ratings, and more. Some specialized databases and their offerings are

ABI/Inform (bibliographies, abstracts, and some full-text articles from more than 800 business, marketing, and management journals)

DIALOG Information Services (the most comprehensive database of all other databases, DIALOG Information Services accesses articles, conference proceedings, news, and statistics from millions of documents in scientific, technical, medical, business, trade, and academic study fields)

BRS (Bibliographic Retrieval Services provides bibliographies and abstracts from life sciences, physical sciences, business, government, and social sciences)

CINAHL (the Cumulative Index to Nursing and Allied Health literature)

ERIC (Education Resources Information Center includes articles on education and related topics)

MEDLINE (articles published in 3,800 biomedical journals)

NEWSBANK (a comprehensive database of national and international newspapers)

LEXIS / NEXIS (bibliographies of law, business, and government publications plus a career center)

Vertical Files

Many libraries maintain a vertical file on timely subjects covered in pamphlets, booklets, bulletins, and clippings that would not otherwise be cataloged. Ask the reference librarian to assist you in locating the file and familiarize yourself with the materials contained within it.

Traditional Card Catalogs

In those libraries that still use traditional card catalogs (an alphabetized compilation of cards) instead of the PAC computerized system for reference and general books, you may locate sources on cards under three separate headings: the author's name, the title of the work, and the subject heading. Thus, a book by Donn B. Parker titled *Fighting Computer Crime* will be cataloged under *P* for Parker, Donn B., under *F* for *Fighting Computer Crime,* and under *C* for Computer Crimes—Prevention. Figure 5.5 shows a typical author card with explanations of the data and Figure 5.6 is the title card for the same book.

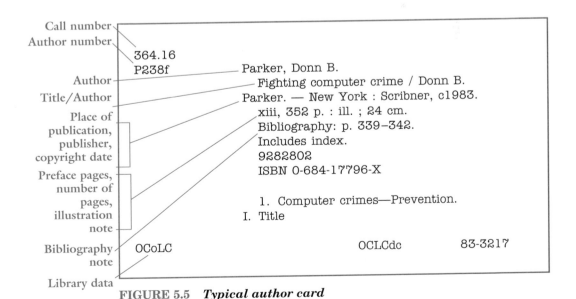

FIGURE 5.5 *Typical author card*

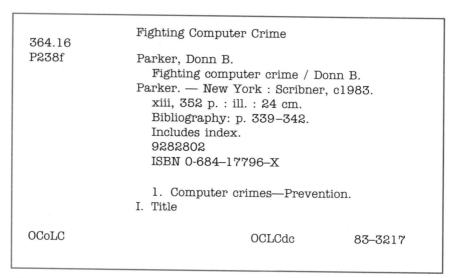

FIGURE 5.6   *Typical title card*

If you do not know the authors or titles of books on your subject, you may locate books by looking up *subject headings*. The subject heading "Computers" will be further divided into subcategories, such as "Computer Crimes—Prevention" and "Computers—Access Control." Figure 5.7 shows the subject card for the same book.

FIGURE 5.7   *Typical subject card*

COMPUTER RESEARCH

You may conduct online research on the library computer system or on your personal computer. Library systems include the free or subscription Web site database indexes to which you may subscribe individually; they tend to be expensive but often well worth it if the database indexes materials in your major field. On the Internet you may access subjects by keywords, specific Internet addresses, or hyperlinks (boldfaced or underlined text, colored words or phrases, pictures, and icons) that when clicked link to other pages in the same Web site or even to other Web sites. Materials that you locate online may be printed or downloaded directly to your computer files.

Search Engines

Search engines are services that index, organize, and often rate and review Web sites that target news, technology, health, science, law, stock market, businesses, and chat groups, to name just a few of the categories. No search engine indexes all Web sites and Web pages, so if your first search isn't productive, try at least one other search engine. If you want to find something specific in a hurry, you might try several search engines, for instance, Yahoo, AltaVista, and Google. Some services rely on people to do the indexing, others use software to identify key information on sites across the Internet, and still others combine both types of service. Those who use people to evaluate sites may lead you to more reliable information. The engines have home pages, such as <http://www.altavista.com>, which can be accessed through browsers such as Netscape Navigator at <http://home.netscape.com> or Microsoft Internet Explorer at <http://www.msn.com>. Some word processing programs offer a search feature that conveniently links to your default browser. Familiarize yourself with a few of the following search engines to determine their categories and which will serve you best. Ascertain what is indexed (whole pages, selected portions, keywords), whether the search techniques allow Boolean AND or OR phrase searches, the speed, whether it retrieves too many or too few titles, and how often it finds what you most need.

Altavista (a huge engine indexing 150 million Web pages, including those in foreign languages from 10 nations and Usenet news.)

Yahoo (another huge global engine using both people and software to list and evaluate the sites.)

Google (this engine searches over1.3 billion Web pages and produces its results in the order of popularity as cyclically determined by the number of links from other popular sites. It includes sites from 10 foreign-language countries.)

HOTBOT (indexes 110 million Web Pages, which can be searched by date, media type, language, or location.)

Excite (uses software to catalog Web sites, but uses people to review a directory—a subsection of its catalogued sites.)

Infoseek (also uses software and people to review Web sites and provides a special directory of sites specified by word order and word, with an emphasis on news.)

Lycos (similar to Excite and Infoseek. It indexes 50 million Web sites. Uses software and people.)

There are many more to investigate.

Source Reliability

CAUTION!! Not all sources are reputable!

It is up to you to evaluate the reliability of your information sources. Libraries contain books, periodicals, and other resources that have generally been well-evaluated and include such information as the author's background, the publisher, and bibliographic material. The Internet, by contrast, has millions of sites where anyone can post just about anything, so you must use extra care when gathering information from the Internet. Consider the following criteria in evaluating all of your information sources.

1. **Authorship.**   Are you familiar with the author because he or she is well-known and well-regarded in your field of study? Has the author been mentioned in a positive manner by another author, professor, or anyone else you trust as an authority? Does the document you are reading give biographical information, including the author's position, institutional affiliation, and address? Can you e-mail the author to obtain professional background information?

2. **Publishing body.**   Is the name of any organization given on the document you are reading and are there indications that it is part

of an official academic or scholarly Web site? Is the organization recognized in the field you are studying and suitable to address the topic? What is the relationship between the author and the publisher or Web site sponsor? That is, was the document prepared as a part of the author's professional duties, or is it of a casual or for-a-fee nature? Realize that companies with services and/or products are using their sites for public relations and marketing; even annual reports are slanted to present the company in the best possible light. Does this Web site reside in an individual's personal Internet account, rather than being part of an official Web site? Personal pages must be approached with the greatest caution.

3. **Point of view or bias.**   Because data is used in selective ways to form information, it generally represents a particular point of view. Is the site scholarly or popular? Could the author have a commercial or sociopolitical bias? Even some respectable sources slant their material. Consider that if you are reading about a political figure at the site of another political party, you are reading the opposition. Consider also that if you are looking at product information, you are probably looking at an advertisement. If you are looking for scientific information on abortion or capital punishment, would you trust a political organization to provide unbiased material? The Internet allows for "vanity" publishing, that is, independent, self-promoting information not subject to any authoritative review.

4. **Referral to and/or knowledge of the literature.**   What is the context in which the author situates his or her work? Can you evaluate the scholarship of the author or do you have knowledge of the trends in an area under a discussion? Is there a bibliography? Are there allusions to or displays of knowledge of related sources with proper attribution (naming of a source)? Does the author display knowledge of related theories, schools of thought, or techniques? Does the author discuss the limitations as well as the value of a new theory or technique? If the author's treatment of the subject is controversial, does he or she know and acknowledge this?

5. **Accuracy and verification of details.**   If you are reading material by an unfamiliar author presented by an unfamiliar organization or presented in a nontraditional way, try to verify the accuracy of facts. Is there an explanation of the research method used to gather and interpret the data? Is the methodology appropriate to the topic and does it allow duplication for purposes of verification? Does the document rely on other sources listed in a bibliography or include links to the other reliable documents? Does the document name individuals and/or sources that cannot be accessed? Can the background information be verified for accuracy? Are the graphics merely decorative or are they factual and ethical?

6. Currency.   Is timeliness of the information important? Does the document include the date(s) at which the information was gathered and/or modified? If no date is given, can you view the directory in which it resides to read the date of latest modifications? Does the document refer to clearly dated information (e.g., "Based on the 2000 Census data")? Does the document include a date of copyright?

Never use information that you can't verify. Be skeptical and become a critical consumer of information in all of its forms.

Other Personal Sources

Review the list of online reference books presented earlier in this chapter. These are readily accessible on your personal computer. You may own other reference materials on CD-ROM. For example, all 43 volumes of the *Encyclopedia Britannica* are on one CD-ROM, which scans a 44 million-word database, indexes more than 400,000 references on more than 65,000 subjects with hypertext links to the actual text articles, provides more than 10,000 links to images and tables, and includes the *Merriam-Webster Collegiate Dictionary, Tenth Edition.* Other CD-ROMs for your personal library include the *New York Public Library Desktop Reference*, the *Grolier Interactive Multimedia Encyclopedia,* and *Microsoft Encarta.*

Articles from your own local newspaper and from major U.S. newspapers are accessible on the Internet. Some other full-text periodicals (newspapers and magazines) not previously mentioned follow. These sites may usually be located through the Web by typing <http:// www.nameofperiodical.com>.

Air Force Times
Army Times
Business Week
Congressional Quarterly
Health Magazine
Nature Conservatory
Navy Times
Newsweek
Scientific American
Smithsonian Magazine

Don't hesitate to search the Internet for other publications; most already have Web sites, and new sites are established every day. You may also access newspapers from Africa, Asia, Australia, Canada, Central America, South America, Europe, and elsewhere.

You may subscribe to electronic "bulletin boards" (interactive specific-subject news groups) on **USENET** or **Listserv.** Your subscription allows you to post questions on a specific subject and retrieve responses from all over the world. Free specific-subject mailing lists are also available for interactive discussions. There are thousands of subjects on which you may post questions and then access e-mail responses. Use keywords to access these services and to subscribe to them.

Informative Web Sites

Although Web addresses are always subject to change or deletion, and others are added constantly, some currently helpful addresses for student researches in the sciences, technology, medicine, and government follow:

<http://www.refdesk.com>. My Virtual Reference Desk site is the bible for hyperlink access to text (uses Google), facts, and graphics on weekly U.S. and worldwide wire services, news services (print, radio, and television), special reference sources (astronomy, finance, politics, biographies, law, science, government, etc.), facts reference sources, and weekly updates on encyclopedia entries, plus access to *Encyclopedia Britannica, World Fact Book, Bartlett's Quotations, Roget's Thesaurus,* and more.

<http://www.elibrary.com>. The Electric Library site is a moderately priced subscription database that is undoubtedly on your library computer system. However, you may subscribe as an individual. Designed for student research, it includes more than 900 full-text periodicals, 9 international news wires, 2,000 classic books, hundreds of maps, thousands of photographs, such as Archive Photos, Reuters, Simon & Schuster, Gannett, *Times, Mirror, U.S. News & World Report,* National Review Publisher, Inc., *New Republic, Psychology Today*, and hundreds more.

<http://www.loc.gov>. The Library of Congress site has resources for research and information professionals, including the catalogs of the Library of Congress and other libraries, plus databases on special topics.

<http://access.gpo.gov>. The United States Government Printing Office site offers links to official documents covering all areas of government.

<http://www.whitehouse.gov>. The White House site has information on the executive branch, cabinet posts, and independent federal agencies including the president's speeches, publicly released documents, executive orders, copies of the budget, and other major publications.

<http://www.nasa.gov>. The National Aeronautic and Space Administration site features text, audio, and visual access to its projects.

<http://www.space.com>. Worldwide news releases on space agencies, shuttles, space stations, space science and astronomy, searches for life on other planets, and more are available on this site.

<http://epa.gov>. The U.S. Environmental Protection Agency Web site has the latest news on acid rain, global warming, cleanups, ecosystems, emergencies, health risks, pesticides, pollutants, hazardous wastes, water, and more.

<http://www.doe.gov>. The U.S. Department of Energy offers information about the newest scientific breakthroughs, technological advances, environmental updates, nuclear security, and more.

<http://www.nsf.gov>. The National Science Foundation site provides research information and press releases on a wide variety of program areas, including biology, computer and information sciences, education, engineering, geosciences, math and physical sciences, polar research, and social and behavioral sciences.

<http://www.ornl.gov>. The Oak Ridge National Laboratory reports on science and technology (the Human Genome Project, robotics, neutron sciences, and more), current energy and matter investigations, newsworthy scientific breakthroughs, and more.

<http://www.nsidc.org>. The National Snow and Ice Data Center reports on polar and cryospheric research, glaciers, polar and atmospheric ice, and more.

<http://www.noaa.gov>. The National Oceanic and Atmospheric Administration site provides information on seasonal and annual climate forecasts, sustaining healthy coasts and fisheries, protected species information, navigation studies, and more.

<http://user88.lbl.gov/NSD_docs/abc/home.html>. This nuclear science site offers information on radioactivity, fission and fusion, atomic structure, half-life studies, cosmic rays, and experiments, and a link to an interactive question/answer site.

<http://www.who.int>. The World Health Organization site contains press releases on worldwide health statistics, hundreds of fact sheets, audio announcement, photos, and videos.

<http://www.medsite.com>. This comprehensive site allows keyword searches for medical topics and diseases and includes a rating system for the linked sites.

You may also link to most colleges and universities in the United States. These sites provide directories for their special scientific and technical departments and schools; they also list research activities and including library links. Usually you need only type **<http://www. college name.edu>.** Try Harvard, Yale, Princeton, Rutgers, Stanford, Massachusetts Institute of Technology, Illinois Institute of Technology, Rensselaer, and so on. William and Mary's address is **<http://www/ wm.edu>.** Johns Hopkins University is **<http://www.jhu.edu>.** You may even connect to the great universities of Great Britain; Cambridge University is **<http://www.cam.ac.uk>,** Oxford University is **<http:// www.ox.ac.uk>,** and the University of Edinburgh is **<http://www. ed.ac.uk>.**

BIBLIOGRAPHY CARDS OR FILES

Each time you locate relevant source material, create a separate bibliography card or add to your computer file of sources. Each entry should contain

1. Author's name
2. Title of the work
3. Publication information
4. Library call number, database or Internet identification code, or other identifier
5. An annotation about the contents of the source (optional, but strongly recommended)

Later you will transform this information into the documentation style suitable to your subject matter. For instance, subjects in the social sciences, biological and earth sciences, education, linguistics, and business require the APA style; subjects about language and literature require the MLA style; subjects in applied sciences (chemistry, computer science engineering, mathematics, and physics) and the medical sciences (biomedicine, health, medicine, and nursing) require the Number System, and so on. Chapter 6 covers the particulars for five documentation styles. Each style has separate forms for books, periodicals, Internet documents, advertisements, art works, broadcast interviews, bulletins, computer software, conference proceedings, dissertations, film or video recordings, personal interviews, letters, microforms, public address or lectures, and recordings on record/tape/disk.

Figures 5.8 and 5.9 show two typical hand-written bibliography cards. You may, of course, key this same information into a computer bibliography file.

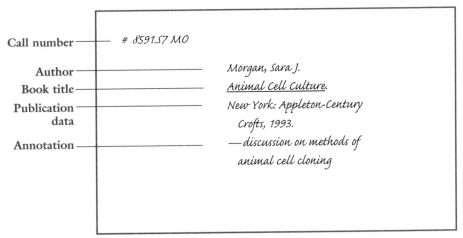

FIGURE 5.8   *Annotated bibliography card for a book*

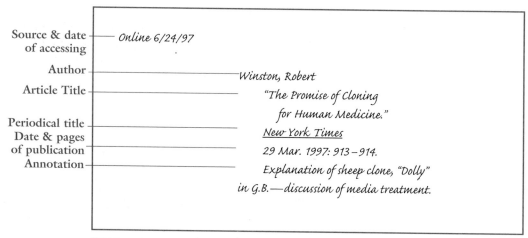

Source & date of accessing — *Online 6/24/97*

Author

Article Title

Periodical title
Date & pages
of publication
Annotation

Winston, Robert
"*The Promise of Cloning*
for Human Medicine."
<u>*New York Times*</u>
29 Mar. 1997: 913–914.
Explanation of sheep clone, "Dolly"
in G.B.—discussion of media treatment.

FIGURE 5.9   *Annotated bibliography card for a periodical located on the Internet*

You will use these cards or file entries to relocate material for note taking and for compiling into the appropriate format for a complete bibliography, works cited, or reference listing at the end of your paper.

NOTE TAKING

With your research purpose firmly in mind and with an outline of your intended paper as your guide, begin now to record the notes from each of your sources. Notes are taken on 3″ × 5″ or 4″ × 6″ index cards or entered into your computer files. If you key your notes into a computer, write each note as a separate temporary file under a common directory so that you may move the notes later as appropriate. Alternatively, write all your notes into a single file, using code words or phrases that can be located by a search command as you write the actual paper. Save all of your note cards or printed copies of your computer files to submit with your final paper, should your professor require them.

Plagiarism

Notes must be recorded carefully to avoid plagiarism. The word *plagiarism* derives from the Latin word *plagiarus,* which means "kidnapper." In effect, plagiarism in writing is kidnapping the words (apt phrases or entire sentences) and ideas (the organization of material, another's argument, or a line of thinking) and then presenting them as if they were your own original thinking, organization, or phrasing. The published words, ideas, and conclusions of an author are protected by law; therefore, the use of the material of others must be acknowledged whenever it appears in your work. Failure to provide acknowledgment may result in a failed paper, failure in the course, expulsion from school, and lasting impairment of your scholarly reputation. To avoid plagiarism, follow five rules:

1. Introduce all borrowed material by stating in the text the name of the authority from whom it was taken.
2. Enclose any exact wording within quotation marks.
3. If you summarize or paraphrase material, be sure that the information is written in your own style and language.
4. Provide a parenthetical reference note for each borrowed item.
5. Provide in your bibliography, works cited, or reference list an entry for every source that is referenced in your text (see Chapter 6).

Following is an original excerpt from the U.S. Global Change Research Program, followed by three note cards. The first two (Figures 5.10 and 5.11) are unacceptable plagiarisms, while the last (Figure 5.12) is an acceptable paraphrased note.

Original

A warming of the global climate is likely to have substantial consequences—for better or worse—around the United States in coming decades, including bumper crops in the heartland, chronic erosion of coasts, summer water shortages, winter floods in the West, and a future New York City that steams in summer like present-day Atlanta. Some scientists are concluding that a buildup in the atmosphere of heat-trapping gases like carbon dioxide has contributed to a warming seen in the last 100 years, and they say the trend is likely to carry on throughout the century.

Revkin, Times   Impact of
USGCRP   Global Warming

The substantial consequences of the global warming around the United States in coming decades will include bumper crops in the heartland, chronic erosions of coasts, summer water shortages, summer floods in the West, and a future New York City that steams in summer like present-day Atlanta. Scientists are concluding that a buildup of carbon dioxide and other heat-trapping gases contribute to the warming seen in the past 100 years.

p. 1

FIGURE 5.10 *Plagiarized note*

Revkin, *Times*   *Impact of*
USGCRP   *Global Warming*

> *Global warming is likely to have substantial consequences in the United States in the future. Bumper crops in the heartland, chronic erosions of coasts, summer water shortages, winter floods in the West, and steamy summer weather in the Northeast will continue in the coming decades. Scientists conclude that heat-trapping gas buildups in the atmosphere for 100 years is the cause.*
>
>   *p. 1*

FIGURE 5.11   *Less obvious, but plagiarized note*

Revkin, *Times*   *Impact of*
USGCRP   *Global Warming*

> *According to the U. S. Global Change Research Program, the greenhouse gas buildup in our atmosphere for over 100 years is going to continue to heat up our global climate. In the next decades, the U. S. will experience abundant crops in the Midwest, chronic coastal erosions, summer water shortages, winter floods along our Western coasts, and "a New York City that steams like present-day Atlanta."*
>
>   *p. 1*

FIGURE 5.12   *Acceptable note employing quotation marks for exact wording of the original*

In the acceptable note: (1) the information is summarized, (2) the language is the note taker's, (3) the author's exact phrases or sentences are in quotation marks, and (4) the source is acknowledged in the text. In addition, a parenthetical reference to the source and a complete "Works Cited" entry will appear in the final draft of the paper.

Note-Taking Guidelines

Use a consistent system to record your information:

1. Enter only one item of information on each card or under each computer file notation so that you may continuously organize and

rearrange the cards or computer notes during all stages of the research.

2. If you use cards, write on only one side of each. A note continued on the back may be overlooked. If an item of information is too lengthy for one card, use two or more and staple them together.

3. Record the topic notation or code words or phrases at the top left or top right of the card or computer note. This topic notation or code may correspond with an entry in your outline.

4. Record the source at the top right or left of the card or file. This notation may be an abbreviated form of the bibliography card, such as the author's last name, the book or article title, or both.

5. Write or type the note in the center of the card or computer list. Refer to the types of notes and their explanations in the next section of this textbook.

6. Record the exact page number(s) or online computer material identifier(s) from which you took the note.

Types of Notes

There are specific strategies for taking notes:

1. **Quotation notes.** Copy the exact words and syntax of an unusually well-expressed opinion from an authority to support your own thesis or topic sentences.

2. **Précis notes.** Copy a review note, abstract, or bibliographic annotations for later development.

3. **Summary notes.** Record factual data or opinions to support your own ideas.

4. **Paraphrase notes.** Use your own words entirely to interpret and express the opinions and explanations of an authority. It is likely that most of your notes will be paraphrase notes.

5. **Personal notes.** Record your own ideas for further development or inclusion as you research your materials.

Sample Note Cards

Most of your notes will consist of summaries or paraphrases, that is, condensations or restatements of the source material in your own words. Following is an excerpt as it appears in the original source material.

Original

Temperatures nationwide rose about 1 degree Fahrenheit in the twentieth century, and it is estimated that by using a "business as usual" assumption in which carbon dioxide levels continue to grow at the rate seen in recent years, global warming will accelerate. At this pace, the average temperature in the country would rise 5 to 10 degrees in the twenty-first century.

Following are three note cards (Figures 5.13, 5.14, and 5.15); the first summarizes and paraphrases the original, the second employs full sentence direct quotation, and the third demonstrates partial quotation. Note the use of the ellipsis (. . .) to indicate omitted words and the brackets ([ ]) to indicate a slight addition by the note taker.

Revkin, *Times*   Temperature
USGCRP   increases

Carbon dioxide levels forced temperatures to rise about 1 degree Fahrenheit in the past 10 years. If the carbon dioxide levels continue to accelerate at the rate measured in recent years, global warming will raise the average temperature 5 to 10 degrees in the next 100 years.

p. 1

FIGURE 5.13   *Summary and paraphrased note*

Revkin, *Times*   Temperature
USGCRP   increases

The USGCRP reports, "Temperatures nationwide rose about 1 degree in the 20th century, and it is estimated that by using a 'business as usual' assumption in which carbon dioxide levels continue to grow at the rate seen in recent years, global warming will accelerate."

p. 1

FIGURE 5.14   *Full sentence quotation note*

Revkin, Times   *Temperature*
USGCRP   *increases*

The USGCRP estimates "[If] carbon dioxide levels
continue to grow at the rate seen in recent years, global
warming will accelerate, . . . [and] the average
temperature . . . would rise 5 to 10 degrees . . . ." in the
next 100 years

p. 1

FIGURE 5.15   *Partial quotation note*

SURVEYS AND QUESTIONNAIRES

Surveys and questionnaires are often used to gather data to guide the writer and to back up conclusions. Essentially, these primary sources serve three purposes:

- Gathering opinions of many people
- Determining solutions to a problem
- Substantiating predetermined positions.

Care must be taken in selecting the test group, motivating a response, constructing the questions, and compiling the results. The survey sheet itself is referred to as the **survey instrument** or the **questionnaire.** Figure 5.16 shows a sample survey instrument.

Advantages

Written surveys have several advantages:

1. Carefully formatted questions require only a check-mark in a box to speed up the survey response process.
2. A large sample of data may be collected in a short time.

AAA's Traffic Safety Survey

Since its founding in 1902, traffic safety has been one of the building blocks of AAA's foundation. Proactive AAA traffic safety programs start in elementary school and continue through mature vehicle operator education.

The School Safety Patrol program supports over 48,000 children in the Florida, Louisiana, and Mississippi public and private school systems. AAA also has proactive alcohol programs that start in kindergarten, along with supplemental printed materials that can be used for vehicle operators of all ages.

Today traffic safety programs include pedestrian, bicycle, school bus, and a variety of driver safety materials. AAA investigates member concerns on local enforcement and traffic engineering issues as well as serves community organizations dealing with transportation and traffic safety issues.

This survey will help us to understand better the traffic safety concerns of AAA members, both now and for the future. Please answer these questions as accurately as possible with the information provided and mail or fax us your answers. Please choose or write in only one answer per question unless otherwise instructed.

1. Should AAA offer a driver improvement refresher program for those over 55?
_____yes
_____no
_____undecided

2. If yes, what would you be willing to pay for this type of course?
_____$10
_____$15
_____$20
_____Other (please specify in $5 increments) $_____

3. Should AAA become involved in a vehicle theft deterrent program, such as Combat Auto Theft?
_____yes
_____no
_____undecided

4. If yes, would you like to see AAA (check all that apply):
_____Provide a brochure on vehicle theft deterrence?
_____Provide vehicle identification number (VIN) etching?
_____Provide decals for a theft deterrent program?
_____Other (please specify)
_____

5. Please indicate the programs you would like to see AAA continue (check all that apply):
_____ Pedestrian safety
_____ Bicycle safety
_____ Alcohol education

_____ Safety Patrol
_____ School bus safety
_____ Safe driving for mature operators
_____ Enforcement investigations
_____ Traffic engineering investigations

6. Should AAA observe major highway construction projects to make sure they adhere to federally established guidelines for motorist safety?
_____yes
_____no
_____undecided

7. Would you like AAA to provide effective reflective materials that can be used by pedestrians, bicyclists, and others to increase their visibility to motor vehicle operators?
_____yes
_____no
_____undecided

8. Would you like local-issue traffic safety articles and/or a column in each issue of *Car & Travel*?
_____yes
_____no
_____undecided

9. Would you like traffic safety information offered on the Member Services page of *Car & Travel* (see page 34)?
_____yes
_____no
_____undecided

10. Sex
_____Male
_____Female

11. Age
_____18–34
_____35–54
_____55–64
_____65 or above

12. Marital status
_____Married
_____Single

13. Annual household income
_____$25,000 or below
_____$25,001–$40,000
_____$40,001–$55,000
_____$55,002–$75,000
_____$75,001 or above

Thank you for your participation in this survey. Please return your completed survey by **Jan. 31** to *Car & Travel*, 1000 AAA Drive, MS73, Heathrow, FL 32746-5080; or fax it to us at 407/444/4140. As always, the information provided in this survey is confidential, but if you would like to be entered in a drawing to win a *Car & Travel* golf shirt, please provide your name, address, and size.

Name_____

Address_____

City/state/Zip_____

Golf shirt size (circle one)

Medium   Large   XLarge

FIGURE 5.16 *Sample survey instrument* (By permission AAA Florida *Car & Travel* magazine [January/February 1997]: 17. Adaptation.)

3. Even distribution (employee mailboxes, targeted mailings) ensures that all relevant people are actually contacted within a given time frame.

4. Printing and distribution are less costly and more efficient than person-to-person inquiries and interviews.

5. Written responses are usually less biased than those garnered by personal interviews.

6. The data can easily be tabulated by counting or computer key-punching.

7. The data are on hand to point out other problems and avenues for future study.

Disadvantages

A survey is not always advantageous. Certain disadvantages or drawbacks may result from the use of written surveys or questionnaires:

1. A *low* response rate (less than 10 percent of those both affected by the study and actually surveyed) will result in unreliable data.

2. A *slow* response rate slows up your findings.

3. Certain members of your group may not respond to written surveys because they are uninformed, disinterested, careless, lazy, prejudiced, or offended.

4. Others may not respond accurately due to inability to read or to understand the precise meaning of your terms.

5. Some people may be reluctant to respond to questions regarding personal data or to offer opinions for fear of repercussions.

Test Group

Selecting the size of the test group and determining the number of responses required to assure validity are problems for the true statistician. For the purposes of accessing information for your reports, try to survey all of the people who will probably be affected by your proposal for change, such as all of the employees in an office or department, all of the users of an inadequate parking lot, all of the residents of your apartment or condominium complex, or all of the neighbors touched by a city project. Remember, you need a 10 percent response rate of those affected and surveyed for validity. Consider whether you wish to collect numbers or percentages, or both, as well as comments, suggestions, and the like.

Familiarize yourself with your survey group. Are they knowledgeable about your subject? Are you looking for new data, substantiation of a position, or opinions and suggestions to determine direction? What prejudices might influence answers (age, sex, race, organization rank, religion, politics, length of involvement, etc.)? Should your tone be formal or folksy?

Motivating Responses

You need a clearly defined goal first. What are you trying to determine? Do your own research of the problems and develop possible solutions before developing your survey. Consider the following motivating factors.

Time.   Do not begin to write your proposal or other type of report until you have your survey results. Consider a schedule for devising, typing, printing, distributing, collecting, and tabulating your data. Distribute forms at an opportune time, not when people are on vacation or occupied with IRS returns, Christmas, and the like. Sometimes distribution at a large meeting of the affected people will guarantee a quick, high return on the spot. Preference and opinion questions do not require research on the part of the respondents. Urge your group to return the form(s) to you quickly by hand, by mailing a preaddressed envelope, or by delivering the forms to a designated spot at a set date and time.

Sponsorship.   Make it clear that you are seeking the information under the auspices of an institution, organization, professor, manager, or the like. Gathering data for a college professional and technical writing course project should lend prestige to your inquiry.

Cover Sheet.   A cover letter, memo, information sheet, or simply an explanatory paragraph above the survey questions may accompany the survey instrument, stating the purpose, nature of the report, and how the data will be analyzed. This is a logical place to state the sponsorship of the survey. Emphasize the importance of the survey and offer to share results of the study if that is feasible. Above all, stress that you need a return as soon as possible.

Anonymity.   Allow for anonymity by guaranteeing confidentiality. Do not ask for names on the forms, particularly if the material covers touchy subjects such as opinions about supervisors or administrators, work schedules, and raises. Ensure anonymity also for questionnaires about highly personal feelings on such issues as abortion, capital punishment, personal habits, and so on.

Format

The overall appearance and arrangement of the instrument should motivate responses. Follow these suggestions:

1.   Select a good quality, possibly colored paper, which suggests care in preparation.

2. Title the survey carefully to clarify the content.

3. Keep your questions as brief as possible.

4. Provide tick boxes for responses and printed lines for comments. Type in lines rather than just leaving white space for the answers.

5. Do not crowd your questions. Allow for plenty of white space.

6. Use language carefully. Such words as *should, might*, and *could* often "lead" answers. Do not ask offensive or very personal questions if possible. (Avoid: *Should employees be blood-tested for AIDS?* [   ] yes [   ] no)

7. Ask for only one piece of information per question. (Avoid: *Do you favor a personnel reorganization of your department or would you be interested in a new position?* [   ] yes [   ] no)

8. Ask for comments, opinions, or other alternatives when it is appropriate to do so.

Types of Questions

There are six conventional types of questions. Each has its own advantages and disadvantages in tabulation.

1. **Dual alternatives.**   This is the easiest type of data to tabulate because there are only two choices: yes or no, positive or negative, true or false. Take care to ensure that there really are only two possible answers to your question.

   Sample: Have you eaten purchased cookies in the past month?

   [   ] yes   [   ] no

2. **Multiple choice questions.**   This type also produces easy-to-tabulate data in that you provide a number of alternatives to indicate a fact, preference, or opinion. Your survey may ask for a single tick or multiple ticks.

   Sample: Check the **one** cookie brand you most prefer.

   [   ] Hydrox   [   ] Fig Newtons
   [   ] Oreo   [   ] Pepperidge Farm
   [   ] Duncan Hines   [   ] Almost Home

   or

   Check **all** of the cookie brands you purchase on a regular basis.

   [   ] Hydrox   [   ] Fig Newtons
   [   ] Oreo   [   ] Pepperidge Farm
   [   ] Duncan Hines   [   ] Almost Home

3. **Rank ordering.**   This format provides respondents with a series of items to rank according to preference, frequency of use, or other criteria. Tabulating this data is easy. However, distortion may occur if the mean (the average) differs from the mode (the most frequently occurring response). The items must be significantly different to make rank choices.

   Sample: Rank (1 highest to 6 lowest) the following cookie brands in order of your preference.

   _____ Hydrox

   _____ Oreo

   _____ Duncan Hines

   _____ Fig Newtons

   _____ Pepperidge Farm

   _____ Almost Home

4. **Continuum scales.**   This format is similar to rank ordering in that it provides a method for respondents to express opinions by rank ordering numerically or verbally on a continuum. Scales are most valid if they have an *even* number of choices (usually 4–6). If a scale has a middle choice, respondents choose it a disproportionately large percentage of the time.

   Sample: Mark the response which best indicates the frequency of your cookie purchases per grocery trip.

   [ ] always   [ ] often   [ ] seldom   [ ] never

5. **Completions.**   This format asks respondents to provide facts and/or opinions in either fill-in or open-ended responses. Questions on age, frequency, or amount are easier to tabulate than are questions that allow for open-ended responses.

   Samples: (Fill-in type) I usually buy _____ cookie products a month.
                                            number

   or

   (Completion type) The quality which most determines my cookie product choice is

   _____

6. **Essays.**   This format asks respondents to fully express opinions or facts. The results may be difficult to tabulate and arrange into groupings. Such questions are more effective if they urge a focus on certain criteria.

   Sample: Suggest criteria which would motivate you to purchase Fig Newtons more frequently. Consider butter content, sugar content, thickness, consistency, and amount of filling.

   _____

   _____

Data compiled from a survey may be presented within a paper in tables, bar charts, circle charts, line graphs, and the like. Figure 5.17 shows a sample survey instrument to determine the popularity of a particular cookie, who buys it, and how the product should be improved. It asks for respondent classification data and employs a variety of types of questions.

PRODUCT PREFERENCE SURVEY

———————————— Classification ————————————

<table>
<tr><td>(multiple
choice)</td><td>1. Please indicate your age bracket.
[ ] 16–20   [ ] 21–30   [ ] 31–40   [ ] 41 or over</td></tr>
<tr><td>(dual
alternative)</td><td>2. Are you [ ] male   [ ] female?</td></tr>
<tr><td>(multiple
choice)</td><td>3. What is your primary occupation?
[ ] student   [ ] housewife
[ ] professional manager [ ] farmer
[ ] operative/laborer   [ ] retired
[ ] foreman/craftsman   [ ] not employed
[ ] clerical/sales</td></tr>
<tr><td>(completion)</td><td>4. Fill in the blank with a number. I generally shop
for groceries_____times a month.
                number</td></tr>
</table>

———————————— Cookie Questions ————————————

<table>
<tr><td>(dual
alternative)</td><td>1. I am the primary cookie shopper in my household.
[ ] yes   [ ] no</td></tr>
<tr><td>(continuum)</td><td>2. In the span of a year I purchase cookie products
[ ] often   [ ] frequently [ ] seldom   [ ] never</td></tr>
<tr><td>(multiple
choice)</td><td>3. Check all of the brands of cookies that you purchase
on a regular (once or more a month) basis.
[ ] Hydrox   [ ] Fig Newtons
[ ] Oreo   [ ] Pepperidge Farm
[ ] Duncan Hines   [ ] Almost Home</td></tr>
</table>

FIGURE 5.17   *Sample survey/questionnaire*

(rank order)

4. Rank your order of preference (1–6) for each brand.
   [ ] Hydrox   [ ] Fig Newtons
   [ ] Oreo   [ ] Pepperidge Farm
   [ ] Duncan Hines   [ ] Almost Home

(completion)

5. Fill in the number. I usually buy _____ cookie products a month.   number

(multiple choice)

6. Which factor most determines your cookie brand choice?
   [ ] packaging   [ ] overall taste
   [ ] calorie content   [ ] cost
   [ ] additives (nuts,   [ ] consistency (chewy,
        raisins, fillings)   crisp, crunchy)

(essay)

7. Please comment on improvements that would motivate you to purchase Fig Newtons more often.

   _____

   _____

   _____

   _____

FIGURE 5.17 *continued*

Figure 5.18 suggests ways that such data may be incorporated into the text of your report.

Female respondents indicate a lower incidence of cookie purchases than do males. Figure 1 shows the frequency of cookie purchases by sex:

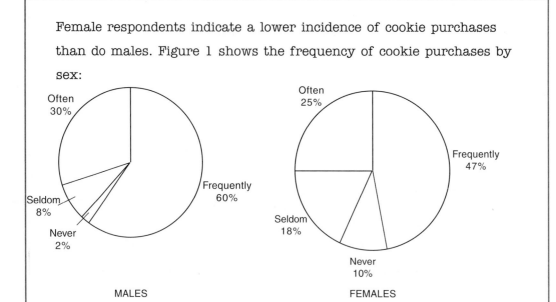

Figure 1   Frequency of cookie purchases per month by sex

Seventy-two percent of female respondents purchase cookies often or frequently, whereas a full 90 percent of male respondents purchase cookies often or frequently. The mean number of cookie sales per month is two for all occupational categories except for students, who indicate a mean number of six purchases per month. Table 1 shows the number of purchases a month by occupational categories:

Table 1   Mean number of purchases per month—600 respondents

Category	Number of purchases
Professional/manager	4
Operative/laborer	2
Foreman/craftsman	2
Clerical/sales	2

FIGURE 5.18   *Samples of data incorporation into text, using graphics*

Housewives	3
Farmers	1
Retirees	2
Not employed	0
Students	6

The 600 respondents indicate that taste is the major factor in determining cookie brand selection. Figure 2 shows the percent of respondents who indicate other determining factors:

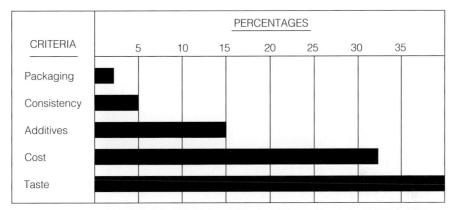

Figure 2   Determining purchase factors—600 respondents

As shown, a full third of the respondents indicated that cost is the major factor. Packaging has the least influence on cookie purchasing.

FIGURE 5.18   *continued*

OTHER PRIMARY SOURCES

Lectures, public forums, television and radio programs, and direct observation may also provide potent material for your purposes. Inquiry letters and their responses (see Chapter 12), interviews (see Chapter 13), and personal testing and field inspections (see Chapter 14) are other valuable primary source research tools. When you incorporate these sources into your documents, include a thorough explanation of public material, the details of your direct observation, copies of the letters, the interview questions and the authoritative background on the interviewee, and the methods and findings of your personal testing and field inspections.

CHECKLIST

Accessing Information

THE SOURCES

❏ **1.** Have I accessed all of the possible sources that pertain to my subject?

❏ General book?

❏ Special reference books (dictionaries, encyclopedias, handbooks, manuals, almanacs, histories, and biographies?

❏ Periodical literature?

❏ Special essays, print indexes, abstracts, and technical reports?

❏ Specialized and/or subscription databases?

❏ Vertical files?

❏ **2.** Have I examined all of the possible media that might contain sources for my paper?

❏ Special databases?

❏ Microforms?

❏ CD-ROMs?

❏ Audiocassettes?

❏ Videocassettes?

❏ Film?

❏ Other?

❑ **3.** Have I accessed the Internet directly or with various search engines for additional sources and text?

    ❑ Specific addresses?

    ❑ Keyword searches?

    ❑ Bulletin boards and mailing lists?

    ❑ Subscription databases?

    ❑ Personal CD-ROMs?

    ❑ Online periodicals?

    ❑ Other?

❑ **4.** Have I assessed the reliability of Internet documents?

BIBLIOGRAPHY CARDS OR FILES

❑ **1.** Have I prepared a separate bibliography card or computer file entry for each source that I will use?

❑ **2.** Have I included all of the necessary information on each card or file listing?

    ❑ Author's name?

    ❑ Title of the work?

    ❑ Library call number, database, Internet address, and source type?

    ❑ An optional personal note for future use?

❑ **3.** Have I alphabetized these cards or files?

❑ **4.** Am I prepared to assemble a final bibliography or works cited page(s)?

NOTE TAKING

❑ **1.** Have I recorded all of my notes with an eye to avoiding plagiarism?

❑ **2.** Have I recorded the information on each card or computer file so that I may access it when I write my paper?

❑ Recorded only one item per card?

❑ Recorded the source at the top left of card?

❑ Recorded a topic notation at the top right of the card?

❑ Used just one side of each card?

❑ Stapled together multiple cards dealing with one item?

❑ Recorded the exact page number or other similar notation of the item?

❑ **3.** Have I included a variety of types of note cards?

❑ Précis notations of summaries abstracts, and listed bibliographies?

❑ Summary notes?

❑ Paraphrased notes?

❑ Direct quotations that are unusually well-expressed?

❑ Personal notes and ideas?

❑ **4.** In taking my notes, have I employed ellipses to indicate omitted words, phrases, or sentences in a direct quotation?

❑ **5.** Have I employed brackets to add clarifying words within a direct quotation?

❑ **6.** Have I incorporated display quotations correctly?

SURVEYS, QUESTIONNAIRES, AND OTHER PRIMARY SOURCES

❑ **1.** If I am devising a survey instrument or questionnaire, have I given careful consideration to

❑ Advantages?

❑ Disadvantages?

❑ Test group selection?

❑ **2.** Have I motivated a response to my instrument?

❑ Date deadline?

❏ Sponsorship authority?

❏ Cover sheet explanation?

❏ Anonymity guarantee?

❏ Other?

❏ **3.** Have I formatted my instrument carefully?

❏ Paper quality concerns?

❏ Clarifying title?

❏ Brevity?

❏ Tick boxes and printed lines for responses?

❏ White space?

❏ Avoidance of leading, offensive, and too personal questions?

❏ One query per question?

❏ Room for comments, opinions, or other alternatives?

❏ **4.** Have I used a variety of types of questions?

❏ Dual alternatives?

❏ Multiple choice?

❏ Rank ordering?

❏ Continuum scale?

❏ Completions?

❏ Essays?

❏ **5.** Have I considered any other primary source research for the paper?

❏ Lectures, forums, television and radio programs?

❏ Letters of inquiry?

❏ Interviews?

❏ Personal testing?

❏ Field inspections?

❏ Direct observation?

❑ **6.** Am I prepared to incorporate the data by a number of appropriate graphics and commentary?

❑ **7.** Am I prepared to include the survey instrument or questionnaire and the circumstances and results of other primary source material into my final paper?

EXERCISES

1. **Subclassifications of a Broad Subject.**   Locate and list at least five periodicals, accessible through your library computer system or through a personal computer search, that are most appropriate to your field of study. Look up a general subject, such as nutrition, medicine, fire science, radiology, electronics, dentistry, bioscience, fiber optics, solar heat, cloning, cardiology, dentistry, pollution, aviation, and so on. List five subclassifications that could be researched for the development of a technical document of about 500 words. Consider whether the material is recent, not too technical, reliable, and available in sufficient amount.

2. **Limiting Your Subject Search.**   In the library or at your personal computer, look up one of the same broad subjects as in Exercise 1. Use three Boolean operators (*and, or*, or *not*) to limit your search and print out the bibliographies located under each of the three searches. Submit these along with the subject headings you used to derive the bibliographies.

3. **Internet Search.**   Through actual Web addresses, keywords, or several search engines, locate Web sites that contain information on a limited subject of your choice within your field of study. List ten or more of the addresses.

4. **Bibliography Cards.**   Prepare five separate bibliography cards on the above subject, one each for a book, a magazine article, a newspaper article, an encyclopedia or other reference book source, and a Web site.

5. **Note Taking.**   Read the excerpt from an article on nuclear wastes (see p. 191):

    a.   Write a summary note of the entire excerpt.

    b.   Write a paraphrase of the sentence: "Until then, they need protection from earthquakes, terrorists, water, infiltration, and un-

---

Sink the Nukes

Leftovers from nuclear power plants remain dangerously radioactive for hundreds of thousands of years. Until then, they need protection from earthquakes, terrorists, water infiltration, and unknowing humans who might stumble across them millennia from now. Fergus Gibb, a geologist at the University of Sheffield in England, has proposed a new way to dispose of the highest-level waste—plutonium and enriched uranium. He suggests dropping the nasty stuff into boreholes three miles deep. Gibb's scheme utilizes heat-resistant waste containers. Thermal energy from the decaying elements would melt the surrounding rock, thereby entombing the containers deep within Earth's crust, where water does not return to the surface. Gibb claims his method would be both safer and less expensive than burying waste in shallower underground mines, such as the trouble-plagued Yucca Mountain site in Nevada.—*Jessica Gorman*

"Sink the Nukes" by Jessica Gorman, DISCOVER Magazine, July 2000. Jessica Gorman/© 2000. Reprinted with permission of *Discover* Magazine.

knowing humans who might stumble across them millennia from now."

   c. Write a note containing a logical partial quotation from the sentence: "Gibb claims his method would be both safer and less expensive than burying waste in shallower underground mines, such as the trouble-plagued Yucca Mountain site in Nevada.

6. Note Taking.   Obtain a copy of an article from your bibliography searches on a subject selected in Exercise 2 or 3 and write three types of note cards: a summary note of the entire article, a paraphrase of some material, and a partial direct quotation. Record the six necessary pieces of information on each card. If the note requires more than one card, staple the cards together. If you enter your notes in a computer file, submit a printed copy.

WRITING PROJECTS

1. Paraphrasing.   In your own words, paraphrase the section on plagiarism in this chapter.

2. Collaborative Project: Survey/Questionnaire.   In groups of four, devise a survey or questionnaire to determine facts about on-campus parking facilities. Use a variety of question types. Determine the number of spaces, hours per day used, peak hours, problems encountered (too few parking spaces, puddles, distance, accidents with car doors, speed bump problems, entries and exits, vandalism, muggings, etc.). List options to improve the facilities and determine a "rating of the options" procedure for the respondents. Seek additional comments. Record statistics, such as whether the respondent is a student, faculty, staff member, or visitor. Consider listing compact, medium, luxury, and recreation vehicle ownership to determine if responses differ for each type of vehicle. What other control data might you include? Title your survey or questionnaire and write a brief purpose statement that will motivate serious and timely responses.

3. **Collaborative Project: Survey/Questionnaire Analysis Paper.** In the groups established for Writing Project 2, administer the survey to three classes other than this one by obtaining another professor's permission to do so. Also, distribute the survey to the faculty and staff in one of the large departments on campus. If possible, distribute the survey to guests on campus, such as attendees at an art exhibit, theater or musical production, or seminar. Use the data that you compile and write a brief paper that analyzes the data. Include at least two graphics and a commentary that interprets your findings.

NOTES

CHAPTER 6

Documenting Research

DUFFY   by Bruce Hammond

S K I L L S

After studying this chapter, you should be able to

1.   Name three functions of documentation.
2.   Name four major systems for professional paper documentation.
3.   Name five documentation strategies that apply to all systems.
4.   Prepare both a working draft and a paper for publication documented in the American Psychological Association (APA) systems.
5.   Prepare a documented paper in the Modern Language Association (MLA) system.
6.   Prepare documented papers in both variations of the Council of Biology Editors (CBE) systems.
7.   Prepare a documented paper in a number system (CBE, ACS, AMS, & AIP) appropriate to a specific field.
8.   Prepare a documented paper in the Chicago style.

INTRODUCTION

Researched scientific and technical documents require formal documentation of the sources that have been summarized, paraphrased, or quoted. Documentation fulfills three functions:

1.   **Ethical acknowledgment of your sources.**   All published material is protected by copyright, so you have a professional and ethical obligation not to plagiarize (see Chapter 5) by citing your sources.
2.   **Establishment of authority and credibility.**   Thorough documentation backs up your scholarship by citing the authorities and adds to the credibility of your original ideas developed in your document.
3.   **Efficiency for further study.**   Documentation allows your readers to locate your sources and to learn more about a subject.

Documentation involves making parenthetical references to the sources cited within your works and preparing a concluding list of references. Prior to 1984 footnotes and endnotes were used in research papers documented in the Modern Language Association (MLA) and the American Psychological Association (APA) systems; however, parenthetical references for those two systems are now preferred. Other systems vary.

For each of your writing assignments from the following chapters, ascertain how much documentation is required by your professor and which style is preferred for any one assignment. Your professor may ask only for a concluding list of references for short papers or for a thoroughly documented paper with references or notes within the document, plus the final list of sources.

There are many accepted documentation styles and systems endorsed by different organizations. The most widely used are

- The **American Psychological Association (APA),** an author/date system used for the social sciences, with similar versions for the biological and earth sciences, education, linguistics, and business.
- The **Modern Language Association (MLA),** a system used for language and literature.
- The **number systems (CBE, ACS, AMS, and AIP),** related systems used for the medical sciences (biomedicine, health, medicine, and nursing), biological and earth sciences, (agriculture, anthropology, archaeology, astronomy, biology, botany, geology and zoology), the applied sciences (computer science, chemistry, engineering, mathematics, physics, and engineering).
- The **Chicago Style,** a system used for some of the humanities (history, philosophy, religion, and theology) and for the fine arts (art, dance, music, and theater).

Comprehensive manuals of these and other discipline-specific systems may be purchased in any bookstore or located online. Information about the APA, MLA, number systems (CBE, ACS, AMS, AIP, and ASCE), and Chicago styles is available on the Internet; use the keywords *American Psychological Association, Modern Language Association, Council of Biology Editors,* or *Chicago Manual of Style.* Some Web sites for a few of the various number system styles are

- **American Chemical Society (ACS)** <http://pubs.acs.org>
- **American Mathematical Society (AMS)** <http://www.ams.org>
- **American Institute of Physics (AIP)** <http://www.aip.org>
- **American Society of Civil Engineers. (ASCE)** <http:// www.pubs. asce.org/authors/index.html>

The bibliography styles, though varied, all include the name of the author, the title of the work, the place of publication, the publisher (or the Internet address or CD-ROM access), and the date. In-text reference citations vary also.

Documentation strategies for all systems are similar:

1. Compiling an annotated working bibliography on cards or computer file (Chapter 5)
2. Taking the notes (Chapter 5)
3. Providing in-text documentation of sources

4. Preparing a final list of references or works cited as determined by your system

5. Presenting the paper in a formal format

APA SYSTEM

The fourth edition of the *Publication Manual of the American Psychological Association* (1994) provides documentation advice for writers in the social sciences, with similar versions for biological and earth sciences, education, linguistics, and business. The APA system is widely endorsed for academic writing. Updates may be accessed online at the APA Web site. The manual is written primarily for authors preparing manuscripts for professional publication in journals, and it covers manuscript content and organization, writing style, and manuscript preparation. An appendix offers advice for student writers.

General Conventions

There are a number of general conventions in the APA system in addition to the internal referencing and final reference list systems:

1. The final paper must contain a title page, an abstract (limited to 960 characters including spaces—approximately 80 to 120 words), the in-text documented paper, and the final reference list. Include an outline only if your professor requires one.

2. The paper should be double-spaced throughout with the pages numbered consecutively in the upper, right-hand corner (with the exception of the title page). The abstract will be on page 2 following the unnumbered title page, and a page number is included on the reference listing.

3. Use last names only within your text (Kline, not Robert J. Kline).

4. The year of the research is primary, so it is featured in the text immediately after the named source ["Kline (1998) . . ." or "Woolsey and his team (1997) . . ."].

5. Because one purpose of a scientific paper is to review the work of previous researchers, the past tense is required when you cite their findings ("Kline (1998) **showed** . . ." or "Woolsey and his team (1997) **demonstrated** . . .").

6. Title your list of sources "References"; do not use all capital letters nor boldface the word.

7. Refer to the manual for the handling of abbreviations, racial refer-ences, capitalization, punctuation, numbers, and other style questions.

Internal References

When you write your paper in the APA style or its variations, several internal referencing practices apply. Essentially you must reference the sources within your text by indicating within parentheses the author and date and sometimes the exact page number. Examples of internal referencing follow:

If the author's name(s) occurs in the text, then all that is needed is the publication date in parentheses:

> According to Campbell (1996) the use of melatonin is . . .

If the author's name(s) does not appear in the text, place it and the date in parentheses at the end of the paraphrased material:

> Studies proved that melatonin is often taken in unnecessarily large doses (Michaels, 1997; Vann, 1998).

If the text contains a direct quotation, you must include the page numbers with "p." or "pp." The pagination note may occur with the date notation or at the end of the direct quotation:

> Campbell (1996, p. 107) stated, "Melatonin, while it may have some efficacy as a sleep or jet lag aid, is totally overrated for its other so-called benefits."

> or

> Campbell (1996) stated, "Melatonin, while it may have some efficacy as a sleep or jet lag aid, is totally overrated for its other so-called benefits" (p. 107).

Write a quotation of 40 words or more as a separate block, which is indented five spaces from the left margin. Introduce the quotation. Do not indent the first line, but do indent the first line of additional paragraphs five more spaces. Omit the quotation marks; the indentation indicates that the material is a quotation:

> Smith (1998) reported the following:
>
> > Melatonin, the much touted herb of the 1990s, has not been found to slow aging as its manufacturers claim. It does stimulate the thymus to release seratonin, but not in any significant amounts.

> Melatonin may be considered a natural hormone, but as such not
> one test has indicated its efficacy for anything beyond a sleep aid (pp.
> 314–315).

Notice that the parenthesized page numbers come *after* the final period
in a block quotation. Paraphrases *may* be handled like quotations. Give
the author's last name in the text, the date in parentheses, and the appropriate page number(s) in parentheses at the end:

> Campbell (1996) suggests that the advantages of melatonin are overrated
> aside from the fact that melatonin may help one get to sleep or overcome jet
> lag (p. 107).

Notice that the period comes *after* the page reference for a short direct
quotation or paraphrase page reference.

Corporate authors must be spelled out in full in the first reference,
followed by the appropriate abbreviation. Use the abbreviation only in
subsequent references:

> One study has questioned that any benefits at all accrue from melatonin
> (American Medical Association [AMA], 1997).
>
> The AMA (1997) concluded that . . .

E-mail, Personal Interviews, Usenet Newsgroups, Bulletin Boards

If you are citing information from an e-mail, personal interview, a news-
group, or a bulletin board, indicate the type of communication and the date
within the parentheses. A sample suitable for all of the above follows:

> Motorola Manager Maria Foster (personal communication, March 28, 2000)
> suggests . . .

Web Sites

If you cite a Web site, it is sufficient to give just the address of the site in
the text; to cite specific parts of a Web document, indicate the chapter,
figure, or table page number as appropriate. For quotations, give the
page numbers (or paragraph numbers) only if they are available as they
appeared in the original printed format; do not use the page numbers
which your printer prints out. A sample follows:

> The American Psychological Association warns that it is possible to send an
> e-mail note under the guise of a different user name (<http://www.apa.org>,
> p. 3).

Specific Documents on a Web Site

Parenthetical references are handled like other books, periodicals, and so on, citing the author or corporate author and date:

> The latest APA (1999) Web Site update on internal referencing states . . .

The foregoing is not an exhaustive coverage of all the requirements, but rather a sampling of the most commonly used internal referencing methods. Figure 6.1 shows some sample text documented in the APA style.

Final Reference List

The APA style manual distinguishes between a reference listing system for a working draft and that for a manuscript for publication in a professional journal. Determine which form your professor prefers. The differences include the following:

1. The working draft will be written in Courier typeface, found on most typewriters, or in Times Roman, 12 points, a standard for computer word processing. A manuscript for publication permits more freedom in the selection of fonts, margins, borders, hanging indentations, and other document design features, in accordance with the publication guidelines and your own preferences.

2. The working draft traditionally requires underscoring of titles, whereas the manuscript for publication traditionally requires italics for the titles, although either underscoring or italics is acceptable providing they are used consistently. If you use italics for journal titles, also italicize the volume numbers (*Biblical Archeologist, 58*). Consult your instructor or editor.

3. The working draft will have an unjustified right margin; the manuscript for publication may use right margin justification.

4. Headings should be used within the document and employ the following capitalization and placement:

Global Warming Evidence ◄— A-level heading, centered with initial capitals

Changes in the Antarctic Ocean ◄————— B-level heading, flush left with major words capitalized

Scientists to conduct tests ◄————— C-level heading, flush left with only first word capitalized

Modern scientific methods, invented in the 16th century, were not only a stunning technical innovation, but a moral and political one as well, replacing the sacred authority of the Church with science as the ultimate arbiter of truth (Grant, 1987). Unlike medieval inquiry, modern science conceives itself as a search for knowledge free of moral, political, and social values. The application of scientific methods to the study of human behavior distinguished American psychology from philosophy and enabled it to pursue the respect accorded the natural sciences (Sherif, 1979).

The use of "scientific methods" to study human beings rested on three assumptions:

> (1) Since the methodological procedures of natural science are used as a model, human values enter into the study of social phenomena and conduct only as objects; (2) the goal of social scientific investigation is to construct laws or law-like generalizations like those of physics; (3) social science has a technical character, providing knowledge which is solely instrumental. (Sewart, 1979, p. 311)

Critics recently have challenged each of these assumptions. Some charge that social science reflects not only the values of individual scientists but also those of the political and cultural milieux in which science is done, and that there are no theory-neutral "facts"

FIGURE 6.1   *Sample documented text in APA style*   (From Stephanie Riger, "Epistemological Debates, Feminist Voices—Science, Social Values, and the Study of Women," *American Psychologist, 57* (June 1992): 78–92. By permission.)

(e.g., Cook, 1985; Prilleltensy, 1989; Rabinow & Sullivan, 1979; Sampson, 1985; Shields, 1975). Others claim that there are no universal ahistorical laws of human behavior, but only descriptions of how people act in certain places at certain times in history (e.g., K. J. Gergen, 1973; Manicas & Secord, 1983; Sampson, 1978). Still others contend that knowledge is not neutral; rather, it serves an ideological purpose, justifying power (e.g., Foucault, 1980, 1981). According to this view, versions of reality not only reflect but also legitimate particular forms of social organization and power and asymmetries. The belief that knowledge is merely technical, having no ideological function, is refuted by the ways in which science has played handmaiden to social values, providing an aura of scientific authority to prejudicial beliefs about social groups and giving credibility to certain social policies (Degler, 1991; Shields, 1975; Wittig, 1985).

Within the context of these general criticisms, feminists have argued in particular that social science neglects and distorts the study of women in a systematic bias in favor of men. Some contend that the very processes of positivist science are inherently masculine, reflected even in the sexual metaphors used by the founders of modern science (Keller, 1985; Merchant, 1980). To Francis Bacon, for example, nature was female, and the goal of science was to "bind her to your service and make her your slave" (quoted in Keller, 1985, p. 36). As Sandra Harding (1986) summarized . . .

FIGURE 6.1 *continued*

<u>Test A</u>. In January 2000 scientists . . . ◄————D-level heading, indented to begin paragraph; underscore or italicize

5. Entries in the final reference list should be double-spaced.

6. A recent change in the system requires that both the working draft and the manuscript for publication have hanging indentation of just three spaces for the final reference list.

All of the following examples employ the hanging indentation and italicized titles. Pay close attention to the capitalization conventions. For a complete explanation of all possible entries, consult the APA style manual.

Book

Burns, M. (1995). *Handbook of psychology* (2nd ed.). Englewood Cliffs, NJ: Prentice Hall.

Two Authors

Fitzpatrick, J., & Swenson, T. (1998). *Studies in sociobiology*. Washington, DC: American Psychological Association.

Book with Editors or Translators

Leonard, R. T., & Coonz, C. R. (Eds.) (1992). *Bilingual education: Teaching English as a second language*. New York: Praeger.

Lura, A. R. (1996). *The mind of a mnemonist* (L. Solotaroff, Trans.). New York: Avon Books. (Original work published in 1965.)

Article in a Book

Gurman, A. S., & Kniskern, D. P. (1995). Family therapy outcome research: Knowns and unknowns. In A. S. Gurman & D. P. Kniskern (Eds.), *Handbook of family therapy* (pp. 742–776). New York: Brunner/Mazel.

Unknown Author

Study finds free care is used more. (1992, April). *APA Monitor*, p. 14.

Encyclopedia

Carey, C. W. (1995). Bonding. In *Encyclopedia Britannica* (Vol. 2, p. 42). Chicago: Encyclopedia Britannica, Inc.

Journal

Gingro, J. (1997). Television in the classroom. *Social Education, 52*, 221–225.

Magazine

Arial, S. K. (1996, August). Computers in the social studies classroom. *Wired*, 37–38, 101–102.

Newspaper

Lublin, J. S. (1996, December 5). On idle: The unemployed shun much mundane work. *The Wall Street Journal,* pp. B14, B18.

Electronic Sources

The system for electronic source documentation continues to be updated periodically; consult your most up-to-date handbooks and online documents. Two excellent Web sites for news of updated formats are the APA site at <http://www.apa.org/journals/webref.html> and Nancy Chase's updates for the University of Vermont at <http://www.uvm.edu/~ncrane/estyles/apa.html>.

E-mail, Personal Web Sites, Personal Interviews

E-mail, personal Web sites, and personal interviews are only parenthetically referenced in the text but are not cited in the reference list. The text should clarify the source. You may include the Web address in the text.

Specific Documents on a Web Site

The format for specific documents on a Web site is similar to that for a print source, with some information possibly omitted and some added, including the retrieval date:

Sleek, S. (1996, January). Psychologists build a culture of peace. *APA Monitor*, pp. 1, 333. Retrieved January 25, 1990 from the World Wide Web: http://www.apa.org/ monitor/peacea.html

Rawlings, A. (1999, August 7). Scientists seek cure for rheumatoid arthritis. *New York Times Online*. Retrieved August 8, 1999 from the World Wide Web: http://search.nytimes.com/search/daily/bi . . .m%29%26%OR%28treatment%29%OR26R%.

Article or Abstract from an Electronic Database

Use the following as a guide:

Kerrigan, D. C., Todd, M. K., & Riley, P. O. (1998). Knee osteoarthritis and high-heeled shoes. *The Lancet, 251*, 1399–1401. Retrieved

January 30, 2000 from DIALOG database (#457, The Lancet) on the World Wide Web: http://www.dialogweb.com

CD-ROM Reference Work

Following are two acceptable methods of citing CD-ROMs:

James, Carol. (2000). *Graphics on-line* [CD-ROM]. New York: RC4
   Symmetric Stream Cipher.
Rheumatoid arthritis [CD-ROM]. (2000). *Compton's interactive encyclopedia*. New York: Softkey Multimedia.

Figure 6.2 shows a sample APA style reference list with three-space hanging indentation, and double spacing, and italicized major titles.

Abstract

You should include an abstract with all APA style papers. An abstract is a quick, thorough summary of the contents of your paper—approximately 80 to 100 words. Abstracts are also covered in Chapter 16, "Producing Professional Papers." The abstract is read first so must be accurate, self-contained, concise, nonevaluative, and coherent. Include it as a second and separate page of your paper, titled *Abstract*.

MLA SYSTEM

The fifth edition of the *Modern Language Association Handbook for Writers of Research Papers* (1995) and the second edition of the *MLA Style Manual and Guide to Scholarly Publishing* (1998) provide documentation advice for writers on languages and literature in the preparation of scholarly manuscripts and student research papers. Some journals carry articles about technical communications but are not truly technical journals; an example is the *Technical Communication Quarterly,* which also uses the MLA style. Your instructor may want you to be able to prepare a research paper in the MLA style as well as in your appropriate discipline style. The *MLA Handbook* includes sections on the purposes of research; suggestions for choosing topics; outline guidance; internal reference and bibliography systems; advice on spelling, punctuation, abbreviations, and other stylistic matters; and formatting of the manuscript.

References

Adams, J. K. (1997). Laboratory studies of behavior without awareness.

Psychological Bulletin, 54, 383–406.

Breuer, J., & Freud, S. (1955). Studies on hysteria. In J. Strachey

(Ed.), *The standard edition of the complete psychological works of*

Sigmund Freud (Vol. 2, pp. 1–307). London: Hogarth Press. (Original

work published 1893–1895).

Hilgard, E. R. (1986). *Divided consciousness: Multiple controls in human*

thought and action (rev. ed.). New York: Wiley-Interscience.

Jacob, L., Lindsay, D. S., & Toth, J. P. (1992). Unconscious influences

revealed: Attention, awareness, and control. *American Psychologist, 47,*

802–809.

Sanderson, K. L. Priming and human memory systems. *Science, 50,*

489–495. Retrieved June 30, 2000 from the World Wide Web:

http://www.sci.com/sanderson/mem/lrs.html.

Tulving, E. (1976). The evolution of intelligence and access to the cognitive

unconscious. In E. Stellar & J. M. Sprague (Eds.), *Progress in psychobi-*

ology and physiological psychology (Vol. 6, pp. 245–280). San Diego,

CA: Academic Press.

FIGURE 6.2   *Sample APA reference list*

General Conventions

In addition to internal referencing and the final works cited listing, a number of general conventions apply in the MLA system.

1. The paper consists of a titled first page with text, followed by the rest of the text, content endnotes (optional), and a works cited page. Include an outline and an abstract only if your professor requires either of them. If you include an outline, abstract, or other introductory matter, use a separate title page as directed by your professor. A non-separate title page includes your name, the professor's name, the course, and the submission date, plus the title of the paper and the introductory text:

    Shawn Sabga

    Professor J. VanAlstyne

    English Comp. 2210.04

    May 8, 2001

    The Phenomenal Facsimile Machine

    Librarian Joan Rutledge helped a student locate a bibliography in a book in another library of a multi-campus university. In the past, Rutledge . . .

    ---

2. The paper should be double-spaced throughout, and the pages (with the exception of a separate title page) must include your last name and the page number in the upper, right-hand corner in Arabic numbers (Sabga 3). If you keyboard the manuscript, use Times Roman, Helvetica, Arial, or other clear and legible fonts.
3. However, if a separate title page is not used, and you include outline pages and/or a separate abstract, include your name and lowercase Roman numerals (Sabga iii.) in the upper, right-hand corner of the outline pages and abstract.
4. If you use an abstract, you may (a) place it on a separate page between the title page and the first page of the text, or (b) place it on the first page of the text one double-space below the title and before

the first lines of the text. If you place it on the first page of the text, indent the first sentence 10 spaces; then indent the rest 5 spaces as a block. Use quadruple spacing at the end of the abstract to set it off from the text.

5.   If you use content endnotes (your own explanatory comments not included within the text), label a separate page *Notes,* place it before the works cited listing, and provide raised superscript numerals within the text to match your superscript numeral notes. Double-space all entries and double-space between them.

6.   Title your compilation list of sources *Works Cited.* Do not use capitals or boldface. Center the title.

7.   Major titles (books, periodicals, and so on) may be italicized or underlined. The MLA suggests using underlining for papers to be graded or edited, but if you wish to use italics, check with your instructor or editor.

8.   Unlike the APA system, which reports on researchers' findings in the past tense, the MLA system uses the present tense to report universal assertions: ("Ernest Hemingway **writes** . . ." or "Patricia O'Malley **proclaims** . . .").

Internal References

When writing your paper in the MLA style, several referencing practices apply. If the author's name or publication title is in the text, then add only the page reference in parentheses after the quotation, summary, or paraphrase:

> McCarroll reports that a fax machine sold for as much as $10,000 and up in 1971 (38).

If the author's name (article or book title if there is no author) is not included in the text, insert both the author and the page number in parentheses directly after the paraphrase, direct quotation, or summary:

> Actually, the origin of the fax machine dates to 1926 when it was invented by Radio Corporation of America (Costigan 2).

> Financial centers in New York, London, and the Far East fax material before and after office hours ("The Fax Revolution" 14).

If a summary refers to an entire work, page numbers may be omitted:

> In his article "The New Fax," Gerald Smith points out that the older fax systems used electromechanical scanning techniques which converted visual tonal variations for transmission to a receiver and were read by scan-

ners, but that modern machines send electronic copies over ordinary telephone lines to the fax machine even at the opposite end of the world.

If quotations are used, the name of the author is included in the text, and the page number is in parentheses at the end with a following period. If the quotation is 40 words or more, indent all of the quotation, omit the quotation marks, and place the parenthetical reference after the period:

> "It's the electronic boom," declares John A. Widlicka, fax marketing manager for Sharp Electronics Corporation (qtd. in Gelfond 59).
> Thomas McCarroll reviews the benefits of the fax machine:
>
> > It eliminates the need for a typist to forward something, as you do when you telex. There are no anxieties about typing or keyboard errors—especially worrisome if you're sending critical financial numbers. You can maintain some degree of confidentiality if the document goes directly from you to a protected receiver. In contrast to e-mail, you don't have to be computer literate to send or receive. (16).

Parenthetical references in the text from works on the World Wide Web are cited just like printed works. Web documents generally do not have fixed page numbers of any kind, so if your source lacks numbering as it would have occurred in an original print edition, you have to omit numbers from your parenthetical references:

> Sometimes two or more early-onset Alzheimer's genes were found (Nash).

If the Web document does contain page numbers or paragraph numbers, give the appropriate abbreviation before the number. Lowercase *p.* and *pp.* are the abbreviations for *page* and *pages*. Lowercase *par.* and *pars.* are the abbreviations for *paragraph* and *paragraphs*.

> Innovation is one of the engines of a market society, a point where economics converges with aesthetics ("The Look of the New," par. 1).

Figure 6.3 shows a sample of an in-text referenced document in the MLA system.

Content Endnotes

Content endnotes are optional. If you need to include notes about research problems, conflicts in the testimony of experts, interesting tidbits relating to your text but not crucial to supporting your thesis, or credits to people or sources not included in the works cited listing, follow these guidelines:

Modern fax machines offer desirable innovative features. Engineer Robert Johns points out that a standard, top-of-the-line fax comes with a photocopy machine, an answering machine, a telephone, and the fax system. Once considered too bulky and costly to be practical, Johns reports, fax machines have shrunk to half their original size. This transformation in size has been taken to the limit. Mitsubishi Electric has introduced a fax unit that fits under a car dashboard and connects to the cellular car phone (McCarroll 38). Medbar Enterprises, Inc. is selling a Japanese-made model targeted for executives on the go. It is battery operated and can be used from a plane or train as long as a telephone line is available (Gelford 69).

The fax has not only shrunk in size, but also in price. In 1971 fax transceivers sold for as much as $10,000 and up (McCarroll 38). Later in the 1970s prices dropped to $3,000 to $6,000 ("The Fax Revolution" 16). In the 1980s faxes the size of a portable typewriter entered the market for as low as $1,500 to $2,000, and the 1990s saw machines available from $500 to $800 (16). By the 2000s fax machines became available from $130 to $250, with such features as answering machine interfaces, collators, caller ID compatibility, size change capabilities, six-second page delivery, and more (*Office Depot: The Big Book* 686–687).

FIGURE 6.3   *Sample paper documented in MLA style*   (Courtesy of student Shawn Sabga)

Sabga 4

If these prices do not attract the buyer's eye, there is still one more cost factor to consider: the amount of money that could be saved by using a fax. Federal Express charges about \$15.00 to deliver a one-page document overnight. The same letter can be faxed in a matter of seconds for the price of the phone call (McCarroll 38). Mark Winther, an electronics analyst at Manhattan-based LINK Resources, says, "The growth of fax is coming out of the hides of Federal Express" and other overnight mail services (qtd. in Costigan 227). As a result of the cost analysis, sales have soared from some 200,000 in 1986 to multimillions in the 2000s (McCarroll 38 and Johnston, par. 3). The fax offers a host of benefits:

> It eliminates the need for a typist to copy something and to send by letter. There are no anxieties about typing errors— especially worrisome if you're sending critical financial numbers. You can maintain some degree of confidentiality if the document goes directly from you to a protected receiver. In contrast to electronic mail, you don't have to be computer literate to send or receive. ("The Fax Revolution" 16)

Fascination with communicating by fax is international. "It's the electronic boom," declares John A. Widlicka, who is in charge of marketing fax machines in the United States for Sharp Electronics Corporation (qtd. in Gelfond 59).

FIGURE 6.3   *continued*

1. Place the content endnotes on a separate page(s) after the last page of the text. Title the list "Notes." Double-space throughout. Indent each separate note.

2. Place superscript numbers in numerical order in your text to refer to each note:

   > The military has long delayed accepting responsibility for the effects of poisons on the soldiers in the Gulf War.[1] An interview with General E. D. Smyth revealed . . .[2]

3. Begin the endnote with the matching superscript:

   > [1]Lately, the military is acknowledging that soldiers have suffered some poison gas reactions, but insists that only those soldiers who did not take proper precautions when destroying our own weapons depots before pulling out can be pinpointed. On this point see Glascoe (4) and Meyers (B8).
   >
   > [2]Funds to travel to Washington to interview General E. D. Smyth were provided by the DCC Foundation.

Final Works Cited List

The essential differences between the MLA style of listing sources and the APA style and its variations are that the MLA requires the full name of authors, rather than initials, and the standard use of capitalization for titles. The date is placed at the end. Center the title *Works Cited*. Each entry employs hanging indentation (usually just one-half inch). Italicize (if you are processing on a computer) or underline book, periodical, and other major titles. Include a page number in the upper, right-hand corner. For listings not included in the following examples, consult the *MLA Handbook*.

Books

Bequai, August. *Computer Crime.* Lexington, Mass.: Lexington Books, 1988.

Hsiao, David K., Douglas S. Kerr, and Stuart E. Madnick. *Computer Security.* 2nd ed. New York: Academic Press, 1979.

Shank, Roger C. *Computer Models of Thought and Language.* Ed. by Roger C. Shank and Kenneth Mark Colby. San Francisco: W. H. Freeman, 1973.

Management Information Corporation. *Computer Privacy.* Cherry Hill, N.J.: Management Information Corporation, 1982.

Cardoza, Juan. *Access Control of Computers.* Trans. by Alan Jameson. New York: New American Library, 1991.

Special Reference Works

Smythe, Charles John. "Computers." *Encyclopaedia Britannica.* 1993 ed.

"Computer Services." *Statistical Abstract of the United States,* 82 (1993):72.

Journal

Barthelmew, Rosemary. "Computer Viruses Increase." *Journal of Communication* 42.2 (1993): 5–24.

Magazine

Taflich, Peter. "Opening the Trapdoor Knapsack." *Time* 25 Oct. 1992: 72.

Newspaper

"Computer-based Home Security System Sales Soar." *New York Times* 15 Dec. 1993, natl. ed.: B7+.

Government Document

United States. Cong. Senate. Subcommittee on DNA Legislation. *DNA and Cloning: A Moral Issue.* 103rd Cong, 2nd sess. S. Hearing 709. Washington, D.C. USGPO, 1997.

Pamphlet (treat like a book)

Radio Shack, a Division of Tandy Corporation. *TRS-80 Model II Micro-Computer System.* U.S.A. 1991.

Interview

Larsen, Robert H. Personal interview. 11 Mar. 1995.

Television

"Should We Get On With Computer Literacy?" *The Firing Line.* Washington, D.C.: PBS-TV, 16 Oct. 1993.

Graphics

Buckley, Charles L. Accounting system flowchart. *Introduction to Accounting.* New York: Oxford University Press, 1997.

Slide Presentation

Marijuana: The Addiction Question. Slide presentation. Developed by Project Cork, Dartmouth Medical School, 7 July 1991. 55 slides.

Electronic Sources

The format is updated periodically; consult your handbook or online sources. Log on to <http://www.mla.org/style/sources.htm> or <http://www.uvm.edu/~ncrane/estyles/mla/html>.

E-mail, Usenet, Newsgroups, Bulletin Boards

Smelzer, Sally. <smelzer@hillspos.hill.af.mil> "Computer viruses." E-mail
to author. 29 Apr. 2000.

Tritt, Merrill. "Virus Hoaxes in E-Mail." Online posting. 12 July 2000.
Computer Viruses. <http://www.mcafee.com/ partner/ MONSTERP
_SPINOFF.txt>.

Personal Web Site

Austin, Paul. Home page. 4 June 2000 <http:// www.refining.org. ca: 7000/
~paustin/index.wtml>.

Specific Documents on the Web (Books, Journals, Magazines)

Nesbit, E(dith). *Ballads and Lyrics of Socialism.* London, 1908. Victorian
Women Writers Project. Ed. Perry Willett. Apr. 1997. Indiana U. 26
Apr. 1999 <http://indiana.edu/~letrs/vwwp/nesbit/ballsoc.html>.

Brockman, R. John. "Considerations of Ethics and the Technical Writer."
Technical Communication Quarterly 2.3 (1999): 17 pars. 22 Feb.
1999 <http://www/hmsu.edu/~gwilson/appl.html>.

Landsburg. Steven E. "Who Shall Inherit the Earth?" *Slate* 1 May 1997
<http://www.slate.com/economics/97-05-11/Economics.asp>.

Article in a Reference Database

"Fresco." *Britannica Online.* Vers. 97.1.1. Mar. 1997. Encyclopedia
Britannica. 29 Mar. 1997 <http://www.eb.com:180>.

CD-ROM Reference Work

Turner, Robert. "EMF Study: Good News." Electrical World Feb. 1999: 10.
ABI/INFORM. CD-ROM. Proquest. Sept. 1995.

Figure 6.4 shows a Works Cited list in the MLA system. Notice that entries by the same authors employ three dashes followed by a period and
that the works are then alphabetized by titles.

NUMBER SYSTEMS (CBE, ACS, AMS, AND AIP)

Number systems are used for the applied sciences and the medical sciences. The systems require an in-text number, rather than the year, and
a list of references that is numbered either alphabetically or in the order
that each is first cited in the text. Writers in chemistry, computer science,
physics, engineering, biomedicine, medicine, nursing, and general health
conform to several general regulations that have been established by the

Janis 11

Works Cited

"Architects." *Encyclopedia of Careers and Vocational Guidance.* 11th
ed. Ed. By William C. Swope. Chicago: I.G. Ferguson Publishing
Company, 1994.

Berger, Karen. <karen@jobweb.org> "Careers in Architecture." E-mail
to recipient. 29 Aug. 1999.

Daley, Thelma T., and others, eds. *4:Construction*, 6th ed. Encino, CA:
Glencoe Press, 1998.

Duckman, Winchell. *A Guide to Professional Careers.* New York:
Julian Messner, 1999.

HDR, Inc. "Careers in Engineering, Architecture, and Corporate Jobs
and Career Opportunities at HDR" 1 May 2000 <http:// www.
hdrinc.com/ careers/hdr_job _opps.html>.

McReynolds. C. M. "So You Want to Be an Architect." *Progressive
Architecture* June 1999: 55–60.

Szerdi, John, President of Szerdi and Associates, Fort Lauderdale,
Florida. Personal interview. 24 June 1999.

U.S. Department of Labor. *Occupational Outlook Handbook.*
Washington, D. C.: USGPO, April 1999.

Wright, John W. *The American Almanac of Jobs and Salaries.* New
York: Avon Books, 1999.

  - - -. "Careers in Architecture." *Architectural Record* May 1999:
200–206.

FIGURE 6.4   *Sample MLA Works Cited page (the three dashes [- - -] indicate the work
is by the previously named author; alphabetization is then by title)*

Council of Biology Editors, the American Chemical Society, the American Mathematical Society, and the American Institute of Physics:

1. Each system requires a list of references with an assigned number to each entry. The list may be alphabetical and numbered consecutively, or the list may be arranged and numbered in the consecutive order in which the references are cited in the text.

2. In-text citations use the appropriate number usually at the end of the summary, paraphrase, or direct quotation. The in-text references vary from numbered superscripts[3] to parenthesized numbers (3) to bracketed numbers [3] to bracketed, boldfaced numbers [**3**]. The name of the authority may, of course, be included in the text, but the number serves as the key reference to the source.

3. Direct quotations require the number of the source and the specific page number in the in-text citation; for example, "The use of photosynthesis in this particular application is crucial to the environment" (8, p. 654).

CBE Style

The sixth edition of the Council of Biology Editor's *Scientific Style and Format: The CBE Manual for Authors and Editors* actually details two systems. The first style is used in general biological and earth sciences (agriculture, anthropology, archaeology, astronomy, biology, botany, geology, and zoology). Refer to the APA system earlier in this chapter for details of the first system; they are very similar, using name and year in-text parenthetical references. The entries in the lists of references are alphabetized; however, the title of the list of references will vary for the specific subjects:

- The references for a paper on agriculture will be titled *References.*
- The references for papers on anthropology or archeology will be titled *References Cited.*
- The references for papers on astronomy or geology will be titled *Literature Cited.*
- The references for biology, botany, and zoology will be titled *Cited References.*

The second CBE style is a number system used for the medical sciences (health, medicine, and nursing) and for the applied sciences (computer science, chemistry, mathematics, physics, and engineering), but variations exist for individual journals. Consult your instructor or specific journals.

Health, Medicine, Nursing (CBE Style)—in-text references. The medical sciences, as a general rule, employ the CBE numbering system or standards established by the American Medical Association's *Style Book Editorial Manual* or specific medical journals, such as *JAMA, Nutrition Reviews, Journal of American College, Health,* and others. References to the source may be cited by superscript number, by parenthesized number, or by bracketed number, according to the specific journal requirements:

> Cocaine use is heavily linked to homicide in New York City.[1]
> Cocaine use is heavily linked to homicide in New York City (1).
> Cocaine use is heavily linked to homicide in New York City [1].

CBE Bibliography Style for health, medicine, and nursing. Label the list *References.* Do not alphabetize the list; rather, number it to correspond to sources as you cite them in the text. Some samples follow:

1. Miner, K. J., Baker, J.A. Media coverage of suntanning and skin cancer: Mixed messages of health and beauty. *J. Health Ed.* 1994; 25: 234–238.

2. Antonovsky, A. *Health, Stress, and Coping.* San Francisco, Jossey-Bass, 1979.

3. Ayman, D. The personality type of patients with arteriolar essential hypertension. *AM J Med Sci.* 1983; 186: 213–233.

4. Scarlatis, George. Optical prosthesis: Visions of the future. http://www.amaassn.org/scipubs/msjama/articles/Vol_283/no_5/jmscool6.htm.

Computer Science (CBE Style)—in-text references. References to the source may be cited by superscript number or by parenthesized number:

> The prices of subscription database searchers is decreasing (4).
> The prices of subscription database search engines is decreasing.[4]

CBE Bibliography Style for computer science. All references are collated in numerical order and listed in the order of the text reference. Title the list *Works Cited* and use numbers followed by periods. Do not indent second and consecutive lines:

1. Gregory, A. "Three ways to insert superscripts." *PC Computing* 33 (July 1994), 54–57.

2. Swenson K. "Future trends in templates." *IEEE Trans. Knowledge and Data Eng.* 2 (2 March 1990), 37–39.

3. Wittmann, Art. "ASPs: A risky bet." Awittmann@nwc.com (accessed July 17, 2000).

4. Schafer, Maria. "Web architect builds a bridge between worlds." http://networkcomputing.com/1113/1113ca.html (accessed 17 July 2000).

Chemistry (ACS Style)—in-text references. References to the source may be cited in three ways: by superscript number, by parenthesized number, or by author name and date:

Oscillation in the reaction of benzaldehyde and oxygen was reported previously.[3]

Oscillation in the reaction of benzaldehyde and oxygen was reported previously (3).

The primary structure of this enzyme has also been determined (Dadel et al., 1984).

Even when references are cited by number, you may also use an author name, directly followed by the reference number:

Jensen (3) reported oscillation in the reaction . . .

In either the number or author/name system, if a reference has two authors, include both:

Recent investigations (Jones and Perez[9]) reveal that . . .

If possible, use a name and page number for a direct quotation:

"The use of genetic removal and replication is crucial" (Victor, 3, p. 44).

ACE Bibliography Style. All references are collated in numerical order if cited by number or in alphabetical order if cited by author. Title the list *References* and number or list each entry in the order of the text reference. If employing numbers, place the numbers in parentheses before the source, omit journal article titles, boldface the date for periodicals, and do not indent second and consecutive lines:

(1) Galidaut, M. et al. *J. Biol Chem.* **1994**, 268, 2166–2167.

(2) Dean, J. L.; Keith. T. T. J. *Organ. Chem.* **2000,** 59, 2745.

(3) Hsu, D. Chemicool Periodical Table. http://www.tech.mit.edu/chemicool/ (accessed Jan. 1999).

(4) Chemistry in the Community Discussion List, discussion of zinc and common cold in archived messages of September 1998, CHEMCOM@listserv.acsu.buffalo.edu.

(5) McCoy, M. Chem. Eng. News. [Online] **1998,** 76, 21–22.

Mathematics (AMS Style)—in-text references. Mathematics papers require the in-text citation number to be placed within brackets in boldface (use a wavy line under the number if you are not keyboarding the text):

Additionally, it is known **[3]** that every D-regular Lindelof space is D-normal. Further results on D-normal space will appear while in preparation **[4–5]**.

Mathematics Bibliography Style. All references are collated in numerical order if cited by number or in alphabetical order if cited by author. Title the list *References.* If not alphabetizing, number each entry in the order of the text reference. For books, the titles are underlined, publisher precedes city of publication, and specific page(s) of books need not be listed. For journals, the title of the article is italicized or underscored, journal titles are not italicized, volume is placed in boldface, followed by year of publication within parentheses, followed by complete pagination of the article. Notice the comma usage in the following samples:

1. H. Colonius and D. Vorberg, *Distribution inequalities for parallel models with unlimited capacity*, J. Math Psy. **38** (1994), 35–58.

2. R. Artzy, *Linear geometry,* Addison-Wesley, Reading, Mass., 1965.

3. O. Solbrig, personal e-mail on evolution and systematics, osolbrig@mit.edu/ (received 22 Aug. 2000).

Physics and Engineering (AIP Style)—in text references. Consult individual journals for differing styles and/or consult the *Online AIP Style Handbook 2000* at <http://www.aip.org>. Usually, in-text superscript numbers are used in the appropriate places.

Physics and Engineering Bibliography Style. All references are collated in the order of text reference, using superscript numbers. For books, titles are italicized or underscored, publisher precedes place of publication, and specific page references *should* be provided. For journals, the title of the article is omitted entirely, the title of the journal is abbreviated and *not* italicized or underlined, the volume is placed in boldface, and the year within parentheses follows the pagination. Samples follow:

[1]T. Fastie, W. G. Rouch, Phys. Today **44**, 37–44, (1991).

[2]C. C. Motchenbacher and F. C. Fitchen, *Low-Noise Electronic Design* (Wiley, New York, 1973), p. 16.

[3]M. Maezawa, T. Yamamori, and A. Shoji, Chip-to-chip communication using a single flux quantum pulse. http://www.ieee.org/pubs/pub_preview/asc_toc.html.

Figure 6.5 shows some selected text and a bibliography for a paper on health documented in the number system.

Venci 2

A brief history of hospice care will further define the philosophy behind the modern hospice care movement. In the Middle Ages,
the hospice was a place of refuge for the traveler. The religious
sects that operated the hospices offered food and shelter to the
poor. In the middle 1800s these shelters developed into retreats for
people who were dying from incurable diseases such as tuberculosis
(2). According to Hamilton and Reid, the advancement of hospice
care came in 1967 when Cicely Saunders opened St. Christopher's
Hospice in London (3). They state that St. Christopher's Hospice was
formed with the purpose of helping the terminally ill patient to remain comfortable and relatively free from pain without any artificial means to prolong dying (3). These same principles are the
foundation of the hospice concept in America today.

The hospice program has taken a different approach to the
treatment of the patient. Although the hospital and hospice are both
committed to medical needs of the patient, the hospital's emphasis is
on the care of victims of acute illness. The goals of the hospital, according to Michael Hamilton and Helen Reid, authors of *A Hospice
Handbook*, are to diagnose and cure disease through modern technology (3). Another hospice expert, Kenneth P. Cohen, states that since
technology is the main source of treatment, it is used to the extent

FIGURE 6.5 *Sample selected text documented and a reference page in number style*
(Courtesy of student Cathy Venci)

Venci 3

of prolonging life, even when there is no known cure (4). He adds that life support systems and aggressive treatment are normal procedures in the hospital setting (4). Health reporter Jane Toot, notes that the treatment of the terminal patient includes

- Intravenous fluids
- Forced feedings
- Nasogastric tubes
- Laboratory examination (5)

Cohen says that the most important treatment the dying patient needs is pain control therapy (4). According to Cohen, hospital procedures have a fixed routine of medication every four hours or so. The patient may be experiencing serious pain before relief is given (4).

There are differences between hospital and hospice attention to the dying patient's family. According to Toot, the hospital is not equipped to deal with the family (5). The patient is often in an acute-care ward that has very strict visiting regulations, which tend to create a sense of isolation for both the patient and the family (5). Toot notes that the family is often confused over the treatment of the patient. She states that this confusion is due to the hospital excluding the family in making decisions concerning the treatment of the patient (5).

FIGURE 6.5 *continued*

Venci 10

References

1. *Mosley's Medical Dictionary*, 1993 ed.

2. Hospice care. *Encyclopedia Americana*, 1993 ed.: 436–438.

3. Hamilton, M., Reid, H. *A Hospice Handbook*. Grand Rapids: William B. Eerdmans Publishing, 1980.

4. Cohen, K. P. *Hospice: Prescription for Terminal Care*. Germantown, Md.: Aspen Publications, 1996.

5. Toot, J. Physical Therapy and Hospice. *Physical Therapy Journal* 64 (1994): 665–670.

6. Coor, C., Coor, D. *Hospice Care Principles and Practices*. New York: Springer Publishing Company, 1983.

7. Venci, C. Survey on Value of Hospice Services. Moore County, 1997.

8. Mudd, P. High Ideals and Hard Cases. *Hastings Center Report* (April 1982): 11–14.

9. Boundy, D. Growth of Hospice Programs Is Cited. *The New York Times* 20 May 1994, sec. 22:6.

10. Friedland, S. Hospice Benefit off to Slow Start. *The New York Times* 22 Nov. 1994, sec. 11: 1, 7–8.

FIGURE 6.5   *continued*

CHICAGO STYLE

The fourteenth edition of *The Chicago Manual of Style* (1993) sets the standards for the fine arts (art, dance, music, and theater) and for some fields in the humanities (history, philosophy, religion, and theology, but not literature). This system employs superscript numerals within the text and places documentary footnotes on the foot (bottom) of the corresponding pages or as endnotes with all of the notes appearing together at the end of the paper. Although no separate bibliography page is usually necessary because it would be redundant, some professors may ask for one at the end of the paper.

In-Text Citations

Use Arabic numeral superscripts. Place the superscript numeral at the end of quotations or paraphrases with the number following immediately without a space after the final word or mark of punctuation, as in this sample:

> Dali advises: "Let us be satisfied with the immediate miracle of opening our eyes, becoming skillful in the apprenticeship of looking well."[11] For Dali photography exemplifies a new method of inquiry into physical appearances through which a simple operation like a change in scale can suggest unusual analogies.[12] Dali explains his position on Surrealism and the automatic process:
>
> > To know how to look at an object, an animal, through mental eyes is to see with the greatest objective reality. . . . To look is to invent. All this seems to me more than enough to show the distance which separates me from Surrealism.[13]
>
> In this way, rejecting Surrealism, automatism, and mimetic painting alike, Dali defines artistic invention as the process through which the physical world is framed as a representation.[14]

Footnotes and Endnotes

Primary Reference Footnotes. Place your footnotes (the corresponding bibliographical reference) at the bottom of the pages where the superscripts occur in the text. Separate the footnotes from the text with triple spacing or a 12-space line beginning at the left margin. Double space between footnotes, but single-space within each entry. Indent the first line five spaces. Your software may have a footnote/endnote feature to arrange your footnotes properly at the bottom of each page or as endnotes. In most instances, the software will first insert the superscript numeral in the text and then skip to the bottom of the page so that you can write the corresponding footnote immediately. Number the footnotes consecutively throughout the entire paper. Italicized book, journal/magazine, and newspaper titles are recommended.

Book

1. Salvador Dali, *Dali on Modern Art: The Cuckolds of Antiquated Modern Art* (New York: Knopf, 1957), 78–79.

Book with Three or More Authors

2. Rafael I Torella et al., *La Miel* (New York: Appleton, 1967), 307–423.

Journal

3. Clement Greenberg, "Avant-Garde and Kitsch," *Art News* 99 (1994), 111.

Newspaper

4. William Rubin, "Dada and Surrealism," *The New York Times,* 18 September 1979, sec. C, pp. C1, C12.

Webpage

5. Valerie Greenstein. "Homepage," 12 May 1999 <http://english.ehu.edu/greenstein/default.htm> (12 June 2001).

Article in a Reference Database

6. Tom Wilson, "'In the Beginning Was the Word. . .': Social and Economic Factors in Scholarly Electronic Communications," ELVIRA Conference Keynote Paper, 1009, 10 April 1995 <http://www.shef.ac.uk/~is/wilson/publications/elvira.html> (23 May 1999).

Special Document on the Web

7. Peter J. Bryant, "Dali's Art," *A History of Art* April 1999 <http://art.bio.uci.edu/~sustain/art65/index.html> (17 July 2000).

Primary Reference Endnotes. With the permission of your instructor, place all your notes together as a single group of endnotes. Begin the notes on a new page at the end of your paper. Title the page "Notes" and use the same format as for footnotes or use superscript numerals placed slightly above the line before each listing. Triple space between the heading and the first note. Double-space the notes and double-space between the notes. Indent each first line of the note five spaces and place second and consecutive lines at the left-hand margin of the page.

Subsequent References. After the first full reference, you should shorten subsequent references to the same work by using only the author(s) last name and the page number. When there are two works by the same author, use the author(s) last name(s), a shortened title, and the page number. In general, avoid Latin abbreviations (*loc. cit.* or *op. cit.*); however, whenever a note refers to the source in the immediately preceding note, use *Ibid.* (abbreviation for the Latin *ibidem* meaning "in the same place") with a page number. Do not italicize *Ibid.* in your note. Notice the footnotes 2, 4, and 5 following:

> 1. Salvador Dali, *Dali on Modern Art: The Cuckolds of Antiquated Modern Art* (New York: Knopf, 1957), 78.
> 2. Ibid., 79.
> 3. Clement Greenberg, "Avant-Garde and Kitsch," *Art News* 99 (1994), 111.
> 4. Dali, *Dali on Modern Art,* 79.
> 5. Ibid., 79.

Your professor may allow you to list all of your notes together at the end of your paper as endnotes because it simplifies your keyboarding of the notes on a computer. If you are allowed to group your notes at the end, entitle the page "Notes," centered and placed two inches from the top of the page. Triple-space to begin the first note. Indent each first line five spaces, and return to the left-hand margin for second and consecutive lines. Double-space the notes and between each note.

Final Reference List

As previously mentioned, a final reference list is redundant because you will have included complete documentation in your footnotes, but some professors may still ask you to prepare one. If you do use one, use a heading that best represents its contents, such as "Works Cited," "Selected Bibliography," or "Sources Consulted." Triple space between the heading and the first entry. Alphabetize the list by the author(s) last name and use the hanging indentation format (unlike your footnotes). Refer to the examples of works cited listings in the MLA system.

Because it is unlikely that you will use the Chicago style for documents written in this course, there are no sample text and reference listings here. Details on the system have been included to provide coverage of the main academic and professional documentation styles.

CHE✔KLIST

Documenting a Research Paper

GENERAL

❏ **1.** Have I selected the appropriate documentation system for the field of research I am pursuing?

❏ **2.** Have I correctly referenced by name and year or number each summary, paraphrase, and quotation?

❏ **3.** Have I provided a bibliography listing appropriate to the style I am using?

APA SYSTEM

❏ **1.** Have I clarified the directions for a working draft or manuscript for publication?

❏ **2.** Have I completed all the parts of the paper?

    ❏ Title page?

    ❏ Outline (optional)?

    ❏ Abstract?

    ❏ In-text documented paper?

    ❏ List of references?

❏ **3.** Have I double-spaced and numbered each page correctly?

❏ **4.** Have I used only last names in the text?

❏ **5.** Have I reported the findings in the past tense?

❏ **6.** Have I correctly referenced by name and year each summary, paraphrase, and quotation (and page)?

❏ **7.** If I am writing a working draft, have I adhered to all of the regulations?

❏ Courier or Times Roman 12 type?

❏ Underscored titles?

❏ Unjustified right margins?

❏ Paragraph indentation for the reference list?

❏ **8.** If I am writing a paper for publication, have I adhered to all of the regulations?

❏ Optional fonts and other document design features?

❏ Italicized titles?

❏ Justified right margin?

❏ Hanging indentation for the reference list?

MLA SYSTEM

❏ **1.** Have I completed all parts of the paper?

❏ Title page?

❏ Outline (optional)?

❏ Abstract (optional)?

❏ In-text documented paper?

❏ Content endnotes (optional)?

❏ The works cited listing?

❏ **2.** Have I double-spaced and numbered the pages correctly?

❏ **3.** Have I reported the research in the present tense?

❏ **4.** Have I correctly referenced summaries, paraphrases, and quotations?

❏ **5.** If I am including content endnotes, have I placed superscripts in the text and before the corresponding notes in the "Notes"?

NUMBER SYSTEMS (CBE, ACS, AMS, AIP)

❏ **1.** Have I distinguished the appropriate system of the two CBE systems according to the subject field?

❏ **2.** Have I documented my paper in the appropriate style for health, medicine, biomedicine, nursing, computer science, chemistry, mathematics, physics, and engineering?

❏ **3.** Have I chosen a consistent in-text documentation system for the discipline?

    ❏ Superscript numbers?

    ❏ Parenthesized or bracketed numbers?

    ❏ Boldface numbers in brackets?

❏ **4.** Have I provided either a numbered and alphabetized or a numbered listing in order of reference in the text as dictated by the subject?

❏ **5.** Have I, at minimum, included a titled first page of text, consecutively numbered pages, and a final list of sources appropriately titled?

CHICAGO STYLE

❏ **1.** Have I used consecutive, superscript numbers within the text for all summaries, paraphrases, and quotations?

❏ **2.** Have I used footnotes to explain each source at the bottom of the page corresponding to the reference, or has my professor allowed a complete list of notes at the end of the paper entitled "Endnotes"?

❏ **3.** Have I used the correct double-spacing for text plus a line to separate text and footnotes allowed by triple-spacing before the first entry?

❏ **4.** Are my footnotes single-spaced and each first line indented five spaces?

❏ **5.** If I use endnotes, is each entry double-spaced and listed with hanging indentation?

EXERCISES

1. Rewrite the content of Figure 6.4 works cited list in the MLA style into the style appropriate for your field of study. Obtain the appropriate manual from your library. If your discipline does not stipulate a particular style, use the APA style for this assignment.

2. **APA Documentation.**   Write the following information in the APA system for a paper on psychological aspects of communication:

    a. An article, "How We Learn to Communicate Feelings," published in the 1997, vol. 83 edition of Psychological Review, written by R. P. Langstroth and M. E. French on pages 41–45.

    b. An Appleton-Century-Crofts publication in New York of a book, Communication Theory, by Theodore Conover, M. D. in 1995.

    c. An article, "Do Men and Women Think Alike?" by M. P. Chadwick in the December 1997 edition of Psychology Today on pages 72–76.

3. **MLA Documentation.**   Rewrite the following information in the MLA style for a "Works Cited" listing:

    a. An article in Technical Communication by C. R. Miller entitled Some Thoughts on Document Design: on pages 108–111 of Volume 37 in July 1997.

    b. Can the Public Understand Scientific Terms? by Dawn Funder, published by Madsen Publishers, Inc. in 1995 in Peoria, Illinois.

    c. A Time magazine article entitled Technical Talk Is Tricky, on pages 37–39 of the February 17, 1996 issue.

    d. An Encyclopaedia Britannica article on CD-ROM published in 1995, written by T. S. Stephan.

4. **Number System Documentation.**   Rewrite the following information in the Number System for an article on computer science:

    a. A book, The Design of analysis of computer algorithms, written by M. E. Grasso, Seth Stephens, and J. D. Ulmann, published by Addison-Wesley in Reading, Massachusetts, in 1995.

    b. An article, Word Perfect 8.0 Sales Soar, by Manning G. Abrams in Computer, on pages 189–199 of volume 27 in September, 1997.

    c. Computers and word processing, on a July 15, 1997 posting on the Internet at <http://www.computers.com/wdprcing/html>.

WRITING PROJECT

1. **Collaborative Project.**   In groups of three, choose one of the suggested listed subjects. Combine your accessing information skills in the library, on a personal computer, through surveys and inter-

views, and so forth (Chapter 5) with your new documentation skills. Research a variety of sources (books, reference materials, periodicals, government documents, Internet materials, CD-ROMs, and so on) on the subject and compile a minimum 10-entry bibliography in the documentation system appropriate to the subject. You may want to extend this work into the development of a professional research paper required in Chapter 16, "Producing Professional Papers."

autism
aardvark mating rituals
new food source
nitric oxide
new surgical glues
new laser surgeries
new fuels
tomography
cloning human body parts
computer scanners
new heart medicine(s)
arthroscopic knee surgery
changes in welfare
changes in life expectancy
facts on UFOs
future of boxing
Mars discoveries
weather satellites
an electronic marvel
future space travel
Hong Kong today
Indonesian economic
    development
new boundaries in Croatia/
    Serbia
low cholesterol diet(s)
homosexuality
e-mail in the business
    world
suction lipectomy

vocal computer problems
zocor
ozone deletion
black holes
robotics
new weapon(s)
digital cameras
new chemotherapy procedures
changes in Medicare
magnetic resonance imaging
religious cults
last year's crime rate in your
    state
effects of weightlessness
trends in transportation
a new Ice Age
new building material(s)
computer-influenced slang
U.S. relations with China
American Catholicism
athletes and steroids
changes in the Democratic
    National Party
an aspect of cyberspace
changes in nursing studies
new tobacco industry
    legislation
abuse treatment centers
digitized TV
911 abuses

other?

NOTES

PART *three*

The Technical Strategies

CHAPTER 7

Preparing Manuals

by Jim Borgman

" WHADDAYA KNOW... AS IT TURNS OUT, WE _DO_ COME WITH AN INSTRUCTION MANUAL ! "

S K I L L S

After studying this chapter, you should be able to

1. Recognize the purpose and uses of various types of manuals.
2. Name the four major writing strategies employed in manuals.
3. Grasp the importance of appropriate language for different manual audiences.
4. Know the procedures for preparing a user manual.
5. Identify and comprehend the appropriateness of the graphics and other visuals in a manual.
6. Recognize the publication options (paper size, quality of paper, section tabs, tiers, etc.).
7. Name at least five other supplements that may be produced in conjunction with a user manual.
8. Evaluate a user guide for content, audience appropriateness, graphics, document design, and publication decisions.

INTRODUCTION

Manuals are written guides or reference materials that are used for training, organizing work procedures, assembling mechanisms, operating equipment or machinery, servicing products, or repairing products. You, no doubt, have any number of user manuals, also called user guides, owner's manuals, or operating manuals, for your computer, printer, fax machine, telephone, cellular phone, CD player, video cassette recorder, and even for your oven, refrigerator, and other appliances. Writing these manuals is the major task of technical writers. Typically, a manual includes

- Precise definitions
- Descriptions of mechanisms
- Step-by-step instructions
- Analyses of processes

In a manufacturing company, professional or technical writers typically prepare manuals for the installation, operation, and repair of products or equipment. Many smaller companies develop training, procedural, troubleshooting, and job review manuals. The preparation, editing, and reviewing of component parts of a manual may be delegated to any competent writer within the organization. Those employees who use such in-house manuals are continually encouraged to review and improve the component parts.

This chapter considers the user guide because it is the most common manual. Chapters 8, 9, 10, and 11 will consider the individual writing tasks of definition, description, instruction, and process analysis, which are the major technical writing components in manuals.

AUDIENCE

In a user guide the language level and technical detail must fit the intended user. Frequently, a novice or layperson is the user of a product or mechanism; thus he or she needs a mechanism description, definitions of terms, operating instructions, and an analysis with suggested solutions to possible problems in the product's operation, as well as many graphic illustrations.

Although the product may be a highly complex mechanism, the manual information must be simple, clear, and accurate. Writers of such user guides often aim their writing at the comprehension level of a seventh- or eighth-grade student. Which of you has not found your computer manual so difficult that you bought a book in the "Dummies" series or relied on the imbedded "Help" tools for a simpler explanation?

The user guide writer must carefully consider the formatting of the message as well as the graphics required. The technical terms, abbreviations, symbols, and mathematical procedures employed must fit the audience's ability to decode and understand the information. The following two examples illustrate language differences in manuals intended for two different audiences. The first example shows the introductory information from an operation manual for a personal pager. The intended audience is the layperson purchaser.

Introduction

Congratulations! You are now using the world's first microprocessor-controlled 900MHz pager. Motorola's advanced technology offers unique features and benefits which provide the ultimate in performance and reliability.

The Dimension 1000 pager is a versatile unit that is designed to provide reliable communications for a variety of applications. To get the full benefit from the pager, please read these operating instructions carefully.

Coding Data Label

Dimension 1000 pagers come in several model configurations equipped with a variety of options which affect the operation of your pager. To determine how your pager operates, refer to the coding-data label located under the belt clip. The pager is capable of one, two, or three calls, depending upon how it was ordered from the factory. The coding-data label indicates the number and type of calls (Figure 2). Tone-only calls are indicated by a

"T," and voice calls are indicated by a "V." If a particular call is not present, that area will be blank.[1]

The next example contains the introductory material from an instruction manual for servicing a walkie-talkie radio. The intended audience is skilled service technicians.

Introduction

The MX300-T "Handi-Talkie" radio described in this manual is the most advanced two-way radio available. Hybrid modular construction is used throughout, reflecting the latest achievements in microelectronic technology. The plug-in modules provide greater flexibility, greater reliability, and easier maintenance.

Each radio contains plug-in hybrid modules. These modules contain over 90% of the electronics—providing faster service and less down-time. Guide pins are provided on the modules to assist replacement and prevent incorrect insertion. Instead of complex wiring harnesses, printed flexible circuits are used in the radio. These durable, thin plastic films eliminate broken, pinched, or frayed wires—with a neat, easy-to-service interior.[2]

The first example is written with a "you" perspective. The language is general and nonspecific. The second example is less personal and contains technical terms such as *hybrid modular construction, plug-in hybrid modules, down-time,* and *flexible circuits,* terms familiar to technicians.

Many major companies market their products internationally. This may necessitate duplicates of user manuals in foreign languages. Shorter guides may include several foreign language translations along with the English version. The company which produces 3M, Scotchbond™, a dental etching gel, prints a two-sided, folded, 13″ × 13 ½″ guide covering storage and use, warranty information, precautions, and instructions in English, French, Spanish, German, Italian, Swedish, Danish, Dutch, Portuguese, Greek, and Russian.

MANUAL PREPARATION

Let us consider the writer's procedural steps for the preparation of a user guide for a technical product.

Step 1—Determining the Audience.   The technical writer must determine if the audience, that is, the potential users of the product, consists of laypersons or skilled technicians. The audience will have a bear-

---

[1]Motorola, Inc., *Motorola Dimension 1000 Binary GSC Pager* (Fort Lauderdale, Florida: Motorola, Inc., Paging Products Division, 1982), a manual.

[2]Motorola, Inc., *Motorola MX300-T Five Channel "Handi-Talkie" Portable Radio* (Fort Lauderdale, Florida: Motorola, Inc., Portable Products Division, 1982), a manual.

ing not only on the language level but also on the complexity of the graphics, the extent and scope of the data, and the manual size and format. If the intended user is a layperson, the technical writer or team must consult with the marketing department to determine through surveys and analyzed sales data who will buy the product, what language level is appropriate, what detail is essential, and what foreign language translations, if any, are necessary. Obviously, the technical writer needs effective interviewing skills.

Step 2—Consulting the Engineering Department. In the past, technical writers were involved only after the product was finished and were given three or four months to complete a guide before the product was shipped. The modern technical writer does considerably more "front end" consultation on product development. The individual writer, or more likely a writing team, consults the engineers to determine how the product works, what instructions and warnings must be stressed, and how much detail should be included for different audiences. These discussions are often tape recorded or even videotaped in order to provide exact documentation for the written materials.

Step 3—Writing the Manual. Bearing in mind the four typical components of user guides (definitions, descriptions, instructions, and process analysis), the writer or team will prepare a content draft; design the document pages, taking note of space considerations, headings, emphatic features, headers and footers, lists, numbering systems, and so on; and determine the need for a table of contents, copyright data, a warranty statement, a glossary and/or other appendixes, an index, and blank pages for notes. Figure 7.1 shows a typical table of contents for an extensive user guide.

Step 4—Preparing the Graphics and Other Visuals. Hand-in-hand with the manual text preparation come decisions about types of graphics and visuals and their execution. Whereas actual photographs and line art may be used to depict the overall product, drawings are used more extensively in manuals. They can emphasize parts and relationships by judicious use of exploded, cutaway, and schematic sketches and eliminate extraneous detail. Lists of items or features can be incorporated into tables for easy reference. Columnar drawings to depict steps in a process are often helpful. Print type and size decisions are important. Color, another important consideration, is usually used sparingly, but covers, borders, headings, and screens often employ color for emphasis and clarity. To protect the company and the user, cautions and warnings are strongly presented, often using graphic icons for emphasis. Figure 7.2 shows typical bold icon warnings. Be aware that the arrow for the "Note" warning is preferable to a pointing finger, which may be offensive in other cultures, for example in Asia.

Table of Contents

FIGURE 7.1   *Typical complex operation manual table of contents*

Step 5—Deciding on the Manual Production and Supplements.
The writer, or writing team must make decisions on the appearance of the
manual, the size and orientation of the paper, the paper quality, fonts, type
sizes, color usage, and the number of manuals to be printed. Other considerations include pagination, section tabs or reference bands for easy accessing, and a comment and evaluation sheet for eliciting feedback. Figure 7.3
shows an evaluation and comments sheet, plus its reverse side, with the
company's address and prepaid stamp. The user needs only to fill out the
form, tear out the page, fold it, staple it, and place it in the mail.

FIGURE 7.2 ***Typical bold warnings***

EVALUATION AND COMMENTS SHEET

FROM:
NAME: ____________ TITLE: ____________

BUSINESS
ADDRESS: ____________

TITLE, PUBLICATION NO., AND DATE ____________

1. BASED ON YOUR EXPERIENCE, RATE THIS MANUAL.

	Good	Average	Poor
Manual Organization			
Manual Readability			
Supporting Illustrations / Diagrams			
Installation Procedures			
Operator's Procedures			
Maintenance / Troubleshooting Procedures			

2. PLEASE WRITE COMMENTS IN THE SPACE PROVIDED BELOW INDICATING ERRORS YOU HAVE FOUND. INCLUDE PARAGRAPH OR FIGURE NUMBER AND THE PAGE NUMBER.

3. WHAT WOULD YOU SUGGEST WE ADD TO MAKE THE MANUAL MORE INFORMATIVE AND USEFUL?

Staple here

No Postage Stamp Necessary If Mailed In The United States

BUSINESS REPLY MAIL
FIRST CLASS PERMIT NO. 5567, MIAMI, FLORIDA

Postage Will Be Paid By Addressee

Racal-Milgo Information Systems
Attn: Technical Writing
8600 N.W. 41st Street P.O. Box 020399
Miami, Florida 33012-8968

Tear Out and Mail

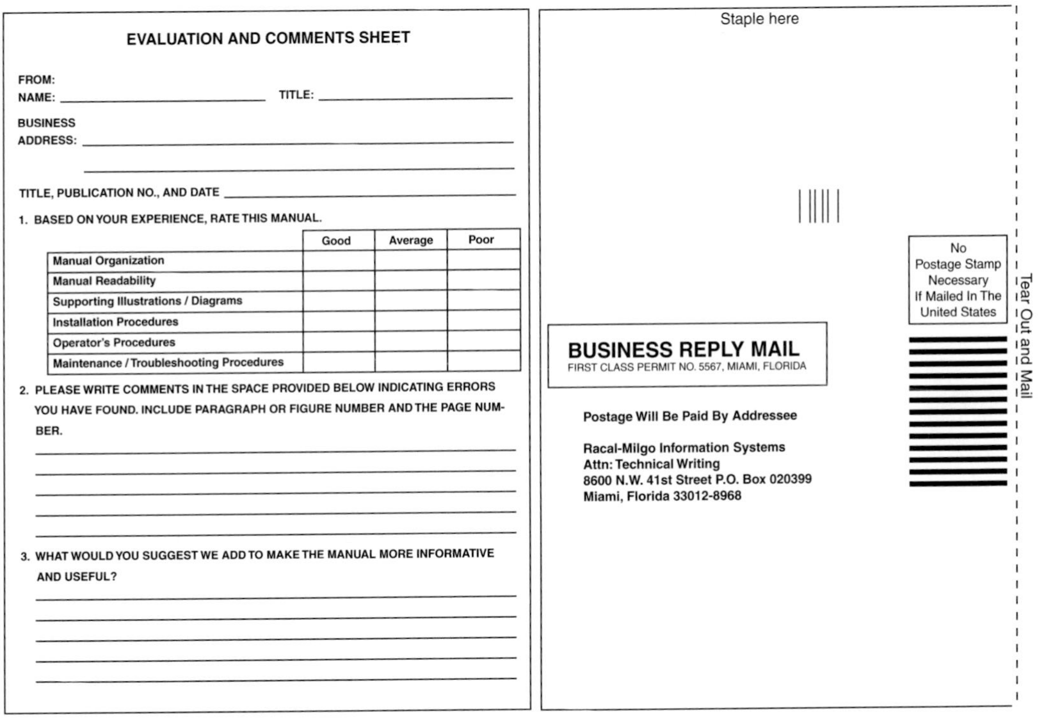

FIGURE 7.3 ***Typical feedback sheet with return address on the reverse side***

Depending on the product and the user, several presentation formats may be considered: fanfolded pamphlets, folders with looseleaf pages, or booklets with a spiral, saddle, or perfect binding.

Currently, technical writers tend to produce **supplements** in the form of, perhaps, three or more **tiers** of manuals: a fanfolded brochure, or a perforated tear-off card for use as quick reference guides; a second-level instruction manual of more detailed material; and a third-level, highly technical and comprehensive manual for the sophisticated user. In addition to print material, technical writers are increasingly providing instructional audiotapes, videotapes, CD-ROMs, computer disks, large posters illustrating basic steps, and even operational information embedded in the product itself (such as the "Help" tools on your computer). Visual aids such as CAD conversions, schematics, and digitized photographs (line art) are increasingly common.

More decisions need to be made on how best to include accessory materials with the print material. Options include end pockets, accompanying plastic bags for computer disks, and boxlike casings for CD-ROMs or videotapes, to name a few.

To reinforce the written and support materials, the technical writer is responsible for a coherent library of posters, flyers, product promotional materials, and multimedia materials. It is the technical writer or team who develops the online materials: the company Web page, product information pages, online manuals, and even quick-line movies showing operations for downloading. All these materials must have the same look and feel. Desktop publishing software, such as Adobe FrameMaker, combines word processing, page layout, graphics, color features, and foreign language translations. It also possesses on-line distribution features, such as disseminating documents directly onto the World Wide Web, inserting hypertext links (highlighted, colored words, phrases, or pictures that can be clicked to skip directly to related documents elsewhere in a collection), and answering product questions submitted online. In fact, the technical writing departments of major manufacturers spend about 60 percent of their time writing and 40 percent developing multimedia presentations and other technical support material.

Figure 7.4 shows one side of a fanfolded brochure, "Four Quick Steps to Everyday Use," produced to accompany a cellular phone. Figure 7.5 shows a "Quick Reference" tear-off page for posting, which is included in the second-tier manual for the same product. In addition to the brochure and the tear-off page second-tier materials, there is an even more comprehensive third-tier supplement manual titled "A to Z Reference," but it is not reproduced here.

To familiarize you with a typical manual, Figure 7.6 shows a complete Motorola pager user guide. It is reproduced actual size, which equates the size of the product and the product packaging. The guide is printed on both sides of the paper and is bound by a staple (saddle binding). The print is landscaped (printed along the length of the paper) to facilitate more efficient and pleasing use of space. Graphics are combined with instructions and analyses on each page. Comments in the left margin explain the content.

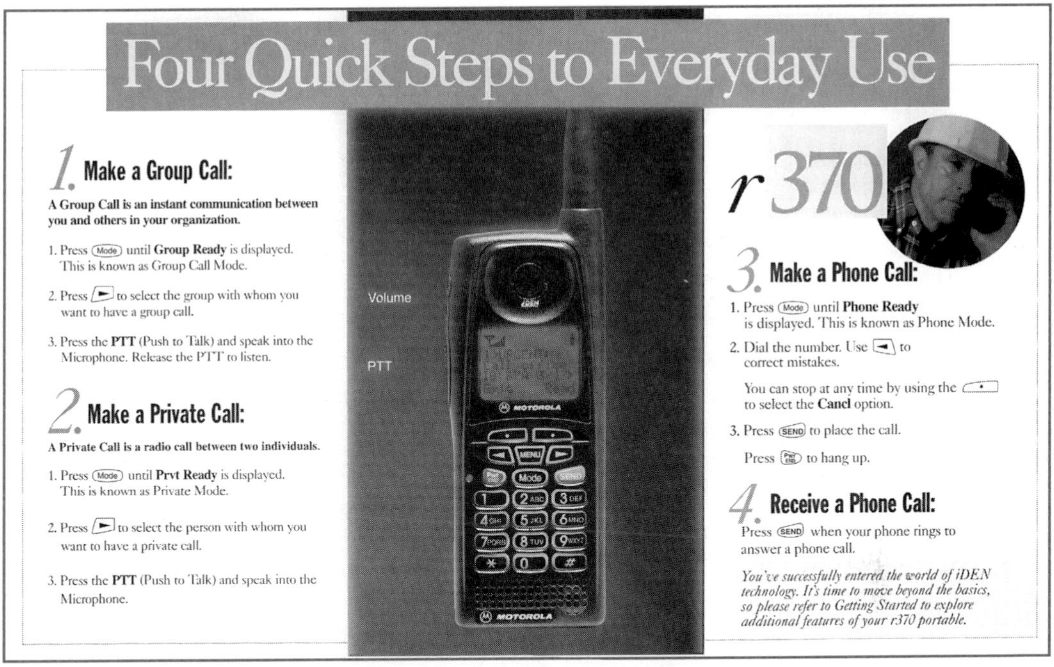

FIGURE 7.4 ***A first-tier, fanfolded user guide for the Motorola i360 cellular telephone***
(Courtesy of Motorola, Inc., Fort Lauderdale, Florida)

PHONE MODE

Making a Phone Call
1. Press (Mode) until you see **Phone Ready.**
2. Enter phone number.
3. Press (SEND) to place call.
4. Press (Pwr/END) to end call.

Answering a Call
Press (SEND) to answer incoming call.

Additional Phone Features
1. Press (MENU) to view other options.

Speed Dial
2. Select **Spd#.**
3. Enter the storage number.
4. Press (SEND) to place the call.

Call Forward
2. Select **Forwd.**
3. Enter the number where you want your calls forwarded.
4. Select **On.**

Call Waiting
2. Select **Wait.**
3. Select **On.**

GROUP CALL MODE

Making a Group Call
1. Press (Mode) until you see **Group Ready.**
2. To initiate the call to the talkgroup shown, press and hold the PTT (Push to Talk) button. Wait for the chirp and then begin speaking.
3. Release the PTT to listen.

PRIVATE CALL MODE

Making a Private Call
1. Press (Mode) until you see **Prvt Ready.**
2. Enter the Private Call ID of the person you want to reach.
3. Press and hold the PTT & speak into the microphone.

Sending a Call Alert
1. Select **Alert.**
2. Scroll to the person to whom you want to send the Call Alert. You may also select the person by using Alpha Search.
3. Press the PTT to send the Call Alert.

FIGURE 7.5 ***A "Quick Reference" tear-off page for posting from a user manual***
(Courtesy of Motorola, Inc., Fort Lauderdale, Florida)

Controls
graphic with
labeled parts

Side bar
references
throughout

Definition of
each button use

Page number
footers
throughout

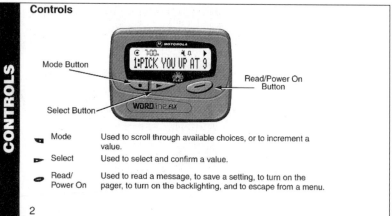

Controls

Mode Button

Read/Power On
Button

Select Button

	Mode	Used to scroll through available choices, or to increment a value.
	Select	Used to select and confirm a value.
	Read/ Power On	Used to read a message, to save a setting, to turn on the pager, to turn on the backlighting, and to escape from a menu.

CONTROLS

2

Descriptive
graphic
showing actual
pager print

Instruction
with brief
analysis

Inserted
boldface,
italic
Note

Turning Your Pager On

MOTOROLA

• 12:35P   4/23

Press ◯ to turn your pager on.

A start-up message is displayed momentarily and your pager activates the currently selected alert.

Note: The start-up alert can be stopped by pressing any button.

When the pager is on and no activity is taking place, the standby screen is displayed. The standby screen displays the power indicator, the time and date, the alert mode, and may display other pager indicators such as alarm status.

GETTING STARTED

3

Description and
analysis

Notice menu
and button
icons to save
words
throughout

Descriptive
graphics

Bulleted and
numbered *in-
structions*
throughout

Menu Icons

The four menu icons on the top row (Ⓜ 🗑 ◀〇) correspond to the four menus: CONTROLS, DELETE ALL, ALERTS, and ALARMS.

The first time you press ◣ these menu icons are displayed. The controls menu icon Ⓜ flashes, indicating that pressing ▶ will enter the controls menu.

Press and release ◣ to display the CONTROLS?, DELETE ALL?, ALERTS?, and ALARMS? menu prompts. Press ▶ to enter the corresponding menus, or press ◯ to return to the standby screen.

Turning Your Pager Off

❶ From the standby screen, press ◣ to display CONTROLS?. The controls menu icon Ⓜ flashes.

❷ Press ▶ to display OFF?.

❸ Press ▶ to turn the pager off. The off screen is displayed without any icons.

GETTING STARTED

4

FIGURE 7.6   *Complete user guide manual incorporating*
definitions, descriptions, instructions, and
analyses   (Courtesy of Motorola, Inc., Boynton Beach, Florida)

Notes, *analysis*
graphics, and
instructions
continued on
both pages

Setting the Time and Date

CONTROLS?

❶ From the standby screen, press ▾ to display CONTROLS?. The controls menu icon Ⓜ flashes.

OFF?

❷ Press ▸ to enter the CONTROLS menu.
❸ Press and release ▾ until TIME/DATE? is displayed.

TIME/DATE?

❹ Press ▸ to enter the TIME/DATE menu. The hour field flashes.

12:00   1/01

❺ Press and release ▾ to adjust the hour.
❻ Press ▸ to move to the next field.

12:35P   4/23

❼ Repeat steps 5 and 6 for the minutes, AM/PM, month, and day fields.
❽ Press ◨ from any field to save and return to the standby screen.

8

Hint: Pressing and holding ▾ scrolls through selections quickly.

Setting the Incoming Message Alert

You can set your pager to alert with a vibrating alert (vibration with no alert tone), one of eight audio alerts, a chirp alert (short beep alert), an alert of increasing volume (Escalert), or no alert (completely silent).

ALERTS?

❶ From the standby screen, press and release ▾ until ALERTS? is displayed. The alert icon ◀ flashes.

VIBRATE?

❷ Press ▸ to enter the ALERTS menu. VIBRATE? is displayed.

AUDIO?

❸ Press and release ▾ until your choice of VIBRATE?, AUDIO?, CHIRP?, ESCALERT?, or NO ALERT is displayed.

CHIRP?

❹ Press ▸ to select the desired alert. The standby screen is displayed with the corresponding alert icon.

9

Descriptive
graphics, hints,
instructions,
and *analysis* on
both pages

Note: If the audio alert mode is selected, your pager automatically displays the audio alert options screen.

Choosing an Audio Alert

If you select the audio alert mode, you can set your pager to alert with one of eight audio alerts.

If you choose no alert or vibrate, your pager emits an audio alert only if a priority message is received, or an alarm sounds.

AUDIO?

❶ From the ALERTS menu, press and release ▾ until AUDIO? is displayed.

ALERT 1

❷ Press ▸ to enter the AUDIO menu.
❸ Press ▾ until your choice of audio alert is displayed. The pager emits a sample of each alert.

12:35P   4/23

❹ Press ▸ to select your choice of audio alert. The standby screen is displayed with the audio alert icon ◀.

10

FIGURE 7.6 *continued*

Analysis,
descriptive
graphics, notes,
and *instructions*
on each page

Setting the Alarm

Your pager has three alarms. Each alarm can be set for either a specific time and date, or for a specific time on a daily basis. If your pager is off when an alarm sounds, it remains off.

Note: An alarm always emits an audio alert.

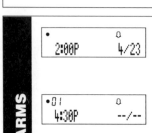

❶ From the standby screen, press and release 🔲 until ALARMS? is displayed. The alarm icon ◻ flashes.

❷ Press ▷ to enter the ALARMS menu. The alarm number 01 flashes.

❸ Press 🔲 to choose alarm 01, 02, or 03.

❹ Press ▷ to move to the next field. The enable ◻ or disable ◻ icon flashes.

❺ Press 🔲 to enable ◻ or disable ◻ the alarm.

❻ Press ▷ to move to the hour field. The hour flashes.

ALARMS

11

ALARMS

❼ Press and release 🔲 to adjust the hour.
Hint: Press and hold 🔲 to adjust the setting quickly.

❽ Press ▷ to move to the next field.

❾ Repeat steps 7 and 8 to set the minutes, AM/PM, month, and day fields.

❿ Press ⬭ in any field to save and exit.

When an alarm is enabled, the alarm icon ◻ is displayed on the standby screen.

To alert on a daily basis, set the month and day to --/--.

When an alarm expires, the alarm icon ◻ flashes, ALARM is displayed with the alarm number, and your pager alerts.

Press any button once to stop the alarm, and again to clear the message.

12

Read Mode

This feature allows you to choose the most comfortable way to view messages.

Setting the Scroll Speed

You can choose the speed at which your messages scroll, or read them line by line.

❶ From the standby screen, press 🔲 to display CONTROLS?. The controls menu icon Ⓜ flashes.

❷ Press ▷ to enter the CONTROLS menu.

❸ Press and release 🔲 until SCROLL? is displayed.

❹ Press ▷ to enter the SCROLL menu.

❺ Press and release 🔲 to choose the scroll speed.

❻ Press ▷ to select the scroll speed.

There are four scroll speeds to choose from: LINE-BY-LINE, SCROLL 1, SCROLL 2, and SCROLL 3 (fastest).

READ MODE

13

FIGURE 7.6   *continued*

Instructions,
analysis,
and *descriptive*
graphics
on each page

MESSAGE FEATURES

Locking Personal Messages

By locking messages, you can save personal messages to prevent them from being replaced when the memory is full. Messages can be locked only while reading them.

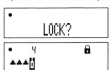

❶ Press ◥◣ while reading a personal message to display LOCK?.

❷ Press ▭▶ to lock the message. The message indicator 🔒 is displayed.

When a locked message is selected or read, the lock icon 🔒 is displayed.

Note: A maximum of eight personal messages may be locked at one time. To lock another message, you must first unlock at least one message.

Unlocking Personal Messages

❶ Press ◥◣ while reading a locked personal message to display UNLOCK?.

14

MESSAGE FEATURES

❷ Press ▭▶ to unlock the message. The message indicator ▲ is displayed.

Private Time

This feature allows you to turn off all pager alerts during a preselected time period. Messages received during this time period are stored. When enabled, private time works on a daily basis. When disabled, no pager alerts are turned off during the private time setting.

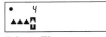

❶ From the standby screen, press ◥◣ to display CONTROLS?. The controls menu icon ⓜ flashes.

❷ Press ▭▶ to enter the CONTROLS menu.

❸ Press and release ◥◣ until PRIVATE TIME? is displayed.

15

MESSAGE FEATURES

❹ Press ▭▶ to enter the PRIVATE TIME menu. The private time enable ⓒ or disable • icon flashes.

❺ Press ◥◣ to enable ⓒ or disable • private time.

❻ Press ▭▶ to move to the next field. The hour field of the start time flashes.

❼ Press ◥◣ to adjust the start time hour.

❽ Repeat steps 6 and 7 to adjust the start time minutes, AM/PM, stop time hour, minutes, and AM/PM.

❾ Press ◖▭ from any field to save and exit. While private time is active, ◀ is not displayed, and the pager does not alert.

16

FIGURE 7.6 *continued*

Instructions

Deleting Messages

Messages may be deleted one at a time or all at once.

Deleting a Single Message

❶ While reading a message press and release ▼ until DELETE? is displayed.

❷ Press ▶ to delete the message.

Deleting All Messages

The DELETE ALL command deletes all read and unlocked personal messages and information services. Locked or unread messages are not deleted.

❶ From the standby screen, press and release ▼ until DELETE ALL? is displayed. The delete icon 🕮 flashes.

❷ Press ▶ to enter the DELETE menu. The delete confirmation DELETE? is displayed.

❸ Press ▶ to delete all personal messages.

MESSAGE FEATURES

17

MESSAGE FEATURES

Storing Messages

Your pager can store up to 16 personal messages. Each stored message is assigned a number, which is displayed when the message is stored. The first message received is 1, the second is 2, and so on.

Automatic Message Deletion

If all message slots are full and a new message is received, the oldest unlocked read message is automatically deleted.

When the message memory is full, MEMORY FULL is displayed. Press any button to return to the standby screen.

If all messages are unread, the oldest, unlocked message is deleted and OVERFLOW is displayed.

Analysis on next two pages

18

Information Services

Information services are typically news or financial reports which provide information that is relevant for a short time (a few hours). Your pager has two information service message slots. You can set each information service message slot to alert you when you receive an information service message. Contact your service provider if you are interested in receiving information services.

INFORMATION SERVICES

19

FIGURE 7.6   *continued*

Analysis,
descriptive
graphics,
and *instructions*
on all three
pages

Reading Information Services

When an information service message is received, ▣ flashes for 12 seconds and the number of unread information services is displayed. After 12 seconds, the standby screen is displayed.

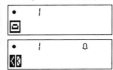

❶ Press 👝 to display the message indicator menu.

❷ Press 👝 to enter the information service menu.
The information service message icons display █ when selected and ▢ when unselected.

❸ Press 👝 to move to the message you want to read. The corresponding message slot number is displayed.

20

❹ Press 👝 to read the message. The time the information service was received is displayed with the first screen of the message.

▶ indicates the message is continued on an additional screen.

Press 🖝 and then 👝 to display the previous screen.

Turning the Information Service Alert On and Off

You can set an information service alert for each message slot.

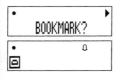

❶ While reading an information service message, press 🖝 until CHIRP ON? or CHIRP OFF? is displayed.

❷ Press 👝 to turn chirp on or off for that information service message slot.

21

Turning the Information Service Bookmark On

You can set a bookmark to hold your place in a lengthy information service while reading it.

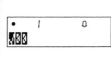

❶ While reading an information service, press 🖝 to display BOOKMARK?.

❷ Press 👝 to activate the bookmark. The message menu screen is displayed.

√ is displayed in the information service menu. The next time you read that information service, the last screen displayed while the message was marked is the starting point of the display.

█ indicates a selected bookmarked message.

22

FIGURE 7.6   *continued*

Analysis of problems

Out of Range

If your pager is equipped with this feature, and if you are outside your paging coverage area, ⊻ is displayed. As long as ⊻ is displayed, your pager cannot receive messages.

Message Error Icons

If there is an error in the message received, the error icon ▮ is displayed at the end of the message.

If this option is enabled, and if ╂ is displayed at the end of the message, either the message was too long, or there was not enough memory to store the message.

Patent information

Patent Information

This Motorola product is manufactured under one or more Motorola U.S. patents. These patent numbers are listed inside the housing of this product. Other U.S. patents for this product are pending.

Cleaning instructions

Cleaning Your Pager

To clean smudges and grime from the exterior of your pager, use a soft, non-abrasive cloth moistened in a mild soap and water solution. Use a second cloth moistened in clean water to wipe the surface clean. Do not immerse in water. Do not use alcohol or other cleaning solutions.

23

OTHER FEATURES

Functionality and Use of Your Pager

For questions pertaining to the functions and use of your Motorola pager please visit our web site at www.mot.com/pagers or call 1-800-548-9954. For questions pertaining to your paging service, contact your paging service provider.

Care and Maintenance

The WORDline and WORDline FLX pagers are durable, reliable, and can provide years of dependable service; however, they are precision electronic products. Water and moisture, excessive heat, and extreme shock may damage the pager. Do not expose your pager to these conditions. If repair is required, the Motorola Service Organization, staffed with specially trained technicians, offers repair and maintenance facilities throughout the world.

You can protect your pager purchase with an optional extended warranty covering parts and labor. For more information about warranties or repair, please contact your paging service provider or retailer, or call Motorola, Inc. at 1-800-548-9954.

Troubleshooting analysis and *instructions*

USE AND CARE

24

Battery Information

Your WORDline or WORDline FLX pager operates with one AAA-size alkaline battery. When the battery is low, the low-battery icon ▯ is displayed between the time and date on the standby screen. Change your battery within five days of receiving a low-battery indication.

Messages are retained when replacing the battery. To retain pager alert settings, turn the pager off before removing the old battery.

Battery analysis, instructions, and descriptive graphic

Replacing the Battery

❶ Turn your pager off.

❷ To remove the old battery, slide the battery door lock towards the top of the pager to unlock the battery door.

❸ While pressing on the battery door, slide the door until the ribs on the battery door align with the ribs on the back cover.

❹ Lift the battery door to free it from the housing.

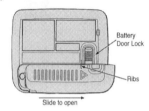

BATTERY

25

FIGURE 7.6   *continued*

Battery
instruction
continued

BATTERY

❺ Remove the battery.

❻ Align the new battery so the positive (+) and negative (-) markings match the polarity diagram in the battery compartment. Insert the battery.

❼ To replace the battery door, align the grooves on the battery door with the grooves on the back cover and slide the battery door closed.

❽ Slide the battery door lock toward the bottom of the pager to lock the battery door.

Double-sided
pages for notes
not reproduced
here

26

WORDline™ and WORDline™ FLX Quick Reference Card

Indicators and Icons

- • On
- ☾ On, Private Time enabled
- ▯ Message, selected
- ▲ Message, unselected
- Ⓜ Controls menu
- ▥ Delete All menu
- ◀ᴵ Audio alert
- ◀ Alerts menu/Chirp alert
- ◀ Vibrate or no alert (silent) icon
- ♪ Alarm menu/Alarm enabled
- ♪ Alarm disabled
- 🔒 Locked message
- ᛉ Out of range

- ▶ Message continuation
- ▭ Low-battery
- ▱ Information service menu, selected
- ▱ Information service menu, unselected
- ▯ Information service message, selected
- ▯ Information service message, unselected
- ▯ Information service, Chirp on, selected
- ♪ Information service, Chirp on, unselected
- ▯ Information service, Bookmark, selected
- √ Information service, Bookmark, unselected
- ▯ Locked message, selected
- 🔒 Locked message, unselected

Control Buttons: Mode   Select   Read/Power On

Next two pages
are the front
and back of
a folded quick
reference
card with
perforations to
tear it from the
manual cover

Setting the Alarm

❶ Press ◢ until ALARMS? is displayed.

❷ Press ▸.

❸ Press ◢ to choose alarm 01, 02, or 03.

❹ Press ▸ to move to the next field.

❺ Press ◢ to enable ♪ or disable ♪ the alarm.

❻ Press ▸ to move to the hour field.

❼ Press and release ◢ to adjust the hour.

❽ Press ▸ to move to the next field.

❾ Repeat steps 7 and 8 to set the minutes, AM/PM, month, and day fields.

❿ Press ⬮ from any field to save and exit.

To alert on a daily basis, set the month and day to --/--.

Locking and Unlocking Messages

While reading a personal message, press ◢ to display LOCK? or UNLOCK?. Press ▸.

Deleting a Single Message

Press ◢ while reading a message until DELETE? is displayed. Press ▸.

WORDline and WORDline FLX Menu Map

Controls	Delete All	Alerts	Alarms
┌OFF	└DELETE	┌VIBRATE	┌ALARM 01
├TIME-DATE		├AUDIO	├ALARM 02
├SCROLL		├CHIRP	└ALARM 03
└PRIVATE TIME		├ESCALERT	
		└NO ALERT	

FIGURE 7.6   *continued*

Standard
copyright
information

6881028B95-O

FIGURE 7.6 *continued*

CHECKLIST

Preparing Manuals

PRELIMINARY

❑ **1.** Have I considered the needs of the audience (laypersons, skilled technicians, international users, etc.)?

   ❑ Appropriate language?

   ❑ Amount of detail?

   ❑ Foreign language translations?

CONTENT

❑ **1.** Have I included definitions, descriptions, instructions, and process analyses as needed?

❏ **2.** Have I designed appropriate graphics?

    ❏ Actual photographs?

    ❏ Overall drawings?

    ❏ Exploded drawings?

    ❏ Cutaway drawings?

    ❏ Schematics?

    ❏ Tables?

    ❏ Warnings and directional icons?

    ❏ Other?

❏ **3.** What other inclusions are necessary?

    ❏ Title page?

    ❏ Table of contents?

    ❏ Copyright data?

    ❏ Warranty statement?

    ❏ Glossary?

    ❏ Other appendixes?

    ❏ Index?

    ❏ Comment sheet for feedback?

    ❏ Blank pages for notes?

DOCUMENT DESIGN

❏ **1.** Have I considered other document design elements?

    ❏ Will it be printed in portrait?

    ❏ Will it be printed in landscape?

    ❏ Spacing?

    ❏ Tabs and columns?

    ❏ Line length, leading, kerning?

    ❏ Headings, fonts, type sizes?

    ❏ Margins, indentations, justifications?

❑ Lists?

❑ Emphatic features (boldface, italics, underlining, numbers and bullets, icons, symbols, reversed type?

❑ Headers and footers?

❑ Borders, fills, watermarks?

❑ Color?

PRODUCTION MATTERS

❑ **1.** Have I considered production matters?

    ❑ Number to be printed?

    ❑ Size of package and manual?

    ❑ Paper quality?

    ❑ Bindings?

    ❑ Pockets for supplements?

    ❑ Section tabs or reference bands?

    ❑ Other?

SUPPLEMENTS

❑ **1.** What other promotional materials should be included?

    ❑ Audiotapes?

    ❑ Videotapes?

    ❑ CD-ROMs?

    ❑ Computer disks?

    ❑ Posters and/or flyers?

    ❑ Imbedded "Help" tools?

EXERCISE

1. **User Guide Feature Recognition.** Collect a variety (five or six) of fairly short user guides for such items as a pager, a telephone answering machine, a portable phone, a calculator, or an alarm clock with multiple features. Include some that you deem weak. Save these for use in this and the next four chapters. Select one for this exercise, photocopy the pages, paste each on a separate blank page, and identify in the margins its parts: title page, table of contents, user identification, definitions, descriptions, instructions, and analysis sections. What else is included?

WRITING PROJECTS

1. **Evaluation of a Manual.** Select the best manual (see the previous exercise) and write a brief report on its effectiveness. Comment on its overall format (size, type of paper, page designs, headings, color, boldface print, numbering system, etc.), its language, the clarity of its mechanism description, awareness of the need for definitions, clear instructions, and the extent of analysis of operational procedures and problems. Use headings in your report and attach the manual, or a photocopy of it, to your report.

2. **Evaluation of Graphs and Other Visuals.** Select another of the manuals you collected for the exercise and write a brief report on the graphs and other visuals. Concentrate on the effectiveness of the graphs and visuals. Are photographs appropriate and helpful or would drawings be more effective? Are the drawings helpful? Are there sufficient numbers of graphs and visuals? Do warnings, notes, and cautions stand out clearly? Are tables used to good effect? Are the fonts and type sizes logical and helpful? Are the graphs and visuals well-labeled and titled? Could other content be presented graphically for beneficial effect? Is color used? Effectively? Other comments?

3. **Collaborative Project—Evaluation of an Ineffective Manual.** With a group of three or four, select a poorly executed manual and write a report on its shortcomings. Comment on the overall format. Is the size appropriate for the mechanism? Is the type of paper appropriate? Too glossy? Too thin? Is the cover design weak? Is the table of contents helpful? Why or why not? Are the page designs logical? Are the headings effective? Are fonts, type sizes, and color used effectively? Are the directions numbered clearly? Are boldface type, cautions, warnings, all capitals, and icons used effectively? Are the graphics, drawings, and other visuals used well? Are there too many graphics? Too few? Is the placement logical and helpful? Is there an

adequate photograph or drawing of the mechanism and its parts? Are the labeling and explanation of parts effective? Are all necessary terms defined? Are the instructions numbered, sequential, and easy to follow? Is there an analysis section (a troubleshooting guide and/or an explanation of the function of the mechanism)? What other weaknesses does the guide have? Utilize all you have learned about document design and writing strategies in your own report. Photocopy or attach the manual to your collaborative report.

NOTES

CHAPTER 8

Defining Terms

ZITS

by Jerry Scott and Jim Borgman

Reprinted with special permission King Features Syndicate.

S K I L L S

After studying this chapter, you should be able to

1. Understand the need for precise definitions.
2. Consider the options for the placement of definitions in a document.
3. Name, explain, and execute the seventeen methods of definitions.
4. Recognize and avoid definition fallacies.
5. Write formal sentence definitions for a list of related terms.
6. Write an extended definition employing at least ten definition strategies.

INTRODUCTION

Because the English language contains the largest, most complex vocabulary of any known language (more than 500,000 official words, excluding the technical, scientific, and most colloquial), it is not difficult to comprehend the need for clarifying terms in every kind of document. This is especially true in professional and technical writing in which jargon, shoptalk, and formality often meet head to head. Computers and related technology have introduced an entire new "language." Words such as *cyberspace, digital cellular networks, gigabyte, hypertext,* and *encryption* bombard us. Even related jargon and argot words, such as *hacker, newbie, booting up, flame,* and *surfing,* baffle the novice.

All medical professionals must be able to recognize words (*melanoma, proteasers, cyclospora*) or a misunderstanding can result in the death of a patient. Engineers must be able to communicate with technicians (*solenoid, circuitry, torque*), and realtors must understand the terminology in their field (*escrow account, liens, easements*). Usually, students find that the most common problem they encounter in their course work is the definition of new terms pertinent to their fields of study. A student who does not fully understand the meaning of the terms *square root* and *equation* will have great difficulty in a math class, and one who cannot define *modifier* or *gerund* will encounter difficulty in English composition.

Moreover, there are two official forms of the English language: Standard English, which spans most of the English-speaking world, and Standard American English, which is used only in the United States. The British and most countries once ruled by Great Britain often spell words differently (*colour* for *color, centre* for *center*) and assign different meanings to American English words (*torch* for *flashlight, jumper* for *sweater*). Thus, a well-developed vocabulary relative to both location and profession is essential to good writing. To define a term requires a solid understanding of what the term actually means in a given context—a difficult task in a language that is filled with words that sound alike, look alike, or are vague in meaning.

AUDIENCE

The audience determines the need for and the extent of any one definition. Writers of manuals, informal or formal reports, professional articles, and the like must define all terms that may be unfamiliar to the audience, words that may have more than one meaning, and those that are used in a special or stipulatory manner. You must determine whether your audience is high-tech, low-tech, layperson, single or multiple, of ethnic or international background, or in a special field.

Consider the simple word *tongue,* which has at least eight distinct meanings to different audiences:

To a biologist	A tongue is a fleshy movable portion of the floor of the mouth of most vertebrates that bears sensory end organs and small glands, and that functions in taking and swallowing food, and in humans as a speech organ.
To a geographer	A tongue is a long, narrow strip of land projecting into a body of water.
To a cobbler	A tongue is a flap under the lacing or buckle of a shoe at the throat of the vamp (the part of the shoe that covers the instep and toe).
To a linguist	A tongue is a spoken language; the manner, or quality of utterance; or the intention of a speaker.
To a belt maker	A tongue is a movable pin in a buckle.
To a carpenter	A tongue is the rib on one edge of a board that fits into a corresponding groove in an edge of another board to make a joint flush (an even and unbroken line).
To a bellmaker	A tongue is a metal ball suspended inside a bell so as to strike against the side as the bell is swung.
To some religious groups	Tongue is the charismatic (divinely inspired) gift of ecstatic (extremely emotional) speech.

The parenthetical definitions of *vamp, flush, charismatic,* and *ecstatic* underscore the need for definition as an essential to meaningful communication.

Definitions may be integrated into a paper, may constitute a major portion of a paper, may be the primary purpose of a paper, or may be included in a glossary, usually at the end of a paper. An instruction manual may include an introductory list of terms to be used within. A policy handbook may begin each chapter with relevant definitions. Legal contracts may define words used within the document.

SOURCES OF DEFINITIONS

The most obvious source of a definition for a term is a dictionary. The standard dictionary for the English language is the *Oxford English Dictionary* (OED), a rather voluminous, but comprehensive, source of definitions providing complete pronunciations, spelling variations, etymologies (the origins and development of words), denotations (concrete, literal meanings), connotations (abstract, implied meanings), and historical perspectives on each different use of the term. The OED is good for an in-depth analysis of a word, but its use may represent an uneconomical use of time for someone in business or technology.

For quicker reference, there are numerous desk dictionaries (e.g., *Webster's New World Dictionary, The American Heritage Dictionary, The American Dictionary of the English Language, The Random House Dictionary*) that are more convenient to use. Though not as comprehensive as the OED, they usually provide the denotation of the term, the most common connotative meanings, some etymology notations, and synonyms. Encyclopedic dictionaries, such as *Webster's Third International Dictionary,* also provide graphics and other data relative to words. Numerous online dictionaries are also available. Try

> *Merriam-Webster Collegiate Dictionary* – <http://www.m-w.com>
>
> *Cambridge International Dictionary of English* – <http://www.cup.cam.ac.uk/elt/dictionary>

There are any number of computer, medical, and scientific dictionaries online, too. Textbooks, technology handbooks, user guides, and documents within a given field may also be valuable sources for definitions.

METHODS

The extent to which a term should be defined depends not only on the audience but also on the complexity of the term itself. Terms may be defined by

- Parenthetical expression
- Brief phrase
- Formal sentences
- Extended sentences or paragraphs

Parenthetical Definition

The simplest way to define a term is to include a synonym in parentheses directly after the term:

> The top half of a drainage map drawing is the plan (aerial view); the bottom half is the profile (horizontal view).
>
> The ring top (round, spoked, carrying handle) of a fire extinguisher corroded (wasted away).

Definition by Brief Phrase

Sometimes a defining phrase will clarify your term:

> If the body temperature is abnormal, which is above or below a range of 97.6°F to 99°F, further diagnostic procedures should follow.
>
> The cause of Reye's Syndrome, a relatively rare but serious disease that appears to be related to a variety of viral infections, particularly chickenpox and influenza, is unknown, although some studies have found an increased risk after the use of aspirin during a viral illness.

Formal Sentence Definition

A specific pattern exists for precise definition of terms. The pattern consists of three parts: the name of the term, the class of the term, and the characteristics of the term which distinguish it from all other members of its class. Some examples follow:

Term	*Class*	*Distinguishing Characteristics*
Arbitration	is a process	by which both parties to a labor dispute agree to submit the dispute to a third party for binding decision.
Sediment	is matter	which settles to the bottom of a liquid.
Assets	are items owned	such as cash, receivables, inventories, equipment, land, and buildings.
Paranoia	is a personality disorder	in which a person feels persecuted or has ambitions of grandeur.
Arson	is a criminal act	of purposely setting fire to a building or property.
A patent	is an inventor's exclusive right	which is granted by the federal government, to own, use, make, sell, or dispose of an invention for a certain number of years.

Down-time   is a period   during which a computer system is
                                        inoperable due to power failure or
                                        hardware breakdown.

Sometimes a formal definition requires more than one sentence in order to read smoothly:

A rifle is a firearm that has spiral grooves inside its barrel to impart a rotary motion to its projectile. It is designed to be fired from the shoulder and requires two hands for accurate operation.

These definitions concentrate on just one meaning of the term. The term *rifle* is also a verb, *to rifle,* which is "to cut grooves," or "to ransack," "rob," or "pillage," or "to search and rob."

DEFINITION FALLACIES (ERRORS)

A fallacy is an illogical or misleading error in a formal sentence definition. A definition fallacy is, therefore, an illogical or misleading definition including terms that are too technical, too broad, too narrow, circular, or that misuse *where* and *when.*

Too Technical.   To be useful a definition should not contain terms that are more technical or confusing than the term itself. Consider which of these two definitions is more useful to a layperson:

Too technical   Dysgraphia is a transduction disorder that results from
                visual motor integration disturbance.

Improved   Dysgraphia is a writing disorder that results from a difficulty in writing what one sees.

The first definition might well convey meaning to a group of learning disability specialists, but the second is more helpful to an undergraduate student.

Too Broad.   The definition writer must avoid using words that are too abstract or broad. Consider the following: "Cerumen (term) is a substance (class) in the internal ear." Is the substance waxy, hard, liquid? How does it get to the inner ear? Is it secreted by a gland or picked up externally? Is it in the canal? Is it in the earlobe? Other examples and their improvements are:

Too broad   Marl (term) is earth (class) that is used in several products
            (distinguishing characteristics).

Improved   Marl (term) is a crumbly soil (class) that consists mainly of clay, sand, and calcium carbonate and is used as a fertilizer and in the making of cement or bricks (distinguishing characteristics).

Too Narrow. Conversely, it is also fallacious to define a term too restrictively. Consider: "A chair is a four-legged seat." Doesn't this restrict the possibility that a chair may have a single pedestal for a base, or have five or more legs arranged in a circle? Other examples and their improvements are:

Too narrow	A catkin (term) is a yellow, tassel-like spike (class) on a birch tree (distinguishing characteristic).
Improved	A catkin (term) is a tassel-like spike (class) that consists of closely clustered, small, unisexual flowers without petals as on a willow, birch, or poplar (distinguishing characteristics).

Circular. It is also important to avoid using any form of the term in the second and third parts of the definition; to do so takes your reader in a circle. Consider this statement: "A radical is a person having radical views concerning social order and systems." Does it define radical? Other examples are:

Circular	Comatose is the state of being in a coma.
Improved	Comatose is a pathological condition characterized by unconsciousness, lethargy, or torpidity (dormant numbness).
Circular	Fertilization is the process of fertilizing.
Improved	Fertilization is the act or process of applying, inseminating, impregnating, or pollinating a plant or cell to assist reproduction and growth.
Circular	A surveyor is one who surveys.
Improved	A surveyor is a trained person who examines the condition, situation, or value of land or other measurable construction.

When and Where. Finally, one must avoid the terms *when* and *where* in formal definition. To write "A crypt is where one is buried" fails to classify the term as "a subterranean chamber" or to distinguish it from a mausoleum or a cemetery, other places where people are buried. Other examples are:

***When* fallacy**	Osmosis is when fluid passes through a membrane into a solution of higher concentration.
Improved	Osmosis is a digestive process in which a fluid passes through a membrane into a solution of higher concentration.
***Where* fallacy**	The Internet is where people can connect to a global network.
Improved	The Internet is a global network of interconnected sites that can be accessed by people with a computer and a service provider.

In summary, when devising formal definitions do not (1) use needlessly technical language, (2) employ classes or distinguishing characteristics that are too broad, (3) employ classes or distinguishing characteristics that are too narrow, (4) use any form of the term itself in the class or characteristics citation, or (5) use the terms *where* or *when* in place of a classification.

EXTENDED DEFINITION

If your intent is to define a term so that your audience has a thorough understanding of it, you may need to extend your formal definition and write what is known as an *extended* or *expanded* definition. Such is the case of engineers or scientists attempting to explain their developments to the less technically oriented, or of inventors seeking a patent for a newly developed invention. The desired definition can be achieved by combining a number of different methods of definition and carefully analyzing the significance of each to the term in question.

The extended definition may be a few sentences, a paragraph, several pages, or even, for the linguist, a lengthy report or thesis. The following methods of definition can be combined to create the extended definition.

Denotation

Denotation is the literal, concrete definition of the word. It is usually represented in dictionaries by the Arabic numeral 1 and often seems far removed from the common usage of the word, as we tend to use language abstractly rather than literally. For example, the *Oxford English Dictionary* tells us that the word *nurse,* when used as a noun, actually means "one who gives suck to a baby." Of course, the term more commonly implies "one trained to care for the sick or infirm." Thus, the denotative meaning of the term is less commonly used than the implied meaning.

Connotation

The connotative, or implied, definition of a word is often the most recognizable to a reader. This is an abstraction from the denotative meaning, sometimes to the point of losing all its original meaning. The connotative usage of a word or phrase is most often used in everyday speech, especially in informal situations. The denotative meaning of the word *grind* is "an action which crushes substances into bits or fine particles between two hard surfaces." A connotative meaning of the word *grind* is "a student who is very industrious in his studies." It is often helpful to understand and explain the denotative meaning of a word in order to comprehend and explain the connotative.

Examples

A second approach to extended definitions is to provide examples. These may be brief or extensive. Two samples follow:

> A parasite is a plant or animal which lives on or within another organism, from which it derives sustenance or protection without making compensation. Some parasites are tapeworms, sheep ticks, lice, and scabies.

> A market is a state of trade which is determined by prices, supply, and demand. In salesmanship the term *market* may refer to a trade or commerce in a specific service or commodity, such as the housing market, the stock market, or the designer jean market. Investors in a market study it carefully before investing. In a stock market a potential investor studies the stock market exchange and current prices of stocks to determine if the "market" is going up or down and to decide whether an investment would be profitable at a particular time.

Description

Following your formal definition, which distinguishes the item from other members of its class, it may be wise to describe the physical parts of the item, if, indeed, the term names an object. The following example begins with a formal definition and includes a brief physical description:

> An otoscope is a hand-held, diagnostic instrument which is used for examining the external canal of the ear and the ear drum. Composed of plastic, aluminum, and glass, it consists of three main parts: a dry-cell battery barrel assembly, a lens assembly, and a speculum (reflector).

To further describe its parts, subparts, dimensions, weight, and method of use would entail other specific writing strategies, which are covered in Chapter 9. A graphic illustration incorporated into the text would help to clarify the term. Figure 8.1 shows the main parts of an otoscope.

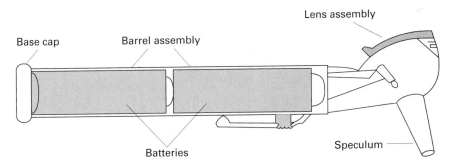

FIGURE 8.1 *The main parts of an otoscope*

Synonym

A third expansion of the basic definition is to list synonyms for the term. When readers are familiar with the synonym, they can better understand the definition. Here are some examples of synonym extension:

> A law is a rule of conduct which is established and enforced by the authority, legislation, or custom of a given community or other group. *Rule, regulation, precept, statute,* and *ordinance* are all synonyms for the word *law.*
>
> A motive is an inner impulse or reason that causes a person to do something or to act in a certain way. The terms *intent, incentive,* and *inducement* are sometimes used synonymously with *motive.*

One must be careful to remember that no two words are ever exactly synonymous. Meanings may overlap, but a careful examination of the differences is what conveys precise meaning.

Contrast/Negation

A fourth writing strategy for extended definitions is to contrast or to negate the term from those terms or items in the same class with which it may be confused. For example:

> In criminal investigation *motive* and *intent* are not truly synonymous. A man who provides a lethal drug to a terminally ill patient has the "motive" to alleviate suffering. Still his "intent" is to kill, making him criminally liable for the death.
>
> A stalactite is an elongated deposit of carbonate of lime which hangs from the roof or sides of a cave. It is not a stalagmite, which is a cone-shaped deposit of carbonate of lime which extends vertically from the floor of a cave.

We often understand better what a thing is by knowing what it is not.

Comparison

Conversely, we can better understand what some terms denote if we can examine similarities to things more familiar. Two comparisons for extending definitions follow:

> A stalactite resembles an icicle in that both are hanging, cone-shaped deposits formed by dripping water. In the case of the stalactite, the water evaporates, leaving the lime deposit; in the case of an icicle, the dripping water freezes.

Resistance, the opposition to the passage of electrical current which converts electric energy into heat, may be compared to friction between any two objects. When a sulfur-coated matchstick is rubbed against an abrasive surface, the friction creates a spark which ignites the sulfur and produces heat.

Analogy

The extended comparison of two otherwise dissimilar things is called an analogy. Although a computer and a thermometer are basically different, the following analogy helps us to understand an analog computer:

An analog computer is an electronic machine which translates measurements, such as temperature, pressure, angular position, or voltage, into related mechanical or electrical quantities. The operating principle of an analog computer may be compared to that of an ordinary thermometer. As the weather becomes cooler or warmer, the mercury in the glass tube falls or rises. The expansion and contraction of the mercury have a relationship to the condition of the weather. The thermometer provides a continuous measurement that corresponds to the climatic temperature. The analog computer makes continuous scientific computations, solves equations, and controls manufacturing processes.

Origin

Examining the source of an item helps us to grasp the meaning of the word. The following passage briefly explains the source, the mining, and the metallurgy of tin:

The earliest known tin is found in bronze (a copper-tin alloy) items excavated at Ur, dated about 3500 B.C. Tin is an element that occurs in cassiterite deposits. The ore is recovered by both opencut and underground mining. The smelting processes include roasting and leaching in acid to remove all of the impurities. The crude tin is resmelted and then refined by further heat treatments of two steps: liquation or sweating, and boiling or poling. Finally, the pure tin is cast in the form of 100 lb ingots in cast iron molds.

Etymology

Etymology is the study of the origin and development of words, and it provides insight to clear understanding of terms. English is an Indo-European language. The majority of our spoken words are derived from the Germanic Old English language of the early Anglo-Saxons. The Norman invasion of 1066 brought the French language to ascendancy in England for roughly 300 years. The majority of words we use when writing formally in English are of Latin and Greek origin. The Renaissance added between 10,000 and 12,000 new words to the English vocabulary. World exploration and settlement added immeasurably to our language. Immigrants and Native Americans in North America introduced their vocabularies, which we adopted into our lexicon. Some examples of these words are

German	*sauerkraut, hamburger*
Chinese	*ketchup, tycoon*
Caribbean	*rodeo, hacienda*
Swedish	*ombudsman, dynamite*
Basque	*bizarre, chaparral*
Spanish	*siesta, coyote*
Hindi	*veranda, calico*
Arabian	*safari, sofa,*
Native American	*skunk, caucus*

Some of our words are acronyms, such as *NATO* (for *N*orth *A*tlantic *T*reaty *O*rganization) and *scuba* (*s*elf-*c*ontained *u*nderwater *b*reathing *a*pparatus). Other words are blends, such as *flurry* (from *flutter* and *scurry*) and *smog* (from *smoke* and *fog*). Still others derive from biographical names, such as *willemite* (a mineral named after King William I) or *magnolia* (a plant named after botanist Pierre Magnol). Many are coined (invented), such as *cyberspace, diskette,* and *astronaut.* New words are constantly entering into our vocabulary.

History

Another way to define a term is to discuss its history. Consider the word *curfew.* A curfew in its modern usage is an order establishing a certain time, usually at night, when certain movement restrictions apply, such as governments forbidding assembly to snuff out civil unrest, or parents imposing a time for teenagers to be home. The word derives from an Old French term. Town criers would order people to put out their fires by a certain time (all the houses were of wood) by shouting *couvre feu,* literally meaning "cover the fire." You can understand how the word came to mean what it does today when you consider its history.

Cause/Effect

To understand some terms completely, an examination of causes and effects is useful. The following example describes the causes and effects of a tornado to amplify the formal sentence definition:

A tornado is a storm characterized by a violently rotating funnel cloud that has a narrow bottom tending to reach to the earth. The cloud may rotate clockwise or counterclockwise at approximately 100–150 mph. The funnel cloud results from the condensation of moisture through cooling by expansion and lifting of air in the vortex. The air outside of the funnel cloud is also part of the vortex, and near the ground this outer ring becomes visibly laden with dust and debris. Although a tornado takes only a minute or so to pass, it results in devastating destruction. Buildings may be entirely flattened, exploded to bits, or moved for hundreds of yards. Straws are known to be driven through posts. The roar of a tornado can be heard as far as 25 miles away.

Process Analysis

A final strategy of expanded definition is to analyze the process in which the item is involved:

A skeleton is the bony framework of any vertebrate animal. It gives the body shape, protects soft tissue and organs, and provides a system of levers, operated by muscle, that enables the body to move. Bones of a skeleton store inorganic sodium, calcium, and phosphorous and release them into the blood. The skeleton houses bone marrow, the blood-forming tissue. Bones are joined to adjacent bones by joints. The bones fit together and are held in place by bands of flexible tissue called *ligaments*.

A pressure cooker is an airtight metal container that is used for quick food preparation by means of steam under pressure. When the lid of the pot is secured by means of a rubber gasket and the container is placed on a heat source, fast-flying molecules of steam constantly bump against each other and the inside surface of the container. The combined blows from all molecules exert heat and pressure, "cooking" the food in approximately one-third of the time required by conventional cooking methods.

Process analysis may be extended into a separate writing strategy and is discussed thoroughly in Chapter 11.

Graphics

Providing a picture, a photograph, or line drawing of the object represented by a word is probably the easiest way to define it. The graphic should be clear, free of all extraneous detail, and may include labeled parts, exploded views, or cutaways.

Figures 8.2 and 8.3 contain two extended definitions. Each begins with a formal sentence definition and utilizes several extended definition strategies. Graphics enhance the definitions. Be prepared to critique these samples in class.

A DEFINITION OF VOLCANO

Formal
definition

     A volcano is a hill or mountain that is formed by lava, ashflows, or ejected rock fragments that come from a central vent. The word *volcano* derives from the name of the

Etymology

little island in the Mediterranean Sea called Vulcano. Many centuries ago the people of this island believed that this lava and ash came from the forge of Vulcan, god of fire and metalworking.

Cause

     In actuality, the lava flows from a magma chamber close to the earth's surface. This chamber is formed from rock that has melted due to increased temperature or re-

Analogy

duced pressure. This second cause, reduced pressure, might best be compared to a pressure cooker. While its lid is on, pressure is maintained, and the food is slowly being cooked. However, when the lid is removed, the pressure drops rapidly and steam rises violently into the atmosphere. If the pressure underneath the earth remains constant, rock will stay in its motionless, solid state, but if the pressure drops, rock will quickly melt and rise toward the earth's surface.

Description

     There are three major types of volcanoes. First, the shield volcano has a gently sloping cone because it is formed from solidified lava flows. The slopes are usually between 2 degrees and 10 degrees. Second, the cinder cone consists of ejected rock fragments, called pyroclasts, such as dust, ash, cinders, and bombs. Since many of the fragments land near the central vent, a peak usually forms. Slopes are generally 30 degrees for this type. Third, the composite volcano is formed by alternating layers of pyroclasts and solidified lava flows. Because of its construction, it erodes at a slower rate than do the other two types of

FIGURE 8.2   *Sample extended definition in expository format*
(Courtesy of student Debra Fuhrhop)

volcanoes and has a steep slope. Figures 1 through 3 show a shield, cinder cone, and composite volcano:

Graphics

Figure 1 Shield Volcano   Figure 2 Cinder Cone

Figure 3 Composite Volcano

Examples

Some examples of composite volcanoes are Mount Rainer and Mount St. Helens in Washington State, Mount Hood in Oregon, Mount Cotaopaxi in Ecuador, and Mount Fuji in Japan.

FIGURE 8.2 *continued*

DEFINITION OF MICROBURST

Introduction

   During takeoff or landing, aircraft can encounter hazardous flight conditions associated with weather storms
known as microbursts. A microburst is a vertical cylindri-

Formal
definition

cal shaft of air between 300 and 100 meters in diameter.
The air within the shaft is colder and thus more dense than
the surrounding atmosphere.

   This denser state causes the shaft air to plunge from
storm-top altitudes downward towards the earth at significant velocities. Upon encountering the ground, the air mass

Description

mushrooms radially outward in a horizontal direction. As
the air diffuses over the ground, it begins to warm, causing
it to become dense. Upon reaching the microburst's radial
edges, the diffusing air and associated winds curl upward
as its less dense state causes it to rise. A portion of the air
mass returns to storm-top altitudes along the outer edges
of the vertical shaft. In this manner, a microburst is able
to sustain itself with varying intensity for up to 30 minutes. Figure 1 illustrates a typical microburst event.

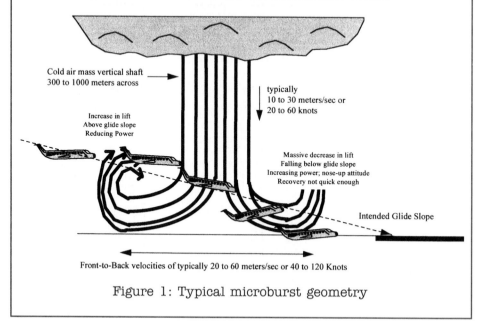

Cold air mass vertical shaft
300 to 1000 meters across

typically
10 to 30 meters/sec or
20 to 60 knots

Increase in lift
Above glide slope
Reducing Power

Massive decrease in lift
Falling below glide slope
Increasing power; nose-up attitude
Recovery not quick enough

Intended Glide Slope

Front-to-Back velocities of typically 20 to 60 meters/sec or 40 to 120 Knots

Figure 1: Typical microburst geometry

FIGURE 8.3   *Sample of a highly technical extended definition*
(Courtesy of Robert Johns, AlliedSignal, Inc.)

Effects

Downward air mass (and hence wind) velocities within the vertical shaft typically range between 10 and 30 meters per second. When the air translates to a horizontal flow at the base of the shaft, the outflow winds typically have front-to-back velocities ranging from 20 and 60 meters per second. Windshear is a defined change in non-turbulent wind velocity and within a microburst. As illustrated in Figure 2, an observer slicing horizontally in range through a microburst would view high velocity oncoming (i.e., *head*) winds with little vertical component diminishing to primarily vertical (i.e., downward) winds followed by high velocity outgoing (i.e., *tail*) winds with little vertical component.

As the air mass plummets toward earth, the rain drops within it mainly have vertical velocity.

Near the ground, air mass mushrooms outward, giving rain drops both horizontal as well as vertical motion; the horizontal rain drop velocity is measured by the radar.

Doppler Measurement Zone

Wind fields

Wind flow horizontal velocity

Range

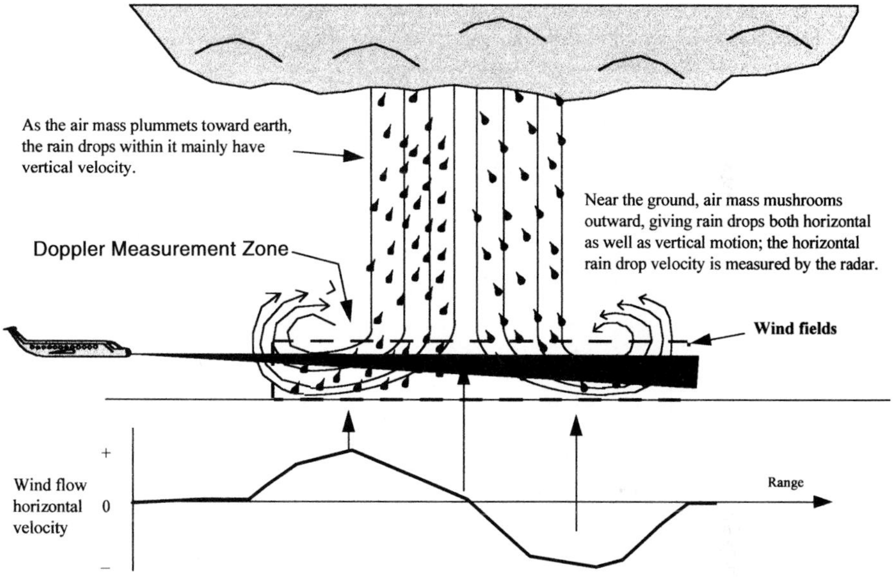

Figure 2: Microburst wind velocity profile

Effects

This change in both wind velocity and direction as a function of range results in a proportional change in the magnitude and direction of aerodynamic forces acting upon the observer. Such a force change is classically defined as a shear; hence the term *windshear* and its characterization as a function of horizontal wind velocity properties.

FIGURE 8.3 *continued*

CHECKLIST

Defining Terms

PRELIMINARY

❑ **1.** Have I considered the audience and field for which the definition is intended?

❑ **2.** Have I checked various dictionary definitions for the term(s)?

❑ **3.** Have I determined where to place my definitions (in the introduction, within the body, or in an appendix)?

METHODS

❑ **1.** Will a parenthetical definition suffice?

❑ **2.** Will a definition by brief phrase suffice?

❑ **3.** Does the term require a formal sentence definition?

   ❑ Have I stated the term?

   ❑ Have I named a class for the term?

   ❑ Have I provided distinguishing characteristics?

❑ **4.** Have I avoided the fallacies?

   ❑ Too technical?

   ❑ Too broad?

   ❑ Too narrow?

   ❑ Circular?

   ❑ *When* or *where?*

❑ **5.** Are methods of extended definition required?

   ❑ Connotations?

   ❑ Examples?

❏ Description?

❏ Synonyms?

❏ Contrast/negation?

❏ Comparison

❏ Analogy?

❏ Origin?

❏ Etymology?

❏ History?

❏ Cause/effect?

❏ Process analysis?

❏ Graphics?

EXERCISES

1. **Fallacies.** There is something wrong with each of the following formal definitions. Rewrite each by providing a precise class and distinguishing characteristics.

    a. A latent image is a prephotographic image on a film that cannot be seen.

    b. Cramming is when a student attempts to learn most of the contents of a course in a short period of time.

    c. Celluloid is a substance that is thin and inflammable and was formerly used for motion picture films.

    d. A bond is when two things, such as concrete and steel, are adhered.

    e. Anxiety is when one is paralyzed with fright and the source of the fear is unknown.

2. **Parenthetical Definition.** Add a brief word or phrase of definition in parentheses after each italicized word.

    a. The culture was studied *in vitro*.

    b. The doctor *sutured* the wound.

    c. The *loess* improved the fertility of the soil.

    d. An *implosion* occurred during the experiment.

    e. The President *vetoed* the bill.

3. **Etymology.**   Look up the etymologies of the following words. How do their linguistic histories help to clarify their current denotations?

anecdote	zero	vandal	talent
candidate	snob	stocks	sock
rain check	meander	graffiti	cobalt

4. **Formal Definitions.**   Write a one-sentence formal definition for five of the following terms. Avoid the five fallacies.

flextime	veto	apogee
kinetics	chiaroscuro	marinade
nanosecond	lift (aviation)	plutocracy
spectrum	steroid	corona
clone	cyberspace	biotechnology

5. **Strategies.**   In the following extended definition, identify each method of definition employed by writing the name of the method in the space provided after each sentence.

A cyclone is a storm that may range from 50 to 900 miles in diameter and that is characterized by winds of 90 to 130 mph blowing in a circle—counterclockwise in the northern hemisphere and clockwise in the southern hemisphere—around a calm center of low atmospheric pressure while the storm itself moves from 20 to 30 miles per hour. _____ Cyclones may be called whirlwinds, hurricanes, and typhoons. _____ The term *hurricane,* however, is properly applied only to a cyclone of large extent and suggests the presence of rain, thunder, and lightning. The term *typhoon* refers to tropical cyclones in the region of the Phillipine Islands or the China Sea. _____ A tornado is not a cyclone. Although a tornado consists of whirling winds, it is characterized by a funnel-shaped cloud which is far smaller in diameter than a cyclone and by winds far exceeding the velocity of winds in a cyclone. _____ The term *cyclone* is derived from the Greek word *kykloma,* which means "wheel" or "coil." _____.

WRITING PROJECTS

1. **Definitions of Terms in a Specific Field.**   Select five related terms from your professional field. For this assignment try to avoid terms that name mechanisms. Title your assignment by stipulating the field of the terms: for example, "Terms Used in Radiation Technology," "Terms Used in Geology," and "Terms Used in Architecture." Develop formal sentence definitions for each of the terms. Each definition must state the term, the class, and the distinguishing characteristics that differentiate your term from all others in its class. Avoid the fallacies. Following are some suggested terms in specialized fields, but select your own if you wish or if your field is not included.

Word Processing

menu
wordwrap
block move
properties
merge

Computers

byte
icons
software
modem
virus

Internet Terms

surfing
host
userid
hypertext
baud

Internet Chat Group Terms

flame
posting
hot chat
smileys
domain

Fashion

godet
grommet
stonewashed
peplum
double-faced linen

Electronics

electron
resistance
frequency
capacitance
Ohm's law

Criminal Justice

larceny
felony
manslaughter
assault
battery

Allied Health

emphysema
atherosclerosis
angina
escemia
vasodilation

Fire Science

arson
pyromaniac
flammable
purple K
cartridge

Psychology

anxiety
psychosis
neurosis
schizophrenia
manic depressive syndrome

Marketing

a good
convenience good
shopping good
specialty good
unsought good

General Business

sole proprietorship
partnership
limited partnership
corporation
conglomerate

Surveying	*Architecture*
azimuth	fascia
stadia	cantilever
transverse	soffit
hub	strut
transit	beam

Political Science	*Astronomy*
democracy	nova
communism	black hole
socialism	albedo
oligarchy	transit
monarchy	solar eclipse

2. **Denotations.**   Look up the denotation of a word in your field of study in four different dictionaries including the OED and compare the definitions in a brief paper. How much etymology is included? History? Denotative meanings? Connotative meanings? Examples? Descriptions? Synonyms? Contrasts or negation? Origins? Causes and effects? Process analysis? Graphics? It is doubtful that the dictionaries will provide all of these extensions of definition. Which dictionary is best? Why?

3. **Collaborative Project—Extended Definition.**   In groups of three, select a broad term (one naming a field of study or a concept or a phenomenon). Write an expanded definition of the term that is suitable for first-year students in the field. Begin with a formal sentence definition and then expand your definition by employing at least ten extended definition strategies. One of the strategies may be a graphic or visual. Use parenthetical or phrase definitions for unusual terms within your extended definition. In the margin indicate the writing strategies that you have employed. Include a brief bibliography listing in the appropriate style for your subject. Some suggested terms are

political science	architecture	speech therapy
psychology	sociology	biology
cyberspace	electronics	plasma sphere
astronomy	accounting	computer science
sunspots	aurora borealis	thermal tide
law	tort	prosecution
semantics	linguistics	jargon
AIDS	hepatitis	cancer

NOTES

Describing Mechanisms

BLONDIE

by Drake Young

Reprinted with special permission of King Features Syndicate.

S K I L L S

After studying this chapter, you should be able to

1. Define *mechanism*.
2. Understand the need for mechanism descriptions.
3. Name the two main purposes of a mechanism description.
4. Name types of publications that present mechanism descriptions.
5. Explain the difference between a *general* and *specific* mechanism description.
6. Write explicit and limiting titles to mechanism descriptions.
7. Understand spatial, functional, and chronological organization.
8. Name the three main sections of a mechanism description.
9. Name four types of graphics employed in mechanism description.
10. Analyze the effectiveness of a well-written mechanism description.
11. Write a general and specific mechanism description.

INTRODUCTION

Written descriptions of the tools, appliances, apparatuses, and mechanisms we purchase or operate are aids to understanding thoroughly their functions. We may call any object—or, for that matter, a system, a location, or a substance—that has functional parts a mechanism. In this chapter, we are considering, then, not just the task of a technical writer, engineer, or other writer of scientific and professional material in describing a simple mechanism (a pocketknife, calculator, or louvered door) or a complex mechanism (computer hardware, an automobile, or an escalator). We are also considering the efforts of an architect, a scientist, an archeologist, or an anthropologist to describe a location (an office space for a new computer work station or the site of a new production plant), body organs and systems (the heart or digestive system), the method of construction of a burial tomb, or the kinship relationship system of a tribal people. Even substances (paint, aspirin, DNA, diesel fuel, etc.) may be described as a mechanism.

Technical writers, marketing specialists, engineers, and other professional writers describe mechanisms to spur sales, to explain assembly, to instruct on operating procedures, to explain functions or composition, and/or to analyze strengths and weaknesses. Mechanism descriptions may appear in textbooks, user guides, service manuals, merchandise catalogs, medical reference materials, specialized encyclopedias, specification catalogs, do-it-yourself trade books, and professional papers.

Except for sales promotion materials that may involve some subjective and persuasive writing, descriptions of mechanisms are characterized by objectivity, specificity, and thoroughness. A well-written description should enable a reader to understand the mechanism and the function of the parts. Further, the description should enable the reader to judge the efficiency, reliability, and practicality of the mechanism.

AUDIENCE

The purpose of the description and the audience for whom it is written will dictate the length and amount of technical detail to be included. Are you describing a general mechanism (a computer, a steering gear, a snake) for an encyclopedia or textbook? Are you persuading a reader to purchase the mechanism by description in a sales catalog? Are you describing a mechanism in a user guide prior to providing instructions for its use? Are you urging a company to consider the feasibility of building or renovating a construction? Figure 9.1 is a portion of an engineer's description of the condition of a condominium's second floor cedar shake and mansard construction to determine if repairs are necessary:

The mansards consist of 2′ × 6′ struts clipped to a sole plate that is bolted onto concrete ledges. The 2′ × 6′s are strapped at the roof level and fastened to the sole plate with three nails through the clips. Wood members (1″ × 4″s) are fastened to the 2′ × 6′s at approximately 8″ centers for fastening the shakes. The shakes are in turn either stapled or nailed to the 1″ × 4″ stripping. The shakes terminate under a copper cap flashing at the roof level and overhang the top of the bottom ledge by approximately 1″. Our investigation of the five mansards found evidence of termites at each location. Live termites were observed at the mansard investigated at Building One. However, the only termite damage observed was in the pine 1″ × 4″ stripping, and only a very small area was affected. No water damage or rotted wood was found. However, all of the shake fasteners and the sole plate anchor bolts were extremely corroded. In some cases, the anchor bolts had no nuts to properly anchor the sole plate to the ledge. With the exception of the anchor bolts and shake fasteners, the overall mansard structures inspected were structurally sound.

FIGURE 9.1   *Engineer's mechanism text*   (Mechanism description on "Mansards" is reprinted by permission of Swaysland Professional Engineering, Plantation, FL.)

More often, a description is lengthier and more detailed.

A *general* description, written for an encyclopedia or a general how-things-work book or article, emphasizes the overall appearance of the mechanism and its parts and explains its purpose, function, and operation. A *specific* description written for a user guide, a service manual, or a proposal emphasizes not only an overall description of the mechanism and its parts, but also includes a detailed description of each part, sub-part, or assembly of parts.

In addition, a description of a mechanism usually discusses its strengths, limitations, cost, availability, and optional equipment and/or similar models to allow the reader to judge the usefulness of the particular brand or model.

ORGANIZATION

Whether your description is general or specific, logical organization will aid your reader. An outline should be developed and followed carefully. There are three major sections to a general description of a mechanism:

1.0 General description, or the mechanism as a whole
2.0 Functional description, or the main parts
3.0 Concluding discussion, or assessment

An outline could be much more detailed. The components of a specific description of a mechanism could be outlined, for example, in the following manner:

1.0 The mechanism as a whole (introduction)
  1.1 Intended audience
  1.2 Formal definition and/or statement of purpose or function
  1.3 Overall description (with graphics)
  1.4 Theory (if applicable)
  1.5 Operation (if applicable, with procedural graphics)
    1.5.1 Who (qualifications)
    1.5.2 When
    1.5.3 Where
    1.5.4 How
  1.6 List of main parts (with labeled graphic)
2.0 The main parts (body)
  2.1 Description of first part
    2.1.1 Definition and/or purpose statement
    2.1.2 List of subparts (if an assembly)
    2.1.3 Shape, dimensions, weight (with graphic)
    2.1.4 Material and finish
    2.1.5 Relationship to other parts and method of attachment
  2.2 Description of second part . . . (etc.)

    3.0 Concluding discussion/assessment (closing)
        3.1 Advantages
        3.2 Disadvantages
        3.3 Optional uses and equipment
        3.4 Other models (with graphics)
        3.5 Cost
        3.6 Availability

This outline is only a guide. The purpose of your description and the intended audience will suggest the amount of detail needed for each report.

Prefatory Material

TITLES

A brief, clear, limiting title is the first writing strategy. The title *BellSouth Cordless Telephone* is not as specific as *A Description of BellSouth's Cordless, 25-Channel Autoscan Telephone, Model 33012. Snakes* will not do when you are writing a full description of a snake's skeleton and internal organs; a better title is *Description of the Internal and External Anatomy of a Snake.* Other examples are

Description of a Turbine Bypass Valve

Description of a Japanese K-D Socket Wrench

Description of a Universal Pressure Cooker, Model 4S

Description of a Hewlett Packard HP Deskjet 600C Printer

INTENDED AUDIENCE

An introductory statement of the intended audience and the purpose of the description may be included.

Examples

This description of a turbine bypass valve is intended for engineering students interested in the general construction, operation, and function of such valves.

This description of an Ace bit brace is intended for a junior high shop class instructional manual.

This description of a K & E pencil-lead holder is intended for a descriptive catalog of architectural, designer, and drafting supplies.

DEFINITION/PURPOSE

As the outline indicates, it is logical to include a formal definition and/or statement of the purpose and function of the mechanism.

Examples

The pressure cooker is an airtight, metal container which is used to cook food by steam pressure at temperatures up to 250°F.

The K-D socket wrench is a hand tool designed to hold and turn fasteners, such as bolts, nuts, headed screws, and pipe lugs.

It is often helpful to compare the mechanism to something similar that is likely to be more familiar to the reader.

Examples

The pressure cooker resembles an ordinary "dutch oven" pot or a large, covered saucepan.

The heart is like a pump in that both draw in liquid and then cause it to be forced away.

OVERALL DESCRIPTION

Next, the physical characteristics of the mechanism are examined. Include a description of the mechanism's shape and/or dimensions, the weight, the materials from which it is constructed, the color, and the finish. Graphic illustration of the mechanism will help the reader to visualize the mechanism.

Example 1

The K-D socket wrench is made of variable grades of steel. The handle is etched to provide a firm grip. The wrench shaft is 6½ in. long, and the head is 2 in. deep. It weighs 13 oz. Figure 1 shows the K-D wrench and its overall dimensions:

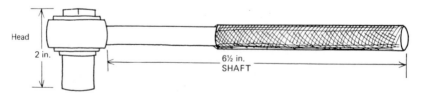

FIGURE 1   *The Japanese K-D socket wrench*

Example 2

The Hewlett Packard Deskjet 600C Printer is made of various grades of plastic with steel and other metal parts plus electric and electronic components. The overall dimensions are 450 mm (16 in) deep, 436 mm (17.2 in) wide, and 199 mm (7.9 in) high. It weighs 5.3 kg (11.6 lb). The exterior color is pale gray with darker gray components. Figure 2 shows the basic parts of the printer.

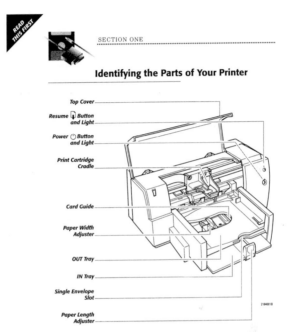

FIGURE 2   *Hewlett Packard Deskjet 600C Printer*   (By permission)

THEORY

If knowledge of theory is essential, it should be included in the introductory or prefatory material.

Examples

The functioning principle of an ordinary mercury thermometer is based on the property of thermal expansion possessed by many substances; that is, they expand when heated and contract when cooled.

The microwave oven cooks food by producing heat directly in the food. As microwaves enter the food, they cause the moisture or liquid in the food to vibrate, and the resulting friction causes the food to heat.

OPERATOR/PROCESS

If the mechanism requires an operator, the qualification or specialty of the operator should be named. If helpful, clarify when and where the process is performed. Next, the process of the mechanism in action should be explained. An explanation of process should not be confused with instructions, which give commands. In explaining the mechanism's process, use third person subjects and present-tense verbs, in either the active or passive voice.

Example 1

Cooks or chefs who wish to extract fresh garlic juice without pulp and skin use the garlic press. The cook *(third person)* places *(present tense, active voice)* the bulb of garlic inside the hollow wedge section of the strainer next to the plate of the press. He *(third person)* squeezes *(present tense, active voice)* the handles together, flattening the garlic and forcing the juice through the small holes of the strainer.

Example 2

The stethoscope *(third person)* is designed *(present tense, passive voice)* to be used by doctors, nurses, and trained paraprofessionals to convey sounds in the chest and other parts of the body to the ear of the examiner. The ear-pieces *(third person)* are placed *(present tense, passive voice)* in the examiner's ears. The bell *(third person)* is held *(present tense, passive voice)* against the area of the body to be examined. The sounds *(third person)* are amplified *(present tense, passive voice)* through the tubing by the diaphragm assembly.

A graphic drawing may help the reader to visualize the mechanism.

LIST OF PARTS

Finally, the main parts of the mechanism are listed. The sequence should have organizational logic. You may list parts *spatially* (from outside to inside or top to bottom as you would logically "see" the mechanism), *functionally* (the order in which the parts are engaged in an operating cycle), or *chronologically* (the order in which the parts are put together).

If a part is complex—that is, it contains a number of subparts, such as nuts, bolts, springs, pins, and so forth—the part may be called an assembly. Use the following sentence pattern:

Sentence pattern	The _____ consists of _____ main parts: the _____, the _____, the _____, and the _____.

Spatial example	The K & E lead holder consists of five main parts: the casing, the push knob, the spring, the tube, and the jaws.
Functional example	The camera consists of six main parts: the housing assembly, the film feed assembly, the viewfinder, the focusing assembly, the lens, and the shutter.

The Main Parts

This section, possibly the lengthiest part of your report, should define and describe each part or assembly in detail. However, a general description of a mechanism will not require as much detail as will a specific description.

DEFINITION/PURPOSE

Each part requires a formal definition or statement of its purpose. Use one of the following sentence patterns to introduce each part or assembly:

Sentence patterns	First, the _____ is designed to _____. The _____, the first main part, supports _____. The first functional part, the _____, connects _____.
Examples	First, the etched handle is designed to provide a firm grip. The base, the first main part, supports all of the other parts. The first functional part, the pedestal, connects the base to the hole punch.

If a main part is an assembly, its subparts should be named.

Sentence pattern	The _____ assembly consists of the following subparts: the _____, the _____, and the _____.
Example	The direction assembly consists of the following subparts: the tension spring, the pin, and the knob.

DESCRIPTION

Next, the shape, dimensions, and weight of the part are described. If the material and finish of a part differ from the overall description, each should be described. The strategy should be to explain how each part is related to the other parts and how each is attached to the overall mechanism. If the part is an assembly, each subpart should be described in the order listed.

It may take practice to handle the punctuation of such words as "the bulb-shaped, 7 in., clear shaft" or "a spring-loaded, S-shaped, trigger." Study the punctuation rules, particularly those for hyphens and commas, in Appendix B, "Punctuation and Mechanical Conventions." Further, to name parts accurately, refer to Figure 9.3, which shows terms used in mechanical descriptions.

A graphic illustration of each part or assembly may be appropriate. Such graphics may include exploded drawings, sections, or schematics.

Example 1

The container, the first main part, is designed to hold and to measure the food to be chopped. It is a round, glass bowl, which is 4½ in. high and 3½ in. in diameter. The container is etched in 2 oz gradients, and it has a capacity of 12 oz (1½ cups). The top rim is threaded to receive the lid (Figure 3).

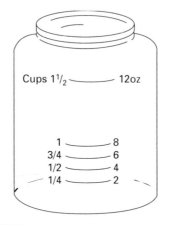

FIGURE 3   *Food chopper container*

Example 2

The plunger assembly, the second main part, consists of the following subparts: shaft, knob cap, shaft housing, shaft housing cap, spring, and blades. When the plunger assembly is depressed, the blades rotate and chop the food in the container.

The shaft is a solid piece of pot metal 8½ in. long and ³⁄₁₆ in. in diameter. It is slightly spatulate at the end where the blades are welded to it. Two stopper tabs protrude 2½ in. from the blade end to secure the shaft in position.

A bulb-shaped, wooden knob cap is pressed securely to the top end. The cap is ⅜ in. long and ½ in. wide.

The shaft housing is a hollow tube 3⅛ in. long and ⅙ in. in diameter. The housing fits over the shaft and contains the spring. A threaded shaft housing cap secures the spring into position.

The steel spring coils around the shaft inside the housing. The spring is 3 in. long.

The two blades, the final subparts of the plunger assembly, are razor-sharp steel. Each is 2½ in. long and ½ in. high. They are bent at a 45° angle and welded to the spatulate end of the shaft.

Figure 4 shows the plunger assembly parts:

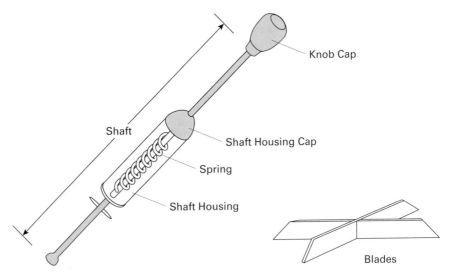

FIGURE 4 *Plunger assembly*

A complete description of a manual food chopper would include description of the other main parts: the lid and the chopping pad.

Concluding Discussion/Assessment

The concluding discussion assesses the efficiency, reliability, and practicality of the mechanism. This assessment may include an examination of the mechanism's advantages and disadvantages, its limitations, its optional uses, the comparison of one model to another, and the cost and availability.

Example 1

The Bostich B8 desk stapler is compact and lightweight, making it easy to store and to transport. The finish is scratch resistant and rust proof. It can be used as a tacker as well as a paper stapler.

Up to 20 pages of copy can be stapled at one time. The Bostich Standard stapler is recommended for larger volumes.

The recommended retail price is $8.95. A box of 5,000 staples is approximately $3.00. The Bostich Standard stapler is sold for $16.75. Bostich staplers are available in most office supply stores.

Example 2

Advantages. The HP Deskjet 600C Printer is compatible with MS Windows, an extensive range of DOS software programs, and OS/2. It will print on all plain, premium, and glossy paper plus transparency film. In addition, it will print standard U.S. and European media sizes plus index cards, postcards, and labels. The built-in feeder will handle up to 100 sheets, 20 envelopes, single envelopes, 30 index cards, and up to 25 sheets of Avery paper labels. It will hold 50 sheets in its out tray.

Depending on fonts, type sizes, color, and graphics, its color speed is from 1 to 4 minutes per page, and its black print speed is also from 1 to 4 pages per minute. It is designed to print 60,000 pages in its lifetime. It will print more than 1,000 fonts at any prescribed size and is capable of both portrait and landscape orientations. It requires 2 watts for power consumption when idle and up to 12 watts maximum when printing.

Limitations. It will not operate well under 41°F or over 104°F. The recommended operating environment is 59 to 95°F within 10 to 80 percent humidity. It is programmed to print out solutions on your computer screen for problems you may have printing.

Cost/Availability. The recommended price is $250.00 but may be less if purchased in a computer/printer package. It is available at all stores that sell computers and their components, through computer mail order catalogs, and directly from the manufacturer.

Graphics

Descriptions of mechanisms should employ ample graphics. Consider an *overall drawing* with the main parts and dimensions labeled. A sketch of the *mechanism in action* also helps your reader to envision its use. As you describe each main part or assembly, picture just that portion of the mechanism with all of its detail and labeling of subparts. Sometimes this entails *exploded* or *cutaway views*. Figures 9.2 and 9.3 show typical drawings of mechanisms (overall, process, exploded, and cutaway views), plus some gears and washers, and terms for configurations, materials, finishes, shapes, and attachments that may help you in your writing options.

Figures 9.4 and 9.5 show sample specific and general mechanism descriptions. Figure 9.4 is a description of a nonmechanical "mechanism."

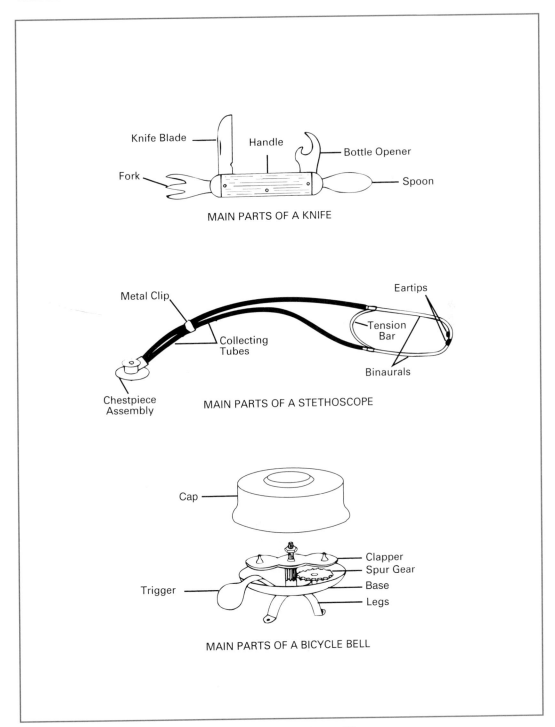

FIGURE 9.2 *Typical drawings of mechanisms (overall)*

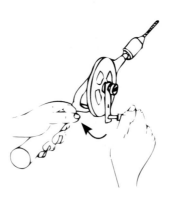

HAND DRILL IN USE

BLOOD PRESSURE CUFF IN USE

FIGURE 9.2   *continued* ***(process graphics)***   (Blood pressure cuff in use from *Coronary Bypass Surgery: A Guide for Patients*. San Ramon, CA: The Health Information Network, 1996. By permission.)

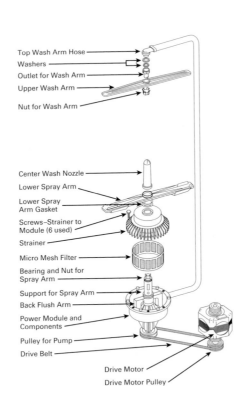

Top Wash Arm Hose
Washers
Outlet for Wash Arm
Upper Wash Arm
Nut for Wash Arm

Center Wash Nozzle
Lower Spray Arm
Lower Spray Arm Gasket
Screws–Strainer to Module (6 used)
Strainer
Micro Mesh Filter
Bearing and Nut for Spray Arm
Support for Spray Arm
Back Flush Arm
Power Module and Components
Pulley for Pump
Drive Belt
Drive Motor
Drive Motor Pulley

EXPLODED VIEW OF JETWASH SYSTEM INCLUDING MOTOR

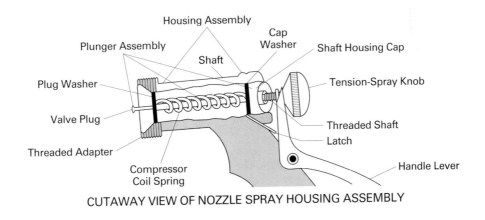

Housing Assembly
Cap Washer
Plunger Assembly
Shaft Housing Cap
Shaft
Plug Washer
Tension-Spray Knob
Valve Plug
Threaded Shaft
Latch
Threaded Adapter
Compressor Coil Spring
Handle Lever

CUTAWAY VIEW OF NOZZLE SPRAY HOUSING ASSEMBLY

FIGURE 9.2 *continued* **(exploded and cutaway views)** (Jetwash system by permission of Maytag Corporation)

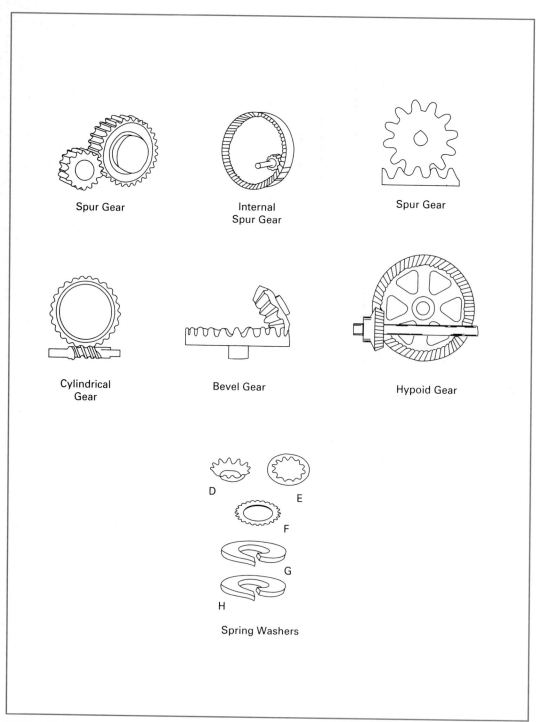

FIGURE 9.2   *continued* (***gears and washers***)

TERMS USED IN MECHANISM DESCRIPTIONS

Configurations

arc
arm
assembly

ball
bar
barrel
bearing
bevels
blade
bolt
bore
bow
brace
bracket
buckle
bushing

calibrations
cap
casing
channel
clamp
clip
coil
collar
cone
cotter pin

diaphragm
disk
dowel

extension arm
eye bolt

face
fin
fitting
flange
frame
funnel

gauge
gear
gradients
groove
guide

handle
hinge
hook
housing
hub

jacket

key

latch
leg
leg ring
lever
lip

marking
matting
mouth

nib
nozzle
nut
   slotted
   square
   wing

O-ring

pad
pin
plate
plug
plunger
pocket clip
point

ratchet
reservoir
ribbing

ring
rivet

screw
   metal
   recess
   wood
shell
sleeve
slot
socket
spline
spool
spring
stem
stopper
switch

teeth
threads
tip
toe plate
tray
trigger
tube

wand
washer
webbing
wedge

yoke plate

Materials

aluminum
copper
noncorroding
  metal
plastic
pot metal
steel
   anodized
   drop-forged

galvanized
stainless

Finishes

brushed
buffed
etched
glazed
lacquered
lustrous
matte (dull)
stained
semigloss

Shape

circular
concave
conical
convex
cylindrical
flared
grooved
hexagonal
hollow
octagonal
rectangular
solid
square
tapered
triangular
u-shaped

Attachment Methods

coiled
compressed
crimped
flange/slot
  attachment
glued
riveted
screwed
soldered
welded

FIGURE 9.3 *Some terms used in mechanism descriptions*

DESCRIPTION OF THE VENOJECT BLOOD COLLECTION SYSTEM

Intended audience

This description of the Venoject Blood Collection System, which is used for obtaining blood specimens for laboratory tests, is intended for medical laboratory students.

General Description

Definition and purpose

The Venoject Blood Collection System obtains blood specimens for laboratory tests. It is designed to obtain multiple blood sample tubes from a patient with only one puncture site required.

Overall description

Assembled, the steel, glass, and plastic parts measure approximately 6 inches long depending upon the length of the selected tube. Figure 1 shows the assembled system:

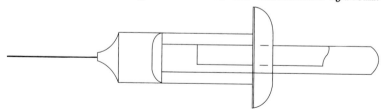

Figure 1 The Venoject Blood Collection System

Theory Process

The system operates on the principle of a vacuum in the collecting tube. First, the stopper on the top of the tube is punctured by the needle, allowing the blood sample to flow into the tube and stop when the tube is full. Second, when the full tube is removed, the needle stops the blood flow until another tube is punctured by the needle. This

FIGURE 9.4 ***Specific description of a mechanism*** (Courtesy of student Joanne Fata)

procedure can be repeated for each tube needed for specific blood tests with no discomfort to the patient. Figure 2 shows the Venoject System in a venipuncture procedure:

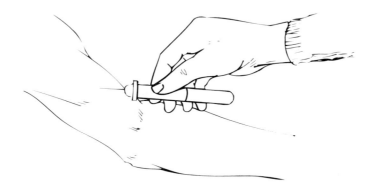

Figure 2   Venoject System in venipuncture procedure

List of parts

The Venoject System consists of three main parts: the double-pointed needle assembly, the adapter, and the blood-collecting tube.

Functional Description

Purpose of first part

The first main part, the needle assembly, functions in two ways: it pierces the skin at the site, and it closes off the blood flow when a collecting tube is not attached. The needle assembly consists of three subparts: the needle, the connector, and the cover. The 2.4-in. sterile needle is hollow steel. A 2.1-in. plastic connector fits securely over the needle at the halfway point. It has threads that screw into the holder and an extended tube to protect the end of the

FIGURE 9.4   *continued*

needle which is inserted into the collecting tube. A plastic cover protects the needle until it is to be used. Figure 3 illustrates the needle and plastic cover:

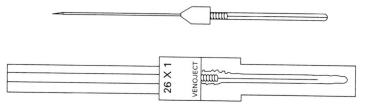

**Figure 3   Venoject double-pointed needle
with plastic cover**

Purpose of
second part

The second functional part, the adapter holder, connects the needle and the collecting tube. The 2.8-in. plastic, cylindrical holder has a diameter of $^3/_4$ in. One end is threaded to receive the needle; the other end is open to receive the collecting tube. Figure 4 shows the holder:

Description

Figure 4   Venoject System holder

Purpose of
third part

The third part, the collecting tube, is designed as a vacuum to collect the blood. It consists of three subparts: the tube, a rubber stopper, and a label tape. Hollow, glass tubes are available in 3-in., $3^1/_2$-in., and 4-in. lengths; the two shorter tubes have a $^1/_4$-in. diameter while the 4-in. tube has a $^1/_2$-in. diameter. A color-coded rubber stopper is inserted into or over the open end of the tube. The color

Description

FIGURE 9.4   *continued*

of the stopper indicates whether the tube contains an anti-coagulant which is necessary for certain blood tests. The collecting tubes in general use are not sterile; sterile tubes are available when needed for bacterial determinations. A plastic label to record the patient's name and date is taped onto every tube. Figure 5 shows three collecting tubes with rubber stoppers and labels in place:

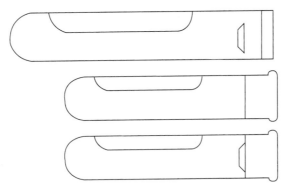

Figure 5   Venoject collecting tubes

Concluding Discussion

Assessment

The Venoject Collecting System is an efficient apparatus for collecting blood for any number of laboratory tests. It allows the technician to obtain multiple samples while preserving the patient's vein. Because it is a disposable system, bacteria and hepatitis cannot be transmitted from one patient to another. It is available at medical supply houses.

FIGURE 9.4   *continued*

DESCRIPTION OF THE
STRUCTURE OF THE HUMAN HEART

Intended
audience

This description of the human heart is intended for a general audience as an introduction to a discussion of diagnostic procedures to determine heart diseases.

GENERAL DESCRIPTION

Definition and Purpose

Formal
extended
definition and
purpose

The human heart is a hollow, muscular organ that is located in the chest, slightly to the left of the body's midline. By means of the heart's pumping action, blood flows through the circulatory system to the various tissues, providing them with the oxygen and nutrients that sustain life and removing waste for elimination through the lungs or kidneys.

Overall Description

Overall
description

The heart may be compared to a large pear about the size of two clenched fists of an adult. It is positioned in the middle of the chest, with its widest portion **(base)** at the top and its smallest portion **(apex)** pointing down and to the left. It lies between the lungs and immediately behind the breastbone **(sternum),** resting upon the diaphragm. It is shielded by the rib cage in front and the spinal column in back. Figure 1 shows two views of the heart: the anterior view of the exterior of heart and a cut view showing the inside:

FIGURE 9.5   *Description of the human heart, a nonmechanical mechanism*

Graphic
description

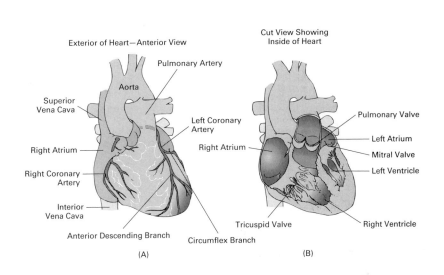

Exterior of Heart—Anterior View

Cut View Showing
Inside of Heart

Pulmonary Artery

Aorta

Superior
Vena Cava

Left Coronary
Artery

Right Atrium

Right Coronary
Artery

Interior
Vena Cava

Anterior Descending Branch

Circumflex Branch

Pulmonary Valve

Left Atrium

Right Atrium

Mitral Valve

Left Ventricle

Tricuspid Valve

Right Ventricle

(A)

(B)

Figure 1   Two views of the normal heart   (From COLUMBIA
UNIVERSITY COLLEGE OF PHYSICIANS & SURGEONS
COMPLETE HOME MEDICAL GUIDE, 4/E, copyright © 1989 by
The Trustees of Columbia University in the City of New York
and the College of Physicians and Surgeons of Columbia
University. Used by permission of Crown Publishers, a
division of Random House, Inc.)

The heart weighs 11 to 16 ounces in the average adult.

Operation

Functional
operation

when

how

The heart begins beating within three months of con-
ception and may continue for 100 years or more. Normal
cardiac rhythm is maintained by the heart's electrical system, centered primarily in the group of specialized **pace-
maker cells.** Blood that is depleted of oxygen and loaded
with carbon dioxide flows into the right ventricle, which
pumps it through the pulmonary artery into the lungs. In
the lungs, the carbon dioxide is removed, and a fresh supply of oxygen is added. The oxygenated blood then travels
through the pulmonary vein into the left atrium and on to
the left ventricle. This chamber is the heart's major pump,
responsible for pumping the oxygenated blood into the

FIGURE 9.5   *continued*

List of
main parts

Description of
main parts

First assembly

Definition
and purpose

Subparts

Description
of second
assembly

List of
subparts

Definition
and purpose

Shape

Description
of third
assembly

aorta (the great artery that forms the main trunk of the arterial system) and eventually to all parts of the body through a vast network of arteries, arterioles, and capillaries before it returns via the venules and veins—a total of about 60,000 miles of blood vessels. A system of valves keeps the blood moving in the right direction through the heart.

The heart consists of five main assemblies: the pericardium, three layers, four chambers, five great vessels, and four valves.

THE MAIN PARTS

The first main assembly, the **pericardium,** is a sac that envelops the entire organ and consists of two layers and the pericardial cavity. The first layer, the **fibrous pericardium,** is a tough, dense, outer membrane that protects the heart and anchors it within the chest cavity. The second layer is a thin, smooth, inner membrane, the **serous pericardium,** that forms the outer surface of the heart. Between these two layers is a slight space known as the **pericardial cavity,** which contains a watery fluid that prevents friction when the heart contracts and relaxes.

The second assembly consists of three layers: the epicardium, the myocardium, and the endocardium. The **epicardium,** the outer layer, is the same as the serous pericardium. The inner layer, the **myocardium,** consists of thick bands of cardiac muscle tissue and forms the bulk of the heart, doing most of the work. By alternately contracting and relaxing, the myocardium draws blood into the heart and then propels it outward again in a steady rhythm. The inner layer, the **endocardium,** is a thin membrane that lines the heart's inner surface covering the cardiac valves and is continuous with the inner lining of the major blood vessels of the heart.

The chambers of the heart, the third assembly, consist of four cavities: two atria (upper chambers) and two ventricles (lower chambers). They are arranged so that one pair sits on top of the other pair. See Figure 7.4a to locate these chambers.

FIGURE 9.5 *continued*

Definition
and purpose

Subparts
description

Shape

Subparts
description

Definition
and purpose

Shape and
dimension
Relationship
of parts

Description
of fourth
assembly
Definition
and purpose

Subparts
description

Purpose

Subparts
description

Shape and
dimension

Subpart
description

The **left and right atria,** two cavities, serve as momentary storage reservoirs for receiving blood from the veins. They act as booster pumps for filling the ventricles. Because they do not need to exert much force, their walls are fairly thin. The right atrium receives blood from all portions of the body except the lungs and is slightly larger than the left atrium. The upper front portion of each atrium features a small conical pouch called an **auricle,** which increases the chamber's surface area.

The two ventricles propel blood out of the heart, so their walls are two or three times thicker than those of the atria. The **left ventricle** has the heaviest workload since it must pump blood at high pressure to all portions of the body except the lungs, and its wall may be as thick as 0.5 in. It is also longer, narrower, and more cone-shaped than the right ventricle. The **right ventricle** receives blood from the tissues and returns it to the lungs to eliminate carbon dioxide and to take on fresh oxygen. The left ventricle receives purified blood from the lungs and pumps it through the arteries to the body tissues.

The five great vessels, the fourth assembly, include the inferior vena cava, the superior vena cava, the aorta, the pulmonary artery, and the pulmonary veins. The pulmonary vessels transport blood to or from the lungs while the other great vessels carry blood to or from the rest of the body. See Figure 7.4a to locate these vessels.

The **inferior and superior vena cava,** the body's largest veins, empty into the right atrium. The superior vena cava returns blood from the head, neck, upper limbs, and thorax. The inferior vena cava returns blood from the lower part of the body.

The **aorta,** the body's largest artery, transports blood to all parts of the body except the lungs. Measuring about 1.2 in. in diameter, the aorta extends upward as the ascending aorta, curves to the back and left to form the aortic arch, and then passes down as the descending aorta.

The **pulmonary artery** originates from the right ventricle and divides into right and left branches, which carry blood to the right and left lungs, respectively. In turn, the

FIGURE 9.5   *continued*

Purpose
four **pulmonary veins** transport blood back to the left atrium.

Description
of fifth
assembly
The four valves, the fifth assembly, keep blood moving in the desired direction by opening and closing in regular sequence due to pressure changes within the cardiac chambers. The atrioventricular valves are the **mitral valve** and the **tricuspid valve.** The mitral valve lies between the left atrium and the left ventricle and has two pointed cusps, or leaflets. It is sometimes called the biscuspid valve. Its counterpart, the tricuspid valve, is located between the right atrium and the right ventricle. The **aortic valve** lies between the left ventricle and the aorta, and the **pulmonary valve** lies between the right ventricle and the pulmonary artery. See Figure 7.4a to locate these valves.

Subparts
description

Shape
Both of these half-moon-shaped valves have three leaflets that spread apart in response to pressure from the bloodstream so that their pointed, free edges extend in the normal direction of the blood flow. To prevent the leaflets from opening backward, the two valves have fine, tendonlike cords called **chordae tendineae,** which anchor to the small muscles on the inner surface of the ventricles. When a valve closes, the free edges of its leaflets normally come together to form a seal preventing blood from leaking backward. The chordae tendineae of the atrioventricular valves stretch tightly against the leaflets, holding them in the desired positions.

Relationship
of parts

Purpose

Relationship
of parts

Concluding
discussion
CONCLUDING DISCUSSION

Efficiency
Efficiency
During an average lifetime of 74 years, the heart beats more than 2.5 billion times. Each minute it beats between 60 and 80 times and pumps about 5 quarts of blood, almost all of the body's blood supply. During exercise, the pumping action automatically increases three- to fourfold in response to the tissues' demand for increased oxygen.

Reliability
Reliability
During fetal development, the heart originates as a tubelike structure that gradually enlarges and twists back upon itself forming ridges that partition it into the four

FIGURE 9.5   *continued*

chambers. The heart is fully formed and beating months before birth although the exchange of nutrients, oxygen, and waste products occurs through the umbilical vessels in the placenta (an organ that connects the fetus with the mother).

The heartbeat depends on electrical impulses generated and transmitted by the heart's conduction system. Because the conduction system does not depend on outside stimuli, generation of the heartbeat is entirely automatic. The rate of the heartbeat is affected by the autonomic nervous system, body temperature, and various hormones and other chemicals, particularly potassium and sodium. During exercise or intense emotion, the heart automatically speeds up. Because athletes have exceptionally strong, efficient hearts, their heart rate is usually slower than that of nonathletes.

Assessment

To sustain its activities, the heart must have a constant supply of blood. This requirement is met by a special vascular system that encircles the heart and reaches deep into its walls, providing the coronary circulation. If a coronary artery becomes obstructed by disease, the flow through that artery will be reduced or completely blocked. The result is that the lack of oxygen to the myocardium will result in chest pain or myocardial infarction (death of the affected tissue), commonly called a heart attack. If a person survives a heart attack, the dead myocardial cells will be replaced by scar tissue that is incapable of contracting. Therefore, the remaining healthy portions of the heart will be forced to work harder than before the heart attack.

Disorders of the heart are due to hereditary, environmental, and infectious processes that damage the heart muscle, the coronary arteries, the valves, or the conduction system. If cardiovascular disease is suspected, there are many diagnostic methods that can help to clarify the type and extent of the disorder and medicines, procedures, and surgeries to control heart disease and disfunction. Controllable risk factors include high blood pressure, cigarette smoking, high blood cholesterol, Type A personality, environmental stress, obesity, diabetes, and a sedentary lifestyle. Uncontrollable risk factors include age, sex, and heredity.

FIGURE 9.5 *continued*

CHECKLIST

Writing a Mechanism Description

INITIAL CONSIDERATIONS

❏ **1.** Have I selected a mechanism with main and subparts?

❏ **2.** Have I determined my audience?

❏ **3.** Does the mechanism warrant a general or a specific description?

❏ **4.** Have I decided upon a spatial, functional, or chronological order of parts?

❏ **5.** Have I provided a precise and limiting title?

PREFATORY DESCRIPTION

❏ **1.** Have I named the mechanism precisely?

❏ **2.** Have I defined and/or stated the purpose?

❏ **3.** Have I provided an overall description?

❏ **4.** Have I discussed the operational theory?

❏ **5.** Have I stated by whom, when, and where the mechanism is operated?

❏ **6.** Have I provided a list of the main parts?

FUNCTIONAL DESCRIPTION OF MAIN PARTS

❏ **1.** Have I described the parts in the order listed?

❏ **2.** Have I defined and/or stated the purpose of each part?

❏ **3.** Have I listed the subparts of assemblies?

❏ **4.** Have I described each part adequately?

    ❏ Shape?

    ❏ Dimension?

    ❏ Weight?

    ❏ Materials?

    ❏ Finish?

    ❏ Relationship to other parts?

    ❏ Method of attachment?

❏ **5.** Have I avoided wordiness?

CONCLUDING DISCUSSION/ASSESSMENT

❏ **1.** Have I assessed the strengths of the mechanism?

❏ **2.** Have I discussed the limitations?

❏ **3.** Have I explained optional uses and equipment?

❏ **4.** Have I compared the mechanism to other models?

❏ **5.** Have I addressed the cost?

❏ **6.** Have I commented on the availability?

❏ **7.** Will the reader be able to judge the reliability, practicality, and efficiency of the mechanism?

GRAPHICS

❏ **1.** Have I used adequate and appropriate graphics?

❏ **2.** Have I referred to the graphics in my text?

❏ **3.** Have I numbered and titled each graphic clearly?

❏ **4.** Have I given credit to the graphics sources if they are not my own?

EXERCISES

1. Reorganize these sentences into a **logical description.**

   a. The air pump consists of three main parts: barrel assembly, the plunger assembly, and the hose.

   b. It is compact, lightweight, and portable.

   c. The hand-operated air pump is designed for inflating bicycle tires and sporting goods, such as basketballs, footballs, rubber rafts, and so on.

   d. It has a 60 psi (pounds per square inch) rating.

   e. The rustproof, steel construction will ensure many years of useful service.

   f. The brass, octagonal barrel cap allows access to the pump mechanism diaphragm. It is threaded to attach to the housing and has a ¼-in. hole in its center to slide over the shaft. A ⅛-in. hole in the side of the cap allows air to enter the housing.

   g. The barrel assembly consists of three subparts: the housing, a barrel cap, and a toe plate.

   h. The operator clamps the hose nozzle on to the filler stem of a tire to be inflated, stands on the toe plate, and pumps the plunger up and down to inflate the tire with air. If the item to be inflated is a sporting good, the supplied filler needle is inserted into the nozzle clamp.

   i. The housing is a hollow, 4½-in. by 17-in. steel barrel. The top end is threaded to receive the barrel cap.

   j. The 18-in., fabric-covered, rubber hose screws into the barrel housing 1 in. above the base with an air-tight brass fitting.

   k. A 6-in., wood handle threads onto the top of the rod.

   l. Welded to the bottom of the barrel housing is a 4½-in. long toe plate base on which the operator stands during the operation of the pump's plunger mechanism.

   m. The plunger assembly consists of three subparts: a rod, a handle, and a diaphragm.

   n. The locking clamp nozzle is inserted into the hose end and is secured with a ⅛-in. metal band. A thumb chuck allows quick release for regular and high-pressure use.

   o. The rod is a ¼-in. by 16½-in. threaded steel shaft.

   p. The plunger assembly fits into the housing and is secured by the barrel cap.

   q. The pump is 16½-in. high and is constructed of steel with rubber hosing and brass fittings.

    r. A diaphragm, a leather washer, is secured to the lower end of the rod by two ¼-in. nuts.

2. Rewrite the following portion of **a mechanism description** to eliminate the instructional commands. Use third person subjects and present-tense verbs in the active or passive voice.

> The operation of a socket wrench is simple. First, select the proper socket size for a specific fastener. Second, lock the socket into place on the driving lug. Third, fit the socket end of the wrench over the fastener. Fourth, set the direction control to the right or left by moving the fastener counterclockwise. Fifth, move the handle in a right to left or left to right motion to twist the fastener. Simultaneously, place your free hand over the wrench and fastener to secure the wrench to the fastener.

3. Correct the **mechanics** of these sentences by adding hyphens and commas.

    a. The frame is nine and one half centimeters long.

    b. The six in long handle connects to the barrel with two 12 mm long copper rivets.

    c. The top portion has a centered 25 centimeter circular cutout.

    d. Figure 9 illustrates a spring loaded L shaped latch.

    e. The base is bolted to the cylinder by 2″ diameter bolts.

4. Examine a simple mechanism (a mechanical pencil sharpener, a pocketknife, a garlic press, a bicycle seat, or the like) and name every part including screws, nuts, bolts, clamps, and so forth. **Classify the parts** into several major categories.

5. Determine a logical order of presentation (spatial, functional, or chronological) for the following mechanisms. Explain why you chose each order.

    a. a flute   **e.** a soccer ball

    b. a basketball team   **f.** a college fraternity

    c. paint thinner   **g.** a paper stapler

    d. a child's toy truck   **h.** aspirin

WRITING PROJECTS

1. **General Description Project.** Write a *general* description of one of the following mechanisms or a mechanism used in your field of study. Do not concentrate on a particular brand. Select a non-electric mechanism that consists of four or five main parts, of which at least three are assemblies. Use graphics wherever possible.

a.	deadbolt lock	**i.**	eggbeater	**o.**	drawing compass
b.	skateboard	**j.**	folding lawn	**p.**	pencil sharpener
c.	paper punch		chair	**q.**	manual can opener
d.	Rolodex file	**k.**	toggle switch	**r.**	kerosene lamp
e.	tape cassette	**l.**	stethoscope	**s.**	sink trap
f.	technical pen	**m.**	butane hair	**t.**	lawn sprinkler
g.	bicycle seat		curler		
h.	hand drill	**n.**	garlic press		

2. **Individual Project.**   Select a mechanism from the following list or choose one of your own and write a *specific* description of it; that is, describe a specific model or brand. Refer to the same outline and samples in this chapter. Do not stint on details, and use graphics wherever possible.

   a. Specific model of a Cuisinart food processor

   b. Specific make and model of a toaster oven

   c. Specific make and model of an automobile jack

   d. Staedtler/Mars drafting compass

   e. Particular make and model of a cellular phone

   f. Other?

3. **Collaborative Project—Nonmechanical "mechanism."**   In groups of three, select a nonmechanical "mechanism" from the following list and write a description of it. Determine and state at the beginning of your document the purpose for your description. Is it to persuade a specific organization to build or establish a new facility? Is it to describe a mechanism to a general audience in an encyclopedia or textbook? Is it to encourage readers to purchase the mechanism? Other purpose? Refer to the sample outline in this chapter. Use copious detail and appropriate graphics for a complete description. Include a bibliography of sources in APA style.

a.	bicycle rack site	**e.**	human ear	**i.**	snake
b.	storage shed site	**f.**	human knee	**j.**	an amoeba
c.	polyurethane glue	**g.**	flatworm	**k.**	plaster
d.	housefly's eye	**h.**	a goldfish	**l.**	aspirin

NOTES

CHAPTER *10*

Giving Instructions

TIGER   by Bud Blake

S K I L L S

After studying this chapter, you should be able to

1. Define *instructions*.
2. Appreciate the need for instructions in all enterprises.
3. Appreciate the legal requirements of instructions.
4. Describe the difference between *instructions* and *process analysis*.
5. Discuss the differences between instructions for a novice, an intermediate operator, and a technical operator.
6. Write precise, limiting titles for instructions.
7. Organize instructions by a variety of numbering systems: Arabic/letters, two decimal systems, and the digit-dash-digit system.
8. Name an intended user and imply or state his or her expected knowledge and skills.
9. State a behavioral or instructional objective for a given set of instructions.
10. Emphasize the importance and benefits of a given set of instructions.
11. Define key terms in instructions.
12. Provide appropriate notes, precautions, cautions, and warnings.
13. List required tools, materials, and apparatus in the order of usage.
14. Sequence the steps chronologically.
15. Use appropriate language (avoid wordiness, use articles *a, an,* and *the,* and eliminate pronouns).
16. Provide visuals and appropriate document designs.
17. Critique sets of instructions.

INTRODUCTION

Instructions are the most common form of technical writing and are seen so often and in so many forms that we often do not realize that they are present. They may be as simple as "Wash your hands before proceeding" or as complicated as those in a computer repair manual. Recipes are sets of instructions as are the contents of the thousands of *how-to* books on the market today. Almost every mechanism that we purchase, from analog computers to food processors to zylophones, includes a set of instructions for assembly and procedural steps as well as tips and warning about the care, potential problems, and improper usage of the item. Which of you has not labored over the instructions for your computer applications, your printer, a programmable telephone, a VCR, and the like? Consider the instructions, advice, and notes and warnings given in the user guide for the pager illustrated in Chapter 7.

Instructions may be oral, such as a request to do something from a supervisor, parent, or instructor, or even gestural, such as those used by a police officer directing traffic. They are not, however, as simple to construct as they may seem.

DEFINITION

Instructions are a chronological set of steps, usually numbered, to be performed in a procedure by a specific operator and include the desired outcomes, notes, precautions, cautions, warnings, and sometimes analyses of outcomes. Instructions often combine with the strategy of *process analyses.* That is, they are the result of a task analysis (a study of steps in a process), but, as you will see in the next chapter, they differ from the formal process analysis in one fundamental way—instructions direct and process analysis informs. Instructions focus on *how to perform a task,* whereas the formal process analysis focuses on *how a task is performed by a third party* and emphasizes *what the results are* for the steps within the process. In other words, instructions require a personal relationship between the writer and the reader whereas the formal process analysis maintains a more impersonal rapport. Though both may be used to define and/or inform, only instructions are expected to be performed by the reader. Whereas a scientific writer, for example, might write a formal process analysis on how a nuclear bomb is developed and assembled, it is more likely the task of a group of nuclear engineers and their technical writers to write a set of instructions on how to build the bomb.

LEGAL CONSIDERATIONS

Because we live in an era of litigation, with lawsuits almost a way of life, it should not surprise you that many of the instructions we read daily are provided primarily to protect manufacturers from lawsuits. Instructions are drawn up with the utmost consideration to warning and caution statements, and some companies even have attorneys working with their technical writing teams to ensure accuracy. Leaving a critical detail out of a set of instructions can lead to loss of time, damage to property, or even death, as in the case of failing to warn about the consequences of improper usage or the dire consequences of noxious and toxic fumes, explosions, electrical shocks, burns, and fires.

Instructions for a procedure or a product's operation constitute a legal contract with the user, guarantee the warranty, and may protect a company from liability. Therefore, all limitations of its operations and adequate warnings and cautions about each appropriate step must be in-cluded. When instructions are incorporated into a user guide, the man-ual should cover pertinent definitions and mechanism description plus

- Assembly instructions
- Operating instructions
- Cleaning and storage information
- Necessary maintenance
- Problem solving discussion
- Emergency steps
- Disposal directions
- Warranties
- Service information

AUDIENCE CONSIDERATIONS

As with all professional writing, the language of instructions should be slanted toward the intended audience. The language in instructions in-tended for novices (beginners) should be simple and detailed and provide definitions wherever knowledge of a term is questionable. Instructions for intermediate users are generally briefer, and those for skilled techni-cians are more cryptic.

For the novice   1. Locate the tray (for storing paper) beneath the copier assembly as shown in Figure 1.

2. Lift the copier assembly and pull open the paper tray.

For the intermediate   1. Lift and pull open the paper tray.

For the technician   1. Disassemble the paper tray at lock A-1.

Titles

Another tactic to aid your audience is to provide a precise, specific, and limiting title. Your title should indicate to the reader exactly what your instructions cover. Consider the differences in the following:

Weak   How to Repair a Circuit Board

Precise   Instructions for Removing a Faulty Transistor from a Printed Circuit Board

Weak   How to Perform a Digital Block

Precise   How to Inject Lidocaine into a Finger to Produce Numbness

Organization

Your audience will not read instructions as a continuous narrative, but as one-by-one steps. Instructions usually require numbering of each separate step to stay on track. In simple instructions, an Arabic numbering system is usually appropriate. Complex instructions may involve main steps divided into substeps. Preliminary outlining with attention to logical groupings will reveal the main steps and their parts. Several numbering systems may be considered: Arabic with alphabetical letters, a variety of decimal systems, or a digit-dash-digit system. The most reliable are as follows:

Arabic/Letters

1. Section
   a. Component
   b. Component
2. Section
3. Section
   a. Component
   b. Component
   c. Component
4. Section

Decimal System A

0.0   Preliminary section
1.0   Section
1.1   Component
1.1.1 Subpart
1.1.2 Subpart
1.2   Component
1.2.1 Subpart
1.2.2 Subpart
2.0   Section

Decimal System B

0.0   Preliminary section
1.0   Section
1.01 Component
1.02 Component
1.02.1 Subpart
1.02.2 Subpart
2.0   Section

Digit-Dash-Digit System

0–0 Preliminary section
1–1 Section
1–2 Component
1–3 Subpart
1–4 Subpart
1–5 Component
2–1 Section

THE INSTRUCTIONS

Well-written instructions require informational preliminary material; attention to notes, precautions, cautions, and warnings; listings of tools, materials, and apparatus required for the procedure; proper sequencing; careful attention to language and visuals; and careful page layout.

Preliminary Material

Because instructions emphasize *what to do* but not *why,* it is helpful to include preliminary statements, such as

- Naming the intended operator
- Delineating the expected prior knowledge and skills of that operator
- Stating the behavioral or instructional objective
- Stressing the importance and benefits
- Defining key terms
- Providing notes, precautions, preliminary cautions, and warnings
- Listing the materials required in the order of use (if the list is short)

Following are two preliminary statements incorporating consideration of the preceding list. These may be included in a numbered or unnumbered introductory statements, such as

For a novice   0–0   These instructions on how to inject lidocaine (a pain-blocking, local anesthetic) into a finger to produce numbness are intended for an allied health student learning minor surgical techniques. If instructions are followed properly, the student can anesthetize (that is, cause loss of pain and temperature sensation in) a finger. Anesthesia is necessary prior to finger surgery or before manipulation of broken finger bones. The anesthetizing of a finger or toe is termed a *digital block.*

Warning   Failure to inject the lidocaine properly may result in undue pain.

Have at hand cotton swabs, alcohol, a packaged sterilized syringe, and the lidocaine.

Instructions for a professional technician would not require as many definitions, warnings, or material assembly.

For a professional   These instructions are to be used by field service personnel to install a repaired DA-1203 antenna. Alignment of the DA-1203 to the aircraft's horizontal position gyro is necessary for proper operation of a stabilized weather radar.

Notes, Precautions, Cautions, Warnings

Notes, precautions, cautions, and warnings differ in intent but are all related to the safety of the operator performing the instructions and the protection of the manufacturer. Because such notations are often not actual steps of the instructions, it is advisable to make them large and easily visible and to place them at appropriate locations, such as in the introduction and at each pertinent step. They should be underlined, boldfaced, capitalized, and/or boxed. Icons help to focus attention on these notations.

Notes. Notes are usually unboxed, supplemental material providing the reader with hints, tips, or information that may be of use, but the information is not necessary to complete the task. Notes are provided wherever needed, especially where a reminder to the reader or a reference to a previous instruction is indicated. Visuals include the following:

NOTE, **NOTE,** or ▶

Precautions. Precautions are light warnings that need to be understood prior to beginning a task or a set of tasks, such as telling the reader to place newspaper on the table prior to applying glue to an object. Precautions do not usually cover health- or life-threatening situations but are simple tasks the reader can perform to avoid later problems. They may be highlighted in the same manner as notes:

PRECAUTION, **PRECAUTION,** or **!**

Cautions. Cautions are used to emphasize actions that may result in damage to equipment or other property. They should be strategically placed either at the beginning of a set of instructions and/or just before the steps that have a potential for trouble. Caution statements are similar to precautions but are considered stronger. They may be highlighted as follows:

CAUTION! or  or

> **CAUTION:** TIGHTENING CYLINDER HEAD BOLTS TO THE IMPROPER TORQUE WILL RESULT IN DAMAGE TO THE ENGINE.

Warnings. Warning statements indicate potential hazards to health or even life. These warnings are very strong and should be placed strategically wherever necessary. The reader must be informed about the possibility of toxic fumes, explosions, fires, and electrocutions. They may be highlighted as follows:

WARNING!! or  or

> **WARNING!!!**
> **THIS PRODUCT PRODUCES NOXIOUS FUMES!**
> **USE ONLY IN WELL-VENTILATED AREAS.**

Figure 10.1 shows a page from a Whirlpool® automatic washer/dryer user guide with safety notes and emphatic warnings. Other warnings about shock and fire are included in the guide next to the appropriate instructional steps.

Operating Your Dryer

The information in this section helps you learn to use your dryer efficiently and safely. Refer to "Laundry Tips" on pages 17–22 for additional information on sorting, loading, and drying most types of washables.

NOTES
- Make sure your dryer is properly vented before using it. See warning and your Installation Instructions.
- Make sure your dryer is properly installed in a well-ventilated room where the temperature is above 45°F (7°C).
- Make sure your dryer is leveled on a floor that can support the weight.

⚠WARNING	⚠WARNING
Fire Hazard	Explosion Hazard
Use a heavy metal vent.	Never place items in the dryer that are dampened with gasoline or other flammable fluids.
Do not use a plastic vent.	
Do not use a metal foil vent.	Do not wash or dry items soiled with vegetable or cooking oils because they may contain some oil after laundering.
Failure to do so can result in death or fire.	Doing so can result in death, explosion, or fire.

Before starting your dryer

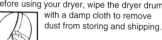

Before using your dryer, wipe the dryer drum with a damp cloth to remove dust from storing and shipping.

FIGURE 10.1   *Notes and emphatic warning in a user guide*
(By permission of Whirlpool® U.S.A.)

Tools, Materials, Apparatus

Logically, the first step in any procedure is to collect the necessary tools, materials, or apparatus. If the list is short, it may be included in a preliminary statement or section (see the previous prefatory material for injecting lidocaine). If the list calls for more precision, list the items under step one or the first task section. Do not forget to include the obvious, such as old newspapers, running water, and the like. Number each item. The following is a list of tools and materials needed to complete a set of instructions on how to develop a black and white photographic print:

1.0   Collect the following equipment and materials:
    1.1   Three developing trays
    1.2   Enlarger and easel
    1.3   Print tongs
    1.4   One liter of developer
    1.5   One liter of stop bath
    1.6   One liter of fixer
    1.7   Black and white photographic paper
    1.8   Negatives to be printed

Figure 10.2 illustrates that tools and materials may be presented by a visual.

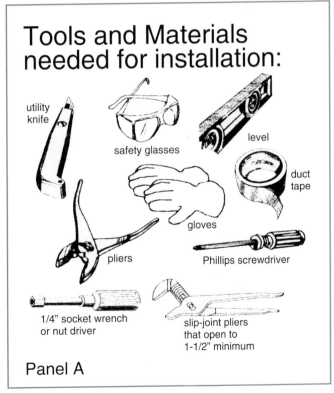

Tools and Materials needed for installation:

utility knife

safety glasses

level

duct tape

gloves

pliers

Phillips screwdriver

1/4" socket wrench or nut driver

slip-joint pliers that open to 1-1/2" minimum

Panel A

FIGURE 10.2   *Visual for required tools and materials for installation* (By permission of Whirlpool® U.S.A.)

Sequencing

Each of the actual steps or commands must be in chronological order. In the following excerpt from instructions on soldering a circuit board connection, Step 9 is obviously out of place; the joint would have to be secured prior to the actual soldering:

6. Preheat the joint to melt the solder.
7. Apply the solder to the joint.
8. Place the soldering iron tip against the solder and the joint for 2 or 3 seconds.
9. Secure the joint with a vise to avoid motion.

The writer of instructions is usually quite knowledgeable about the procedure, but the user is not. Therefore, not only is careful chronological order essential but also the inclusion of all steps is a must for effective operation. It is just as important to instruct users to turn on a word processing computer CRT as to instruct them on how to perform a global search of a text.

Language

Instructions demand precision, clarity, parallel construction, simplicity, and thoroughness. Each step usually begins with a command word, which is an active-voice verb stated in the imperative mood, such as *switch, disconnect, lift, depress.*

Incorrect	The following tools *should be collected.*
Correct	*Collect* the following tools.
Incorrect	*You cut* out the premarked damaged section.
Correct	*Cut* out the premarked damaged section.

Further, the command verbs should be precise.

Vague	Remove the bolt.
Precise	Unscrew the bolt by rotating the wrench in a counterclockwise motion.
Vague	Turn on the computer.
Precise	Depress the ON/OFF key to the ON position.

Occasionally you must precede the action command with explanatory words, such as:

While depressing the RECORD button with your left forefinger, push . . .
Using a straightedge, outline the damaged portion . . .

Similarly, avoid all other vague terms.

Vague	Check the patch to ensure that a good bond has been obtained.
Precise	Pack the edges of the patch with stiff spackling compound to eliminate wobbling.

Vague	Allow the glue to dry adequately.
Precise	Allow the glue to dry for six hours.

Vague	Screw the woodscrews only partially into the anchors.
Precise	Place a stack of three nickels (approximately $\frac{7}{32}$ in.) against the mounting surface next to one of the screw locations and turn the woodscrew in until the head touches the coins.

In a set of simple instructions that are not numbered, words—such as *first, second, next, following,* and so forth—mark time and sequence.

Short sentences are easier to understand and to execute than are wordy commands.

Wordy	Insert the mounting post of the breaker arm into the recess in the cylinder so that the groove in the mounting post fits the notch in the recess of the cylinder.
Short	Fit the mounting post groove into the notch of the cylinder's recess.

Component parts of your instructions should be expressed in parallel (identical) grammatical form.

Nonparallel	1.0	Remove the damaged wall board.
	2.0	Prepare the patch.
	3.0	The patch should be spackled and painted.
Parallel	1.0	Cut out the damaged wall board.
	2.0	Fit a wall-board patch into the hole.
	3.0	Spackle and paint the patch.

In your effort to eliminate wordiness, do not eliminate articles *(a, an,* or *the)*. Also, do not use pronouns *(it, them, that)*.

Poor	Push them through circuit board holes.
Improved	Push the leads A and B through the circuit board holes.

VISUALS AND DOCUMENT DESIGN

No set of rules covers where and when to place visuals into a set of instructions; common sense is the key to success. Whenever and wherever a drawing or other visual would make the procedure clearer, include one.

Visuals are usually placed below each step or to the left or right of each step. Use plenty of white space to aid the reader and eliminate crowding and confusion. Figure 10.3 presents three basic page layouts for the same instructions.

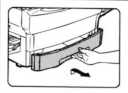

1. Lift and pull open the paper tray.

1. Lift and pull open the paper tray.

2. Push down on the shiny plate until it locks into position.

2. Push down on the shiny plate until it locks into position.

3. Adjust the paper guides to the desired paper size.
   • Squeeze the side guide.
   • Lift and insert the rear guide.
   • When adding 14" paper, remove the rear guide and store it in the pocket in front of the side guide.

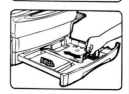

3. Adjust the paper guides to the desired paper size.
   • Squeeze the side guide.
   • Lift and insert the rear guide.
   • When adding 14" paper, remove the rear guide and store it in the pocket in front of the side guide.

1. Lift and pull open the paper tray.

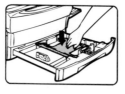

2. Push down on the shiny plate until it locks into position.

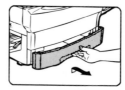

3. Adjust the paper guides to the desired paper size.
   • Squeeze the side guide.
   • Lift and insert the rear guide.
   • When adding 14" paper, remove the rear guide and store it in the pocket in front of the side guide.

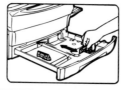

FIGURE 10.3   *Three basic page layout options for loading a copier tray* (By permission of Xerox Corporation)

Figures 10.4 and 10.5 present two sets of instructions. Figure 10.4 uses a simple Arabic numbering system and right-column graphics. Figure 10.5, because it involves several main steps and several substeps, illustrates a complex numbering system.

Memory Dialing

Your AT&T Cordless Telephone 5455 can store nine different phone numbers that you can dial just by pressing [MEM] and one of the number buttons.

Programming a Number into Memory

The handset must be turned OFF.

1. Press [PROG] (Figure 1). The PHONE Light will blink to show that you are in the programming mode.

2. Dial the phone number you want to store. The number can be up to 16 digits long.

3. Press [MEM].

4. Press any number button from 1 to 9. This assigns the phone number to the memory location you selected.

After pressing the number button, you will hear a three-part tone that means the number was stored properly. If you hear a long buzzing tone, or nothing at all, press [OFF], then follow the steps above to program the number again.

Follow the steps above for each phone number you want to store, assigning each one to a different number button.

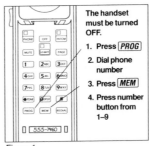

The handset must be turned OFF.

1. Press [PROG]
2. Dial phone number
3. Press [MEM]
4. Press number button from 1–9

Figure 1

NOTE: The numbers stored in memory may be lost when you change the handset batteries, or if the batteries run down completely. Follow the steps above to store the numbers again.

FIGURE 10.4   *Instructions for using the Memory Dial feature of the AT&T Cordless Telephone 5455*   (By permission of Lucent Technologies, Inc.)

Memory Dialing *(continued)*

Directory Cards

1. To use the directory card concealed in the back of your handset, press the arrow above the word DIRECTORY (Figure 1) and slide the door toward the top of the handset. The directory card has an erasable surface. If you write in pencil, you'll find it easy to change names when necessary.

> **NOTE:** If the door slips off the handset, slide it onto the track and back in place.

2. To write on the directory card on the Portable Handset Cradle (Figure 2), remove the plastic cover by inserting a pointed object in the hole and gently prying the cover up until it pops out.

Dialing a Number Stored in Memory

1. Press [PHONE] to get dial tone (Figure 3).
2. Press [MEM].
3. Press the number button (1-9) you assigned to that phone number.

For example, to dial the phone number you assigned to button "6," press [PHONE] [MEM] [6].

Figure 1

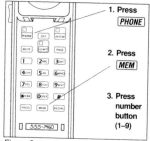

Figure 2

1. Press [PHONE]
2. Press [MEM]
3. Press number button (1–9)

Figure 3

FIGURE 10.4   *continued*

INSTRUCTIONS FOR TAKING AN ORAL
TEMPERATURE WITH AN ELECTRO:THERM

0.0 These instructions are intended for a medical assistant who is working in a doctor's office. If instructions are followed properly, the assistant will obtain an accurate temperature reading in less time than if a mercury type thermometer were used.

> **PRECAUTION:** <u>DO NOT</u> take an oral temperature with the electro:therm if the patient
> 1. has a mouth injury,
> 2. is an infant or young child who is not old enough to hold his lips closed when told to do so, or
> 3. has had something hot or cold in the mouth in the last five minutes.

1.0 Collect the following apparatus and place on the counter:
   1.1 an electro:therm thermometer
   1.2 a box of sterile plastic covers that are made to be used with the electro:therm
   1.3 a box of clean tissues
2.0 Familiarize yourself with the electro:therm by studying the parts as diagrammed in Figure 1:

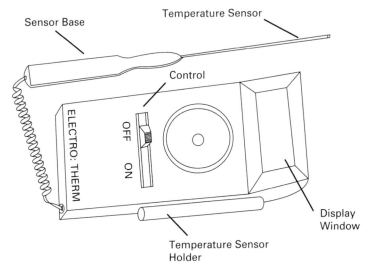

Figure 1. Parts of an Electro:therm

FIGURE 10.5 *Sample instructions with complex numbering system*

3.0　Prepare the electro:therm for operation.

    3.1　Remove the temperature sensor from the control base by pulling the temperature sensor back toward the cord with your left thumb and first finger.

    3.2　With your right hand, pick up one sterile plastic cover by its paper wrapping.

> **WARNING:**　Your hands should NOT come in contact with the sterile plastic covering.

    3.3　Insert the temperature sensor into the sterile plastic cover until the plastic cover is completely engaged over the narrow sensor rod.

    3.4　Gently pull off the paper wrapping, leaving the sterile plastic cover on the temperature sensor.

Figure 2 shows what the temperature sensor looks like with the plastic in place:

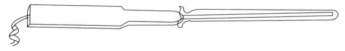

Figure 2.　Temperature sensor with plastic cover in place.

    3.5　Switch the base of the temperature sensor to your right thumb and first finger.

    3.6　Pick up the control base with your left hand.

    3.7　With your left thumb, turn the control base to the ON position.

    3.8　Observe the electro:therm's flashing numbers in the display window.

> **NOTE:**　**The numbers are flashed every second and give the temperature in degrees to the nearest tenth of a degree, as illustrated:**
>
>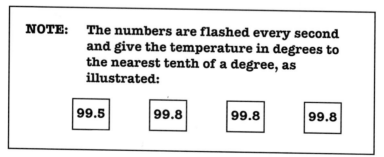

FIGURE 10.5　*continued*

4.0 Take temperature reading.

> **CAUTION:**   **Unlike the mercury thermometer, the temperature sensor should be held continually in the patient's mouth while taking the temperature. DO NOT remove the sensor from the patient's mouth until the same degree displays at least two times in a row. The electro:therm is so sensitive that a patient taking a breath can lower the reading.**

    4.1 Insert the sterile, plastic-covered temperature sensor into the patient's mouth, making sure it is under the tongue.

    4.2 Instruct the patient to keep lips closed tightly and not to talk.

    4.3 Leave the temperature sensor in the patient's mouth for one minute.

    4.4 After the minute has lapsed, observe the temperature as displayed on the control base.

    4.5 When the same numbers have flashed for two times, remove the temperature sensor from the patient's mouth.

5.0 Deactivate the electro:therm.

    5.1 Turn the control base switch to the OFF position with your left thumb.

    5.2 Place the control base on the counter.

    5.3 Switch the base of the temperature sensor to your left thumb and first finger.

    5.4 With a tissue in your right hand, pull the used plastic cover off of the temperature sensor and discard the tissue and the plastic cover.

    5.5 Slide the temperature sensor back into its holder on the side of the control base.

6.0 Record the patient's temperature on the chart.

FIGURE 10.5   *continued*

CHECKLIST

Writing a Set of Instructions

PRELIMINARY CONSIDERATIONS

❏ **1.** Have I determined if the audience is a novice, intermediate, or technical operator?

❏ **2.** Have I provided a precise, limiting title?

❏ **3.** Have I chosen an appropriate numbering system?

   ❏ Arabic/letters?

   ❏ Decimal system A?

   ❏ Decimal system B?

   ❏ Digit-dash-digit system?

THE INSTRUCTIONS

❏ **1.** Have I provided an appropriate preliminary statement?

   ❏ Does it name the intended user?

   ❏ Does it state or imply the desired knowledge and/or skills of the operator?

   ❏ Does it include a behavioral or instructive objective?

   ❏ Does it stress the importance and benefits of the instructions?

   ❏ Does it define key terms?

   ❏ Does it include necessary notes, precautions, cautions, and warnings?

❏ **2.** Is it appropriate for me to include storage information, maintenance tips, problem-solving discussion, emergency steps, and/or disposal directions?

❏ **3.** Have I provided a list of necessary tools, materials, and apparatus?

❏ **4.** Have I organized each step chronologically?

❏ **5.** Have I used proper sequencing?

❏ **6.** Have I used appropriate language?

    ❏ Have I eliminated wordiness?

    ❏ Have I use the articles *a, an,* and *the?*

    ❏ Have I avoided pronouns?

VISUALS AND DOCUMENT DESIGN

❏ **1.** Have I provided visuals for notes, precautions, cautions, and warnings?

❏ **2.** Have I used underlines, boldface, capitals, and/or boxes to make my visuals easily visible?

❏ **3.** Are my cautions and warnings strategically placed?

❏ **4.** Have I presented the instructions in the best possible document design?

EXERCISES

1. **Analysis of Instructions.** Locate both a simple and a complex set of instructions for the assembly, procedures, or care of a mechanism. Make a copy of both for each member in your class. Discuss whether the directions are effective or not. Consider titles, organization format and numbering, prefatory material (intended operator, desired skills of the operator, final objective, discussion of importance and benefits, key term definitions, lists of required materials, cautions and warnings), chronology of steps, sequencing, language considerations, visuals, and page layouts.

2. **Cautions and Warnings.** Photocopy at least five cautions and warnings from sets of instructions. Bring them to class. Be prepared to discuss their uses, importance, and effectiveness.

3. **Edit Instructions.**   Edit the following set of instructions for language problems. Look for verbs that are not expressed in the imperative mood or active voice. Also look for vague verbs and other terms, overlong sentences, nonparallel constructions, eliminated articles and questionable pronoun references.

INSTRUCTIONS FOR THE HOLGER-NEILSON
(BACK PRESSURE-ARM LIFT) METHOD OF
MANUAL ARTIFICIAL RESPIRATION

1–1 Positioning the victim

   1–2 Check the victim's mouth for foreign matter and wipe it out quickly.

   1–3 WARNING: CHECK it for obstructions every 30 seconds.

   1–4 Place the victim face down, bend elbows, and hands are upon the other.

   1–5 The victim's head should be turned slightly to one side with head extended and chin jutting out.

2–1 Administering the respiration

   2–2 Kneel at his head.

   2–3 You place your hands on the flat of victim's back.

   2–4 Rock forward until your arms are vertical to his back.

   2–5 The weight of the upper part of your body should now be forced down to exert a steady, even pressure downward upon your hands which are already placed on the back of the victim.

   2–6 Slide your arms to the arms of the victim just above elbows and draw the arms upward and toward you.

   2–7 NOTE: Enough lift to feel resistance and tension in his shoulders should be applied.

   2–8 The victim's arm must be lowered to the ground.

2–9 Repeat this cycle 12 times a minute.

4. **Text for Instructions.**   Convert the following paragraph into a set of instructions.

   To assemble an ABC vacuum cleaner connect the hose by inserting it into the opening of the machine by lining up the largest projection on the hose with the largest notch of the opening. Turn the hose to the right to tighten it. By depressing the latch and turning the hose to the left, you may disconnect the hose. To attach the extension wands, cleaning tools, and nozzles, the other end of the hose connects by turning the plastic latch ring on the right hand grip until the outer slot lines up with the inner slot. Pushing your hand down hard onto the wand or tool will push the button projection into the slot. The latch ring must be turned to lock it into place. To remove the wand, tool, or nozzle, the procedure is reversed.

WRITING PROJECTS

1. **Simple Instructions.** Select one of the following subjects and write a simple set of instructions of only 10 to 15 steps, using an Arabic number/letter system. Use visuals wherever appropriate. Consider your document design carefully. Refer to the chapter checklist.

   a. How to carve a turkey

   b. How to tune a guitar

   c. How to extinguish a campfire

   d. How to cuff pants

   e. How to back up data on a zip disk

   f. How to wax a car

   g. How to take an oral temperature

   h. How to wash windows

   i. How to select a tennis racket

2. **Collaborative Project—Complex Instructions.** In groups of three or four, write a complex set of instructions from the following list. Use one of the decimal numbering systems or the digit-dash-digit system. Use visuals where appropriate. Consider your document design carefully. Refer to the chapter checklist. Include a bibliography in MLA style.

   a. How to operate scuba gear

   b. How to clean a toaster oven

   c. How to care for in-line skates

   d. How to check out a book from the library

   e. How to apply for financial aid

   f. How to replace a lost driver's license

   g. How to apply for a passport

   h. How to dispute a grade

   i. How to make size reductions on a copy machine or scanner

   j. How to perform a complex operation related to your field of study

NOTES

Analyzing a Process

BLONDIE   by Drake Young

Reprinted with special permission of King Features Syndicate.

S K I L L S

After studying this chapter, you should be able to

1. Understand the purposes of a process analysis.
2. Understand for whom process analyses are written.
3. Distinguish between a process analysis and a set of instructions.
4. Name and identify the five types of process analysis.
5. Use indicative mood verbs in the active or passive voice to describe the steps in analysis.
6. Write a precise, descriptive, and limiting title for a process analysis.
7. Name the logical divisions of a process analysis.
8. Employ suitable graphics and visuals in a process analysis.
9. Write a process analysis embodying logical organization, significant detail, and appropriate language.

INTRODUCTION

Process analysis is a method of explaining how something occurs, how it is accomplished, or how it is organized by separating the process into its parts to examine their natures, proportions, functions, and relationships. Such an analysis concentrates not on instructions but on an ordered sequence of events such as occurs in glass blowing, wastewater management, human digestion, court decision appeals, tornadoes, or the enactment of laws. The processes may be **linear** (enacting laws), **cyclical** (human digestion), **independent** (glass blowing), or **interdependent** (court decisions appeals).

A process analysis may be complete in itself or part of a longer technical document. An analysis may be interspersed with instructions and mechanism descriptions in a user guide (see the sample in Chapter 7) to explain how a mechanism functions or what will happen under different situations. Actually, process analyses are everywhere: in textbooks explaining how diseases are transmitted or how bills are passed into law, in our newspapers and magazines explaining the process of digital television or how broadcast television bounces off relay satellites and into our homes, and in special reports to explain what will occur if a proposal is adopted or recommendations are accepted.

Scientists and engineers read analyses to understand the processes vital to their research and development. Technicians and operators read them to understand better the procedures they perform. People in business read them to promote their products or services and to perform their duties effectively. Students read them to learn the fundamental processes relevant to their fields of study.

A process analysis may contain elements similar to instructions or mechanism descriptions but should not be confused with either. Instructions emphasize *what to do,* and mechanism descriptions emphasize *how something is put together;* neither often explains *why*. An effective analysis of a process emphasizes

- *What* occurs
- *Where* it occurs
- *When* it occurs
- *How* it occurs
- *To what extent* it occurs
- *Under what conditions* it occurs
- *Why* it occurs

Perhaps the *why* is the most important element. The reader of an analysis must be able to judge the reliability, the practicality, and the efficiency of the process. The reader should be able to estimate with great accuracy the difficulty of the process, the problems likely to occur, and successful solutions to these problems.

TYPES OF ANALYSIS

An informational process analysis informs the reader about a particular process for the purpose of increasing his or her general knowledge. There are three basic types of process analysis:

1. The *historical analysis* explains how and why an idea or event occurred or an institution originated. Historical analysis explores subjects such as how the microcomputer was developed or how American teachers unionized.

2. The *scientific, mechanical,* or *natural analysis* explains how such processes occur or should occur. Subjects appropriate to scientific, mechanical, or natural analysis include how chemotherapy cures cancer, how a computer printer functions, or how smog occurs.

3. The *organizational analysis* explains the steps, pitfalls, and methods of efficiently performing a human process. Organizational analysis examines subjects such as how a plant is propagated or how a manager motivates a staff of workers.

PRELIMINARY CONSIDERATIONS

Preliminary considerations include audience analysis, titles, and language appropriateness.

Audience

The reader of any given process analysis is usually a novice seeking to understand the particulars of a process; therefore, the language should not be highly technical. For such an audience, provide ample background material, definitions of all terms, simple language, and graphics and visuals. However, trained technicians, engineers, and scientists also read and review process analyses in order to increase their knowledge about the advances in their fields. When an analysis is written for such audiences, the language may include far more technical terms, equations, and so on.

Titles

The title should be precise, descriptive, and limiting. Avoid a *how-to* title that implies that instructions, rather than an analysis, will follow. Some examples are:

> The Process of Taking Blood Pressure with a Sphygmomanometer
>
> Preparing an Income Statement for a Small Business
>
> How a Congressional Bill Becomes a Law
>
> How Fossil Fuels Are Created

Language

Instructions are written in the imperative mood to give a series of commands (*apply* the alcohol, *spread* the fertilizer, *smooth* the layer of resin). A process analysis is written in the indicative mood to explain steps (the technician *applies*, the gardener *spreads*, the layer of resin *is smoothed*). These indicative mood verbs may be active (the gardener *spreads*) or passive (the resin layer *is smoothed*). Avoid the words *you* and *your* in a process analysis. Some sample statements follow:

Incorrect	All banks recommend when you receive your monthly bank statement that you reconcile your records immediately.
Correct	All banks recommend that their depositors reconcile their records as soon as they receive their monthly bank statements.

or

> All banks recommend checking the monthly statements immediately to reconcile personal records.

Although metaphors, analogies, and personifications often help to make the complicated seem clearer, the technical writer should avoid these devices that may be interpreted as sexist, racist, classist, or ethnocentric, all unethical linguistic constructions. Heather Brodie Graves and Roger Graves explore unethical constructions in a number of technical manuals. In a manual for a nailgun, the fastener is nicknamed a "studgun," and the manual writer discusses the "net penetrating capability" of the fastener and covers its "ramrod devices" to achieve "different intensities of thrust." This discussion may be interpreted as sexist with sexual innuendoes, which might irritate the female operator of the nailgun.

In another example, the writer of a microprocessor application handbook discusses a "master-to-slave exchange" in an analysis of how data transfers in a microprocessor. Clearly the master/slave analogy is classist.

In a manual on the proper use and maintenance of an automatic steam vaporizer, the writer states "excessive mineral content *water will shorten* the product life" of the vaporizer, which personifies the water and avoids stating that the user is the real agent of the action. You must vigorously avoid such literary devices in technical documents.

Organization

A process analysis consists of three parts:

- Prefatory material
- Analysis of the steps
- Concluding discussion

This organization is similar to that used in a description of a mechanism (see Chapter 9). The following in a typical process analysis outline:

1.0   Prefatory material (introduction)
   1.1   Intended audience
   1.2   Definition/purpose
   1.3   Background and theory
   1.4   Who, when, where
   1.5   Special considerations
   1.6   Tools, materials, supplies, apparatus (if applicable, with graphics)
   1.7   List of chronological steps (or graphic flow chart)
2.0   Analysis of the steps (body)
   2.1   First main step (with appropriate graphics)
      2.1.1   Definition and/or purpose
      2.1.2   Theory (if applicable only to the specific step)
      2.1.3   Special considerations (if applicable only to the specific step)

        2.1.4   Substeps (if applicable)
        2.1.5   Analysis of the step
     2.2   Second main step . . . (etc.)
 3.0   Concluding discussion/assessment (closing)
     3.1   Results evaluation
     3.2   Time and costs (if applicable)
     3.3   Advantages
     3.4   Disadvantages
     3.5   Effectiveness
     3.6   Importance
     3.7   Relationship to larger process

A process analysis is usually written in a narrative format, although it may employ outline, decimal numbering, or digit-dash-digit techniques (see Chapter 10).

Prefatory Material

AUDIENCE STATEMENT

You must consider the purpose and audience of your process analysis. Your audience will determine the extent of your analysis and the degree of technicality. State exactly what the purpose is and for whom the analysis is intended.

Examples

The process analysis of taking a patient's blood pressure is designed for a beginning nursing student with no prior experience.

This analysis of how fossil fuels are created is intended for laypersons who are interested in geology.

This analysis of digital television is designed for technical engineers.

DEFINITION/PURPOSE

A formal definition of the process or a clear statement of the purpose of the process is necessary before an analysis of the parts. To avoid later interruption, the prefatory material may include definitions of other key terms to be used within the body of the report.

Example 1

Blood pressure is the force exerted by the blood against the walls of the blood vessels. It is created by the pumping action of the heart. The greatest pressure, known as *systolic pressure,* occurs during the contraction of the heart. The lowest pressure, known as *diastolic pressure,* occurs during the relaxation or rest period of the heart. The purpose of taking a patient's blood pressure is to relate it to other health factors, to determine if the patient is healthy, or to determine the cause of illness.

Example 2

Fossil fuels are energy producers, such as coal, oil, and gas, that have been formed deep underground by heat and pressure on dead vegetation and plankton (plant and animal organisms that float or drift in fresh or salt water) over millions of years.

Example 3

The NECAR4 uses onboard fuel cells to enable hydrogen to combine slowly with oxygen at moderate temperatures and produces H_2O and electricity with zero carbon dioxide to pollute the atmosphere.

BACKGROUND/THEORY

It may be necessary to explain the historical or scientific background, theory, or principle of a process. Such discussion also belongs in the prefatory material.

Example 1

A brief understanding of the conditions under which blood circulates in the body is necessary. Blood passes from the heart throughout the body by way of a system of vessels that eventually return the blood to the heart. This journey is so rapid that a single drop of blood usually requires less than one minute to move from the heart through the body and back to the heart.

A single tube leading from the heart divides into smaller and smaller vessels, the arteries. The smallest arteries branch into capillaries, the most minute blood vessels. Through the thin walls of the capillaries, the blood supplies the body with oxygen from the lungs and collects the waste products of the body for subsequent removal by the kidneys and other excretory channels.

Beyond the capillaries, the branching process is reversed. The capillaries join to make slightly larger tubes, which next unite to form larger vessels, the veins. Eventually the blood is returned to the heart by two large veins. Therefore, pressure is greatest in the arteries and least in the veins.

One records blood pressure as a fraction, such as 120/80 mm Hg (the chemical symbol for mercury). The normal blood pressure for a healthy, resting adult ranges from 100 to 140 mm Hg systolic and from 60 to 90 mm Hg diastolic.

Example 2

As dead vegetation in swamps or decaying microorganisms on the floors of oceans are compressed by increasing layers of earth and sediment, such vegetation and microorganisms are transformed by heat and pressure into fossil fuels.

Example 3

Fuel cells were invented in the 1800s and adopted by NASA for generating clean power in space in the l960s. Only in the 1900s were fuel cells made small enough to fit inside a car's engine.

WHO/WHEN/WHERE

If your analysis involves an operator, the qualifications of the operator should be specified in the prefatory material. When and where the process is performed should also be explained.

Example 1

Nurses, doctors, medical assistants, and other paraprofessionals trained in the use of a sphygmomanometer determine blood pressure in the daily routine of patient care for diagnostic purposes. Because blood pressure will vary at different times of the day and because readings are usually taken for the purpose of comparison with previous readings, readings should be taken at the same time every day. Generally, only a doctor is qualified to evaluate blood pressure readings in relation to a patient's sickness or health. The readings are taken in a doctor's office, a hospital, at the scene of medical rescue operations, or in other clinical settings.

Example 2

The highest grade and most abundant coal comes from forests that grew in hot swampy areas during the Permian and Carboniferous periods 280 million to 360 million years ago. Crude oil and natural gas began as decaying microorganisms on the floors of oceans in the late Tertiary and Cretaceous periods between 30 million and 180 million years ago.

SPECIAL CONSIDERATIONS

Special conditions, requirements, preparations, and precautions that pertain to the entire process, and not just to one step, should be indicated in the prefatory material.

Example 1

It should be noted that many factors influence blood pressure. A patient should be asked if any of the following factors could be influencing his or her blood pressure at the time of the reading:

1. Increasing blood pressure factors
   a. eating
   b. stimulants
   c. exercise
   d. emotional stress

   2. Decreasing blood pressure factors

   a. rest

   b. fasting

   c. depression

Other factors that should be considered are pain, climate variation, tobacco, bladder distension, hemorrhage (blood loss), blood viscosity (thickness), and the elasticity of the arteries.

Example 2

For an oil field to form, the oils must be trapped between layers of impermeable rock such as shale.

TOOLS/MATERIALS/SUPPLIES/ APPARATUS

A precise, detailed list of all tools, materials, supplies, and apparatus used in the performance of the process should be provided. It may be necessary to write a brief description of an unusual mechanism used in the process (see Chapter 9).

Example

The following supplies are used to take an accurate blood pressure reading:

   1. Stethoscope, an instrument used to magnify the sounds of arterial pulse.

   2. Sphygmomanometer, a three-part instrument consisting of a mercury pressure gauge, an arm band with an inflatable rubber bladder, and a pressure bulb to control the flow of air going through connecting tubes in and out of the bladder.

   3. Alcohol to clean the earpieces of the stethoscope.

   4. Cotton balls to apply the alcohol.

LIST OF CHRONOLOGICAL STEPS

After dividing the process into its main steps, each based on completion of a stage of work or action, the steps should be listed in chronological sequence: first, second, third, and so on. Use the following form or a flow chart:

Pattern　The process consists of _____ main steps; first,
_____ing the _____; second,
_____ing the _____; third,
_____ing the _____; and finally,
_____ing the _____.

Example 1

The process of taking a blood pressure reading consists of nine main steps: first, preparing the patient; second, assembling the sphygmomanometer; third, attaching the arm cuff to the patient; fourth, placing the stethoscope over the brachial artery; fifth, closing the pressure control valve; sixth, inflating the cuff; seventh, opening the control valve; eighth, taking the reading; and ninth, removing the apparatus.

Example 2

The process of coal formation consists of four main steps, as shown in the following flowchart:

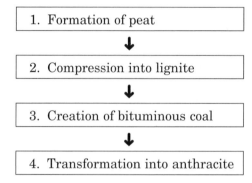

Analysis of the Steps

The body of the process analysis thoroughly examines and analyzes each step. Your emphasis here should be to explain why the process is performed in a particular manner. Analyze what would happen if the process were not performed in the correct manner.

DEFINITION/PURPOSE

Identify each step and write a formal definition and/or a statement of purpose of the step.

Example 1

The fourth step, placing the stethoscope over the brachial artery, is done so that the clinician can hear the rhythmical, thumping sounds of the blood. The brachial artery is the large artery of the arm at the inner crease of the elbow.

Example 2

The first main step, the formation of peat (partially carbonized vegetable matter used as fertilizer and fuel), is necessary before peat can be compressed into lignite (a low-grade, brownish-black coal).

THEORY

If a particular step is based on a theory, explain how the theory applies to the step.

Example 1

The brachial artery is used because it is near the heart, large enough for specific recognition, and near the surface of the skin.

Example 2

For an oilfield to form, the oils must be trapped between layers of impermeable rock such as shale.

SPECIAL CONDITIONS

Describe in detail any special considerations, requirements, apparatus, preparations, and precautions that apply to this step only.

Example 1

In most patients, the brachial artery is found quite simply. If there is any difficulty in locating it, the opposite arm may be more yielding. An injured arm or one that contains an intravenous injection should not be used.

Example 2

Bacteria and fungi are necessary to attack dead vegetation in order to break down the vegetation and form a mixture rich in hydrocarbons.

SUBSTEPS

If a main step contains substeps, list them chronologically prior to explaining each.

Example 1

The third step, attaching the arm cuff to the patient, consists of three substeps: checking the cuff bladder, positioning the cuff, and attaching the cuff to the arm.

Example 2

The first step, the formation of peat, consists of three substeps: bacteria and fungi attacking dead vegetation, the vegetation forming into a mixture rich in hydrocarbons, and the squeezing out of water.

ANALYSIS OF STEPS

Finally, explain each step and its substeps with attention to the reasons each is performed in a specific manner.

Example 1

The cuff bladder should not contain any air at the time of positioning the cuff. If it does, a secure fit will not be affected. The armband is wound around the arm above the elbow at a level with the heart, allowing room beneath it for the stethoscope bell. The band is fastened by means of the hooks, snaps, or Velcro material provided for this purpose. If no means of fastening are provided, the end of the band may be tucked securely under the top of the band. If the band is wound too tightly, it will bind the arm, create extra pressure, and cause discomfort to the patient. If the band is wound too loosely, the sounds will be deadened by the cushion of air required to tighten the band sufficiently to compress the brachial artery.

Example 2

Coal buried deeper than three miles beneath the surface, where the temperature is 400°F, is transformed into anthracite. It requires between 25 tons and 75 tons of plant matter to make one ton of anthracite.

Each main chronological step should be analyzed by the same writing strategies.

Concluding Discussion/Assessment

One or more of the following factors should be discussed in the conclusion:

- Time and cost
- Advantages
- Disadvantages
- Effectiveness
- Importance
- Relationship to a larger process

Some of these points may already have been covered in the introduction. Keep in mind that your reader is seeking to assess the efficiency of the process.

The following concluding discussion explains time and cost, advantages, importance, and relationship to the process.

Example 1

Taking a blood pressure reading requires only a few minutes. Because such readings are routine in regular medical check-ups and patient care, the cost is included in the consultation fee. Many health associations provide free blood pressure readings to the public in such places as shopping centers, libraries, and schools.

A blood pressure reading is the one sure method of detecting hypertension, the silent killer. Early detection can prevent strokes, coronaries, and kidney failures.

It is important to have an ongoing record of readings so that variations can be detected. By itself, a reading will not tell the doctor what is wrong, but along with other diagnostic procedures, it will help to determine a patient's condition.

Example 2

All of the energy produced in the United States in 1991 totaled 67.5 billion BTU. One BTU equals the energy released in burning a wooden match. The breakdown of U.S. energy sources in 1991 was coal at 32 percent, natural gas at 31 percent, crude oil at 23 percent, nuclear electric power at 10 percent, hydroelectric power at 4 percent, and other sources at 0.3 percent.

Allowing for population growth, known coal reserves should last about 300 years. Without a major switch to alternative fuels, known petroleum reserves probably will last about 40 years.

GRAPHICS AND VISUALS

Graphics, such as a flow chart or pictograph of the steps in a process, drawings or schematics of individual steps, and charts and tables of costs and/or time requirements, are helpful. Figure 11.1 shows a variety of sample graphics that will help to organize and analyze a process. Notice also the graphics in the sample process analyses in Figure 11.2 and Figure 11.3. Figures 11.2 and 11.3 show a scientific and organizational process analysis, respectively. Figure 11.2 shows a scientific analysis of a tidal wave/tsunami written for a science encyclopedia. Figure 11.3 shows an organizational analysis of a procedure (stopping a felony suspect's vehicle) designed for a police trainee manual. Either margin notes or headings within the text of each explain and identify the analyses' particulars. Be prepared to critique the examples in class.

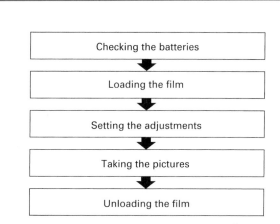

Figure 1. Steps in the process of operating a camera.

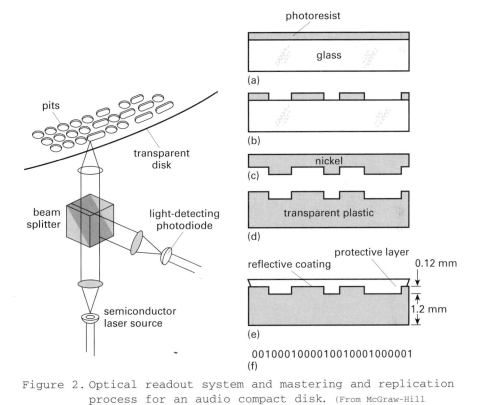

Figure 2. Optical readout system and mastering and replication process for an audio compact disk. (From McGraw-Hill Encyclopedia of Science and Technology, 7th ed., 1992. By permission of The McGraw-Hill Companies.)

FIGURE 11.1 *Sample process analysis graphics (typical flow chart of steps, pictograph, schematic, and entire process)*

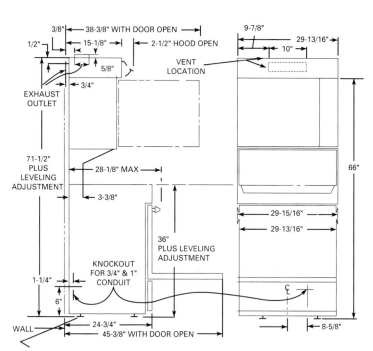

Figure 3. Schematic for positioning a cabinet
microwave oven in a wall

Making electricity and steam together

Exxon's cogeneration systems typically use a
gas turbine to generate electricity. Exhaust from
the turbine heats water in a boiler to make steam
for use in refinery and chemical plant processing.

The hot exhaust gases from the turbine
are passed through a boiler which
generates steam.

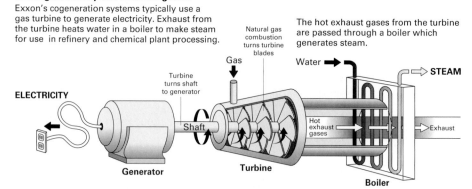

Figure 4. Process of Exxon's cogeneration system
(By permission of Carol Zuber-Mallison, Brook and Company, Dallas, TX)

FIGURE 11.1   *continued*

Tsunami

A radially spreading, long-period gravity-wave system caused by any large-scale implosive sea-surface disturbance. Being only weakly dissipative in deep water, major tsunamis can produce anomalous destructive wave effects at transoceanic distances. Historically, they rank high on the scale of natural disasters, having been responsible for losses approaching 100,000 lives and uncounted damage to coastal structures and habitations. Because of their uncertainty of origin, infrequency (10 per century), and sporadicity, tsunami forecasting is impossible, but the progressive implementation, since 1946, of the effective International Tsunami Warning System has greatly reduced human casualties. Present efforts, aided by advances in the fields of geomorphology, seismicity, and hydrodynamics, are directed toward an improved understanding of tsunami source mechanisms and the quantitative aspects of transocean propagation and terminal uprush along remote coastlines. These efforts are abetted by the ever-increasing coastal utilization for nuclear power plants, oil transfer facilities, and commercial ports, not to mention public recreation. SEE GEOMORPHOLOGY; SEISMOLOGY.

Tsunami generation. While minor tsunamis are occasionally produced by volcanic interruptions or submarine landslides, major events are now recognized to be generated by the sudden quasi-utilized dislocations of large fault blocks associated with the crumpling of slowly moving sea-floor crustal plates, where they abut normally against the continental plates. Such sources are predominantly confined to techtonically active ocean margins, the majority of which currently ring the Pacific from Chile to Japan. As opposed to secular creep, tsunami-producing block dislocations are invariably associated with major shallow-focus (< 18 mi or 30 km) earthquakes of intensity greater than 7 on the Gutenberg-Richter scale, followed by swarms of aftershocks of lesser intensity that decay in frequency over a week or two. Independent lines of evidence indicate that the aftershock perimeter defines the dislocated area, which is usually elongate, with its major axis parallel to the major fault trend. That only about 20% of large earthquakes fitting the above description produce major tsunamis raises the additional requirement

that the dislocations have net vertical displacements, a view supported by seismic fault-plane analysis and confirmed by direct observations in the case of the 1964 Alaskan earthquake. SEE EARTHQUAKE.

Because the horizontal block dimensions are characteristically hundreds of miles, any such vertical dislocation (a few yards) immediately and similarly deforms the water surface. The resulting tsunami whose energy is concentrated at wavelengths corresponding to the block dimensions and whose initial heights are determined by the local extent of vertical dislocation. SEE FAULT AND FAULT STRUCTURES.

Deep-sea propagation. Having principal wavelengths much longer than the greatest ocean depths, tsunamis are hydrodynamically categorized as shallow-water waves; to good approximation, their propagation speeds are proportional to the square root of their local water depth (400–500 or 600–800 km/h in the Pacific). Thus, after determination of the source location by early seismic triangulation, real-time warnings of wave arrival times can be complied from precalculated travel-time charts (**Fig. 1**).

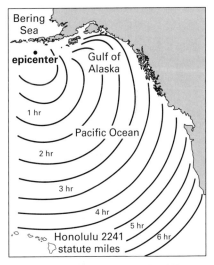

Fig. 1. Advance of tidal wave of April 1, 1946, caused by an earthquake with epicenter southeast of Unimak Island. (After L. D. Leet and S. Judson, Physical Geology, 2nd ed., Prentice-Hall, 1958)

FIGURE 11.2 *Scientific process analysis of tidal wave/tsunami* (William G. Van Dorn, *McGraw-Hill Encyclopedia Of Science And Technology,* 6th Ed., 1987. By permission of The McGraw-Hill Companies)

Concluding
discussion

Because the initial wave energy imparted by the source dislocation is spread ever thinner as the wave pattern expands across the ocean, the average wave height everywhere diminishes with travel distance, amounting to only a few centimeters (virtually undetectable in deep water) halfway around the globe. Beyond this point, energy converges again toward the antipole of the source, and wave heights increase significantly. This convergence accounts, in part, for the severity of coastal effects in Japan from Chilean tsunamis, and conversely. Additionally, azimuthal variations in local wave height are caused by source orientation and eccentricity because, as with a radio antenna, the energy is radiated more efficiently normal to the longer axis. Lastly, further variations of wave height arise from refractive effects associated with regional differences in average water depth. *SEE OCEAN WAVES; WAVE MOTION IN FLUIDS.*

Coastal effects. After the waves cross the continental margins, the average wave height is greatly enhanced, partly by energy concentration in shallow water, and partly by strong refraction which, like a waveguide, tends to trap and further concentrate energy against the coastline. Ultimately, the shore arrests further progress. Here, the tsunami is characterized by swift currents in bays and harbors, by inundation of low-lying areas as recurrent breaking bores, and by uprush against steep cliffs, where watermarks as high as 60 ft (20 m) above sea level have been observed. *SEE NEARSHORE SEDIMENTARY PROCESSES.*

Prediction methods. While most of the above behavior has been predicted qualitatively in theory, the increased accuracy required for engineering design and hazard evaluation has led to the development of numerical modeling on large computers. Taking as input a representative time- and space-dependent source dislocation, as inferred from seismic or observational evidence, and the oceanwide distribution of water depths from bathymetric charts, the computer generates the initial surface disturbance **(Fig. 2),** propagates it across the sea surface, and yields the time history of water motion at any desired location. Where the source motion is known (Alaska, 1964), computer simulations agree quite accurately with observations at places where the linearized equations of motion can be expected to apply. But localized nonlinear effects, such as breaking bores, are more realistically modeled hydrodynamically, using the computer-generated wave field offshore as input.

William G. Van Dorn

Bibliography. L-S. Hwang and D. Divoky, Tsunamis, *Underwater J.,* pp. 207–219, October 1971; T. Iwasaki and K. Ilida (eds.), *Tsunamis: Their Science and Engineering,* 1983; National Academy of Sciences, The great Alaskan earthquake of 1964, *Oceanography and Coastal Engineering,* 1972.

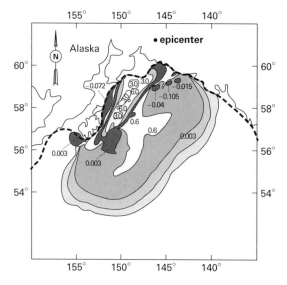

Fig. 2. Computer simulation of tsunami of March 24, 1946, in Gulf of Alaska, 18 min after earthquake. Surface contours in centimeters; wave height in meters; t = 1100 s. 1 cm = 2.5 in., 1 m = 3.3 ft. (*This figure was first published in Underwater Journal, October 1971, pp 207–219, and is reproduced here with the permission of Butterworth-Heinemann, Oxford, UK*)

FIGURE 11.2   *continued*

STOPPING THE FELONY SUSPECT'S VEHICLE

AUDIENCE

This analysis is designed for police academy trainees with no prior police experience.

Purpose/Definition

The purpose of this analysis is to familiarize police academy students with the correct, efficient, and safe way to stop a felony suspect in a vehicle. A felony is any crime, defined by law, for which the culprit could receive imprisonment in a penitentiary or the death sentence.

Background

During a tour of duty a police officer may observe thousands of vehicles. The officer may observe a license number of a wanted or stolen vehicle, or he may recognize a wanted criminal inside a vehicle. In any such encounter, the police officer must immediately distinguish between proper procedures and carelessness. At times, officers have failed to make full use of the advantages that proper procedures offer. Fear of being called "overcautious" or of "crying wolf" have led some officers to disregard the responsibility of self-protection.

Precaution

Before stopping a suspected felon in a vehicle, the officer should ascertain that a supporting or back-up officer will be available for assistance.

Who/When/Where

This process could be performed by any police officer in any city, county, or state when a felony suspect is spotted in a vehicle.

Tools

The officer requires a patrol car, a two-way radio, and a service pistol.

Steps

Stopping a felony suspect in a vehicle consists of five main steps:

FIGURE 11.3   *Sample process analysis* (Courtesy of student Madalyn McGown)

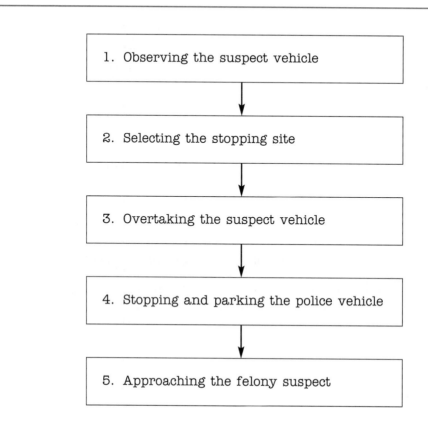

1. Observing the suspect vehicle

2. Selecting the stopping site

3. Overtaking the suspect vehicle

4. Stopping and parking the police vehicle

5. Approaching the felony suspect

ANALYSIS OF STEPS

The first step, observing the suspect's vehicle, requires the officer to notify the radio dispatcher at once. The officer gives the following information:

1. Identification of the police unit
2. Location of the contact
3. Description of the vehicle and license number
4. Description and number of occupants
5. Direction of travel and name of the last cross street passed

The last cross street is broadcast at intervals to aid the dispatcher in predicting the suspect's course of travel, thereby hastening the arrival of supporting police units. The officer writes the license number, description of the car, and the number of occupants on a memo pad inside the police car. This is necessary in case the

FIGURE 11.3   *continued*

dispatcher did not receive all of the officer's radio transmission. The suspect's vehicle is trailed until the supporting units are close. The officer should be alert for sudden stops, turns, or other evasive action on the part of the suspect's vehicle.

The second step, selecting a stopping site, requires the officer to select a location with which he is familiar. Familiar surroundings will be to the officer's advantage as he will be able to direct additional assistance to the location more quickly. The officer will also be in a better position to make an apprehension if the suspect attempts to flee. Stopping near alley entrances, openings between buildings, vacant lots, and other easy escape should be avoided. At night, a well-lighted area will enable the officer to observe whether or not the suspect is disposing of any evidence or a weapon.

In the third step, overtaking the suspect's vehicle, the officer maintains constant vigilance to guard against any evasive action on the part of the suspect. At this time, the officer operates the siren and vehicle emergency lights. The police car is driven directly to the rear of the suspect's vehicle. Overtaking the suspect's vehicle is illustrated in Figure 1:

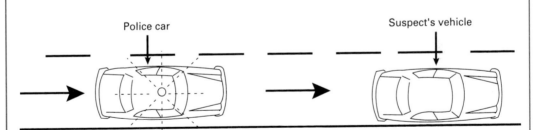

Figure 1   Overtaking the suspect's vehicle

The distance between the two vehicles will vary with the speed. Usually a distance of one car length for every ten miles per hour provides a safety zone for the officer. The officer should constantly consider the safety of other motorists and pedestrians in overtaking and stopping the suspect's vehicle. The suspect's vehicle should be stopped as far as possible to the right side of the roadway or off of the roadway altogether.

FIGURE 11.3   *continued*

In the fourth step, stopping and parking the police car, the officer notifies the dispatcher of the location of the stop. The officer should park his vehicle about 10 feet behind the suspect's vehicle with the front of the police vehicle pointing toward the center of the street. The police vehicle should be on a 45° angle to the suspect's vehicle. The police vehicle position is illustrated in Figure 2:

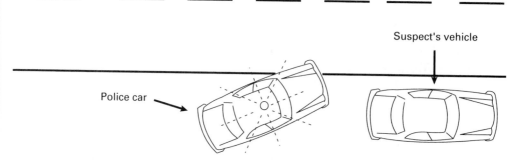

Figure 2   Parking position of stopped vehicles

The police vehicle emergency lights are left on so that other motorists will be warned of danger and the supporting police units will be able to find the officer quickly.

In the fifth and final step, approaching the suspect, the officer making the stop should take complete command of the situation. The officer should get out of the police vehicle and with gun drawn proceed to the left front fender of the police vehicle. Using the body and engine of the vehicle as protection, the officer identifies himself in a loud, clear voice: "Police officer! You are under arrest! Turn off your motor and drop the keys on the ground." The officer may order the suspect to place both of his hands out the driver's window or to place his palms against the inside of the windshield. The officer should stay by his police vehicle until a supporting or back-up officer arrives. The back-up officer should park his police vehicle to the rear and a little to the right of the first police car. The back-up officer should proceed to a position off the right rear of the suspect's vehicle with his gun drawn. The position of the police vehicles and officers are illustrated in Figure 3:

FIGURE 11.3 *continued*

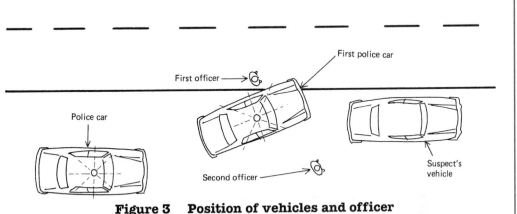

Figure 3   Position of vehicles and officer

The suspect should be made aware of the second officer's presence by the first officer to reduce the possibility of sudden attack from the suspect. The suspect is ordered out of his vehicle by the first officer. The first officer then orders the suspect to assume a search position.

CONCLUSION

Searching a suspect is another process and is not discussed here. No two felony stops are identical in nature. The unpredictable actions of the felony suspect make the officer rely on his training and judgment in each case. By following the steps in this process, the danger to the officer is minimized.

FIGURE 11.3   *continued*

CHECKLIST

Writing a Process Analysis

PRELIMINARY CONSIDERATIONS

❑ **1.** Have I determined what type of process analysis I am writing?

    ❑ Historical?

    ❑ Scientific?

    ❑ Mechanical?

    ❑ Natural?

    ❑ Organizational?

❑ **2.** Have I determined my audience?

    ❑ Laypeople?

    ❑ High-tech engineers and scientists?

    ❑ Low-tech workers in specific fields?

❑ **3.** Have I provided a precise, descriptive, limiting title?

❑ **4.** Will the language I use be appropriate?

    ❑ Indicative mood with consistent active or passive verbs?

    ❑ Elimination of *you* and *your?*

    ❑ Elimination of metaphors, analogies, and personifications that may be sexist, racist, classist, or ethnocentric?

ORGANIZATION AND CONTENT

Preliminary Considerations

❑ **1.** Have I included an intended audience statement in my prefatory material?

❏ **2.** Have I provided a definition and/or purpose statement of the process?

❏ **3.** Have I explained the historical or scientific background, theory, or principle of the process?

❏ **4.** Have I specified the who, when, and where of the process?

    ❏ If there is an operator, have I stated who that operator is and what his or her qualifications should be?

    ❏ Is there a specific time to perform the process or a time period when the process occurs?

    ❏ Have I stated where the process takes place?

❏ **5.** Have I spelled out the special conditions, requirements, preparations, and precautions that pertain to the entire process?

❏ **6.** Have I provided a precise, detailed list of all the tools, materials, supplies, and apparatus that are required for the process?

❏ **7.** Have I divided the process into main steps based on a completion of a stage or work or action and listed them in chronological order?

Analysis of the Steps

❏ **1.** Have I analyzed each step in the correct order?

❏ **2.** Have I divided a main step into substeps if necessary?

❏ **3.** Have I provided a definition or purpose statement for each step?

❏ **4.** Have I explained the theories that pertain to any one step?

❏ **5.** Have I described any special considerations, requirements, apparatus, preparations, and precautions that apply to any one step only?

❏ **6.** Have I analyzed each step and substep with attention to the reasons *why* each is performed in a specific manner?

Concluding Discussion/Assessment

❏ **1.** Have I discussed, if appropriate, the time and cost involved in the process?

❏ **2.** Have I discussed, if appropriate, the advantages of the process?

❏ **3.** Have I discussed, if appropriate, the disadvantages of the process?

❏ **4.** Have I discussed the effectiveness of the process?

❏ **5.** Have I discussed the importance of the process?

❏ **6.** Have I explained the relationship of this process to a larger process?

❏ **7.** May I expect my reader to be able to judge the reliability, practicality, and efficiency of the process?

   ❏ Will my reader be able to estimate the difficulty of the process?

   ❏ Will my reader be able to foresee the problems that are likely to occur?

   ❏ Will my reader be able to ascertain successful solutions to the preceding problems?

Graphics and Visuals

❏ **1.** Have I provided graphics and visuals to clarify the analysis?

   ❏ A flow chart of steps?

   ❏ A drawing, picture, or photo of required equipment?

   ❏ A schematic?

   ❏ A diagram of the substeps within a step?

   ❏ Tables of costs, time requirements, or advantages/disadvantages?

   ❏ Other?

❏ **2.** Is my document design appropriate?

EXERCISES

1. **Types of Process Analysis.** Classify the following process analysis subjects as historical (H), scientific (S), mechanical (M), natural (N), or organizational (O).

   a. How earthquakes occur

   b. Changing a tire on a hill

   c. How the liver functions

   d. Making pennies

   e. How AIDS is treated medically

   f. How (your state) became a state

   g. How intravenous fluids are administered

   h. Evaluating a college teacher

   I. How digital television developed

   j. How labels are printed on a computer

2. **Language.**   The following excerpt is from a process analysis of developing black and white film. Rewrite it to eliminate commands, the pronouns *you* and *your*, and literary language. Use third person subjects and indicative mood verbs in a consistent active or passive voice.

   > In the first step, loading the film into the roll, you attach the film to the developing reel. Place the end of the film into the notch, like a little tooth, at the end of the reel. Wind the entire roll onto your reel. Next, you place the reel into the developing tank. Snap the top securely into place. You may now turn on your lights.

3. **Organization and Naming of Steps.**   Using the pattern in this chapter to name and list chronological steps of a process, devise and number the steps in the following organization processes.

   a. Balancing my checkbook

   b. Cooking a favorite dish

   c. Learning to drive a car

   d. Scheduling my classes for next term (or this term if you will be graduated next term)

   e. Planning a spring break vacation

4. **Visuals.**   Locate or devise five or more visuals from the following list.

   a. How a digital camera, digital cell phone, or digital television operates

   b. How an organ of the body functions

   c. How to avoid damage or injury in a given procedure or assembly

   d. How an earth's phenomenon (tides, new moons, seasonal changes, night/day, etc.) occurs

WRITING PROJECTS

1. **Audience.**   Locate a brief process analysis written for a specific purpose and audience. Look in your textbooks, medical or engineering journals, or other sources that publish materials in your field of study. Rewrite the analysis for a different purpose and audience. For instance, change a medical textbook analysis of AIDS progression to a student newspaper account to be included in an article on the perils of the disease.

2. **Process Analysis.**   Select a subject from the following list and write a *historical* or *organizational* process analysis. Use the checklist to ensure that all requirements are covered. Include a bibliography in the MLA style.

    a. Making clay tiles

    b. How the President of the United States is elected

    c. Tuning the motor of a particular car

    d. Recording a program on a specific VCR

    e. How women in the U.S. won the vote

    f. Maintaining a salt water aquarium

    g. How North and South Korea settled their cold war

    h. How social security was enacted in the United States

    i. Selecting an automobile

3. **Collaborative Project—Scientific, Mechanical, or Natural Process Analysis.**   With a group of three or four students, write a *scientific, mechanical* or *natural* process analysis of one of the following subjects. Use the checklist to ensure that all of the requirements are covered. Include a bibliography in the style appropriate to your subject.

    a. How the Appalachian Mountains evolved

    b. How gum disease develops

    c. How ulcers develop

    d. How a cellular phone works

    e. How e-mail is transmitted

    f. How the lungs function

    g. How osmosis occurs

    h. How a body metabolizes sugar

i. How sound waves are transmitted

j. How a compact disc is made

k. How a fire extinguisher works

l. How a plant propagates

m. How icicles form

n. How rust forms

o. How salmon procreate

NOTES

PART *four*

The Professional Strategies

Composing Correspondence

DILBERT

by Scott Adams

TO: ALL USERS	TO: NETWORK ADMIN	HAVE YOU NOTICED THERE'S TOO MUCH COMMUNICATION IN THE WORLD, DOGBERT?
FROM: NETWORK ADMIN	FROM: DILBERT	
	CC: ALL USERS	
PLEASE REFRAIN FROM FRIVOLOUS E-MAIL. IT BOGS DOWN THE NET-WORK.	I AGREE!	YEAH, EVERY DAY AT ABOUT THIS TIME.

E-mail: SCOTTADAMS@AOL.COM

5/10 © 1995 United Feature Syndicate, Inc. (NYC)

DILBERT reprinted by permission of United Feature Syndicate, Inc.

S K I L L S

After studying this chapter, you should be able to

1. Determine whether a conventional letter, electronic mail, or faxed mail is acceptable for a recipient.

2. Understand the advantages of electronic mail and facsimile machines for rapid and cost-efficient correspondence.

3. Recognize a variety of memorandum formats, e-mail headings, and fax cover pages.

4. Implement the International Standard for dating correspondence.

5. Write a brief, focused memorandum in an appropriate format.

6. Write a brief, correctly formatted electronic memorandum or letter (e-mail).

7. Understand and adhere to all the particulars of "Netiquette," Internet etiquette as it pertains to e-mail.

8. List, define, and write an example of the regular and special elements of letters.

9. Design letters in full block and modified block formats.

10. Head a second page of a letter correctly.

11. Write effective positive (good news) letters: an inquiry or request, an invitation, an order, a congratulatory or thank-you letter, and a sales or service offer.

12. Write effective negative (bad news) letters: a negative response, a complaint, and a series of collection letters.

INTRODUCTION

All professionals need to perfect the writing strategies required for effective correspondence. Memorandums, electronic mail (e-mails and faxed material), and letters are records that promote action, transact business, and maintain continuity.

A one-page letter costs as much as $20 to produce and deliver. Consider the equipment needed, such as a computer, word-processing software, a printer, possibly a photocopy machine or a fax machine, paper and envelopes, and postage. Therefore, mastering the strategies of effective correspondence is crucial for cost-effective transactions.

Memos and letters can be delivered by traditional mail, by e-mail, or by facsimile machine. The latter two methods allow writers to correspond throughout the world in just seconds. Business is now conducted daily on a global scale, so an understanding of international practices, such as dating, language differences, and approaches, has become essential for the professional writer (see Chapter 2, "Executing the Writing Process: Theory and Practice"). Security can be broached and must be a consideration for the type of correspondence you select. As always, you must avoid sexist or "politically incorrect" nuances or word usage.

Overview of Types/Security

A **memorandum,** usually called a **memo,** is used primarily for internal (within the company or organization) correspondence as a rather informal request for information, confirmation of a conversation, brief announcements, congratulations to a coworker, a meeting summary, or to transmit longer documents and reports on day-to-day activities. They are usually placed in an employee's mailbox directly or distributed through interoffice envelopes. Security depends on the ethics of coworkers and the reliability of the deliverer.

Electronic mail (e-mail) is used for the same casual correspondence as memos, but may also contain longer and more formal correspondence. The advantages of e-mail are, of course, the speed in which a message may be sent anywhere in the world, the speed at which it can evoke reaction, and its ability to be printed, saved, or deleted, thereby not necessarily cluttering up someone's desk with needless paper. E-mail is faster and more cost efficient than deliverable interoffice mail, telephone conversations, and posted correspondence; however, it may not provide the same security in that it can be lost in transmittal, the original layout may not transfer exactly, and it might be tampered with or read by others with access to the system.

Letters are used for external correspondence and are usually characterized by more formality than memos and most e-mail. A letter may be used to request information, respond to inquiries, promote sales and services, raise funds, register complaints, offer adjustments, collect payment, provide instructions, offer thanks and appreciation, report on any other activity, or seek employment (see Chapter 13). Letters are the most secure correspondence, because the U.S. Postal Service is reliable, the page will not change its format, and privacy laws protect letters' contents.

Faxes can quickly transmit hard copy of letters, spreadsheets, and other documents at the cost of a regular phone call, faster and often cheaper than regular post. The security of a fax depends on the address accuracy of the sender and the ethics of coworkers with access to the receiving fax machines.

Each type of correspondence commands its own format. Standard formats and text may be created by developing templates with a word-processing program, removing some of the drudgery of setting up a format or repeating text each time for new correspondence. Each type of correspondence should be characterized by a "you" perspective as well as by brevity, clarity, accuracy, and attractiveness.

Types of **memos** may vary from a brief, handwritten message that will probably be discarded by the reader after the appropriate action is taken to a carefully typed message that will be filed as part of the permanent record.

Memorandum and Fax Formats

Many companies use a preprinted communications memorandum for telephone and caller messages, as shown in Figure 12.1:

To _____

Date _____ Time_____

WHILE YOU WERE OUT

M _____

OF _____

PHONE _____
                Area Code   Number   Extension

TELEPHONED		PLEASE CALL	
CALLED TO SEE YOU		WILL CALL AGAIN	
WANTS TO SEE YOU		URGENT	
	RETURNED YOUR CALL		

Message _____

_____

_____

_____

                                    _____
                                              Operator

Campbell 09301

FIGURE 12.1   *Preprinted communication memorandum*

Some organizations also use preprinted memo forms on 8½″ × 11″ or half-page paper. These usually include the company name and logo plus the guide words *TO, FROM, SUBJECT,* and *DATE.* Other organizations use a "speed memo," one with carbon copies that allow the reader to send the message, retain a copy, and solicit a reply, as shown in Figure 12.2.

If your company does not have preprinted forms, the model shown in Figure 12.3 is the most common format. Notice the capitalization,

R
RYDER   RYDER TRUCK RENTAL, INC.   DATE _____

TO _____ A+ _____   SUBJECT: _____

FROM _____ A+ _____   LOCATION CODE: _____

MESSAGE _____

▶

_____ Signed _____

DATE _____

REPLY _____

_____ Signed _____

S10-01 (1-79)

ADDRESSEE TO RETAIN WHITE COPY—RETURN GREEN COPY TO ORIGINATOR
ORIGINATOR TO RETAIN BLUE COPY—SEND WHITE AND GREEN COPIES WITH
CARBON INTACT TO ADDRESSEE

FIGURE 12.2   *Preprinted speed memorandum*   (By permission)

MEMORANDUM

TO: John Snyder, President

FROM: Mary O'Neill, Personnel Specialist MO

DATE: 14 January 200X

SUBJECT: Fire in Reception Area, 13/1/0X

_____

_____
_____
_____

MO:jv

Enclosures

FIGURE 12.3   *Standard memorandum format*

margins, and spacing in Figure 12.3. The message paragraphs are usually single-spaced with double spacing between paragraphs. The writer's
signature or initials appear after the typed name in the *FROM* line for
authorization and for possible legal purposes, too.

   You may design a template for a facsimile cover sheet. Fax cover
sheets tend to be designed as memos but include logos or company slo-
gans for easy recognition. Figure 12.4 shows a typical fax cover sheet.

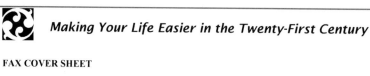

Making Your Life Easier in the Twenty-First Century

FAX COVER SHEET

TO: John Budge, Racal Electronics
         Media Communications Director
            Fax: 954-201-3700

FROM: Wendy Mills, Motorola
         Technical Writing Specialist
            Fax: 561-323-7890

DATE: 2 April 200X

SUBJ: Line Art for Product 5009D - Completion

FIGURE 12.4   *Sample fax cover sheet on a letterhead*

Addresses, Subject Lines, and Dates

Although frequently written in haste, a memo message requires careful consideration. The format requires identification lines of the sender, the intended receiver, the subject, and the date by using TO:, FROM:, SUBJECT: (or SUBJ:), and DATE:.

1. **Address.**   The address line of a memo should contain the full name and title of the receiver. Most employees wear several hats within an organization. An employee may be the director of human resources, chairman of the comptroller search committee, and a member of the ad hoc committee on leave policies. The receiver of a memo wants to know immediately in which capacity he or she is being addressed.

   Example

   TO: Dr. Glen R. Roth, Director of International Sales

   Mr. John Bass, Director of Multimedia Development

   Ms Mary Sue Johnston, Director of Legal Affairs

2. **Sender.**   The sender should also include his or her own name and title in a memo, indicating the position from which the memo is written.

   Example

   FROM: Susan, Long, Policy Committee Chair

3. **Subject line.**   The subject line should be precise. It may even convey the most important part of the entire message.

   Imprecise

   SUBJECT: Committee Meeting

   Precise

   SUBJECT: Resume Review Committee Meeting—18 May 2001

4. **Dates/Format.**   An exact date on all correspondence is important to both the sender and the receiver of a written communique. It allows both parties to maintain records of correspondence, encourages promptness by discouraging procrastination; and ensures mutually accepted times for meetings, deliveries, and other time-bound events. The date is usually spelled out rather than written numerically (7/19/01 or 19/7/01).

Example

DATE: 19 July 2001

Notice that the date is given here in the **International Standard of Dating:** day month year (with no commas) as opposed to the American Standard of Dating: month day, year (January 10, 2001). In America we often use numbers in haste in place of the month, such as 1/10/01. This may present a problem when corresponding with any country outside of the North American continent because almost all other countries use the International Standard. You should understand that someone in Italy receiving an order to be delivered to New York on 1/10/01 is likely to deliver the merchandise on October 1, 2001, causing quite a delay in the desired January 10 delivery. Thus, the U.S. military and all U.S. government agencies have officially adopted the International Standard to avoid confusion. Likewise, the Modern Language Association, the watchdog of writing practices, suggests that the general public should also adopt the International Standard.

Memo Content

A memo is a tool, not a piece of literature. Five conventions guide the content of most memorandums:

1. **Brevity.**   Most employees read numerous memos each day and must react to each quickly. Keep your memos succinct. One of the most effective memos ever written was by the famous commanding general of the Marine Corps', David Shoup. When Shoup completed his first inspection upon becoming the Marine Corps commandant at Parris Island, he noticed that officers, from noncoms to full colonels, were carrying swagger sticks, an affectation adopted from the British Royal Marines. Shortly, he posted the following memo:

   TO: All Personnel
   FROM: Commanding General

   _____

   If you need one, carry one.

   The next day not a swagger stick was to be seen. Although I do not advise you, necessarily, to be as brief as Shoup, remember that reading and writing take time.

2. **Immediacy.**   Usually, the in-house reader is somewhat informed about the general subject. Get directly to the point and avoid lengthy background. Just state your message and exclude irrelevant detail.

3. **Unity.**   Even though you may have several messages to convey to the same person, do not confuse your reader. Cover only one topic in each memo.

4. **Attractiveness.**   Pay attention to document design features, such as space, headings, margins, list options, emphatic features, and so on (see Chapter 4).

5. **Timeliness.**   Announce meetings several days in advance. Respond as quickly as possible to memos you receive. Interoffice correspondence is designed for efficiency.

Figure 12.5 is a sample memo that not only covers a single subject but also is brief, to the point, attractive, and timely.

MEMORANDUM

TO: Members of the Long-Range Planning Committee

FROM: Kenton Duckham, Chairperson *K.D.*

DATE: 27 April 200X

SUBJECT: Change in May meeting date

The regularly scheduled 15 May 200X meeting of the Long-Range Planning Committee is rescheduled for 22 May 200X, at 3:00 P.M. in the Board Room because President Jean Laird must attend a marketing conference in Atlanta.

Please notify me at extension 6647 if you will be unable to attend.

KD:jv

FIGURE 12.5 *Sample memorandum*

ELECTRONIC MAIL

It is probable that you will compose and send more e-mail memos and letters than print memos and letters in your career. Not even counting the approximately 1 billion e-mails sent daily by home computer users, employers and employees within companies send approximately 11 billion messages a month, and the number keeps rising. You must master current standards for format and etiquette, called Netiquette, for all e-mail.

Unfortunately, many people send out sloppy notes, infuse their messages with socio-emotional content, and treat their messages as electronic phone calls, reacting immediately and writing without thinking about organization, grammar, and punctuation. These spontaneous and unedited messages lead to misunderstandings and overreactions. You must remember that there is no privacy on the Internet. Management may legally intercept your message, and not even deleting messages ensures that they are not retrievable from your hard drive. "Cookies," files placed by certain Web sites onto your hard drive to monitor your online movements, can allow others to include information about you on a database available to direct mail catalogs and retailers. The federal government and the e-commerce industry have not yet fully agreed on privacy standards. Further, your messages may inadvertently be sent to the wrong person, skewed in the transmission, and received by many more people than intended.

E-mail Format

E-mail is not effective for long or formal messages, for permanent records, or for confidentially. Use e-mail for short, informal messages, singular or plural readers, immediate response elicitation, and to send attachments.

E-mail messages should resemble memos by including the elements of the heading (To:, From:, Date:, Subject:) although in a different order.

1. **Subject Line.**   Do not leave it blank. Type in the appropriate subject box a brief informative phrase to indicate the subject of your e-mail. Most incoming e-mail box listings include the sender's name and subject, allowing the recipient to scan the list and then select the mail in his or her preferred order. A subject line may be a noun phrase, contain the entire message, or indicate urgency.

   Examples

   Subject: Weekly Meeting Dates for August
   Subject: Friday's meeting will be moved to Room 207.
   Subject: URGENT! Today's 11:00 a.m. meeting is canceled.

2. **Date.**   The date and time of the message, the second line of received e-mail, are automatically generated. Ensure that your system's clock is correctly set or either your message may be misdated or all of your messages may contain the same date. Contrary to the preferred International Standard of Dating, e-mail programs use numbers and virgules (/) in the order of month/date/year.

Example

Date: 7/28/00 11:22.49 AM Eastern Daylight Time

3. **Sender's Address.**   E-mail software generates the "From:" line automatically, but your e-mail address may be difficult to interpret. If your address does not contain your full name and title, set up your system to include it in parentheses.

Example

From: Ladykay@aol.com (Katherine Martin, Director of Technical Writing)

Also set up your system to include a *signature block* containing your name, title, company, address, phone number(s), and e-mail address to attach automatically at the conclusion of each message. Each software system handles this differently, but your manual or systems expert will instruct you how to accomplish this task. A couple of clicks will delete the signature block from personal mail if you desire.

Signature Block Example

John Hansen
Director of Business Development
RST Television Ventures
200 Green Street
San Francisco, CA 91045
Office: (416)555-9999
Mobile: (416)555-7100
E-mail: John.Hansen@RSTTV.com

4. **Recipient's Address.**   The fourth line of e-mail messages begins with a "To:" line, which includes your selected e-mail address(es) and, desirably, the full name of the recipient. Make this letter-perfect or the message will either be returned to the sender as undeliverable or end up in the e-mail box of the wrong person. If the wrong person receives the mail, it could embarrass you, or worse, the receiver could just delete it, and you will not even know that the message did not reach your intended audience.

Example

Fred Dice@cs.com   (Fred Dice, Director of Business Development)

Use judgment when sending carbon copies (cc:) and blind carbon copies (bcc:). Overuse of these options not only creates junk e-mail but also may indicate that you, the sender, don't trust the intended recipient.

5. **Salutation.**   In personal e-mail, the sender tends to omit a salutation or just types, "Hi," or "Dan,". In a professional communication, however, use a more formal greeting, including certain titles and last names. Use an informal tone only if the recipient has indicated that it is appropriate.

Examples

Dear Dr. Jordan:

Dear Janet,

Janet,

6. **Message Length.**   Write a concise message focused on a single subject. Most researchers suggest that messages not exceed 25 lines, the amount that fills the typical screen. To send long messages, consider using "attachments," a separate file to be sent along with the concise message. However, not all recipients' software can access every kind of attachment. Remember, also, that your message will not always appear on the recipient's screen in the exact format that you sent it.

7. **Quoting.**   To respond to previous e-mail, explicitly repeat the question or content to which you are responding. The initiator will not necessarily remember what he or she queried. Use a >, a relatively standard practice, before quoted material followed by your response.

Examples

> Will you present the budget at Friday's meeting? (initiator's question)

Yes.   (your response)

> Do you need suggestions for the Responsibility Statement?

No, the committee discussed these particulars at our last meeting.

Netiquette

E-mail etiquette has developed over the past several years through on-line forums, computer-related magazines articles, and general news articles. Adherence to the rules will mark you as a considerate and professional communicator.

1. Don't forward private e-mail without the permission of the original sender.

2. Don't create or forward chain e-mail.

3. Don't "spam" (send the same message to hundreds of e-mail addresses) in hopes of hitting a few interested people. It is rude and a waste of resources.

4. Do not use all capitals (e.g., MEET ME ON TUESDAY TO DISCUSS SALARIES.) All capitals are considered the equivalent of screaming.

5. Use careful judgment when sending copies to other recipients.

6. Consider privacy issues. The Electronic Communications Privacy Act of 1986 and individual state tort laws make it a legal offense to read someone else's e-mail. The penalty may be up to $250,000 and imprisonment. However, as already mentioned, the prohibition does not apply to employers or systems operators of a company. A number of court cases involving the privacy rights of employees have found for the employer. The courts hold that a company's interest in preventing inappropriate and unprofessional comments or even illegal activity using its e-mail system outweighs any privacy interests of the employees. If you send e-mail with inappropriate sexual or racial comments or material, you may be found guilty of harassment. Pretty Good Privacy (PGP) is just one trademarked program for encrypting messages so that only the intended recipient can read the message. You may download this program at the Massachusetts Institute of Technology Web site <http://web.mit.edu/network/pgp.html>.

7. The use of slang and emoticons, punctuation marks typed to resemble sidewards smiling, winking, or frowning faces, are inappropriate in professional e-mail. Common emoticons are

> :-)   ;-)   :-(

Also avoid color print, color backgrounds, and fancy or oversize fonts.

Figure 12.6 shows a typical e-mail response letter.

LETTERS

Effective letter writing, whether electronic or print, is the key to successful business operation. Letters are the essential link between you and your business connections and between one organization and another. Letter content promotes and confirms most business transactions.

Subj: **Thank you for your time this afternoon.**
Date: 8/3/00 2:14:38 PM Eastern Daylight Time
From: marks@l90.com (Mark Smelzer)
To: Ladyvanj@aol.com

Judy:
Thank you for your time this afternoon. It was a pleasure learning more
about your organization's online advertising objectives.

As we discussed, L90 provides custom online advertising solutions to meet
our client's objectives. I look foward to brainstorming further to
determine solutions for your company.

Our solutions range from low CPM, blind banner buys across our network,
targeted channel banner buys, and site specific banner buys, to
beyond-the-banner solutions such as microsites, content-integration,
database names rental, opt-ins, and "tell-a-friend" email marketing.

Please take a moment to review our website, http://www.L90.com and our
online media kit, http://www.L90.com/external_site/emediakit/ for additional
information.

I plan to be in South Florida next month, and will call next week to arrange
an appointment. Thank you again for your interest and consideration.

Best regards,

Mark Smelzer
VP, Entertainment
L90: Internet Advertising Solutions
2020 Santa Monica Blvd, Suite 400
Santa Monica, CA 90404
email: marks@L90.com
Tel: 310-449-6508

http://www.L90.com

Los Angeles . New York . San Francisco . Chicago . Detroit . London

———————— Headers ————————
Return-Path: <.marks@l90.com>
Received: from rly-za03.mx.aol.com (rly-za03.mail.aol.com [172.31.36.99]) by air-za02.mail.aol.com (v75_b3.11) with
ESMTP; Thu, 03 Aug 2000 14:14:37 -0400
Received: from latitude90exch.l90.com (mail.l90.com [216.35.186.3]) by rly-za.03.mx.aol.com (v75_b3.9) with ESMTP;
Thu, 03 Aug 2000 14:14:09 -0400
Received: by LATITUDE90EXCH with Internet Mail Service (5.5.2650.21)
     id <.QC92ZLP6>; Thu, 3 Aug 2000 11:14:42 -0700
Message-ID: <.D3F1FF86C434D3119A4800902760DDAF01BF2D24@LATITUDE90EXCH>
From: Mark Smelzer <.marks@l90.com>
To: Ladyvanj@aol.com
Subject: Thank you for your time this afternoon.
Date: Thu, 3 Aug 2000 11:14:33 -0700
MIME-Version: 1.0

FIGURE 12.6 *Sample electronic letter* (Courtesy of Mark Smelzer)

Print Format

This chapter has already covered electronic format and mechanics. For printed letters, several acceptable formats are in use today. Individual companies tend to adopt a uniform format for use on preprinted letterhead stationary.

If letterhead stationery is not used or if you are writing business letters as an individual, select white, unlined, 8½″ × 11″ bond paper of about 20-lb. weight. Cheap, flimsy paper neither feels nor looks important. The quality of the paper may seem like a small point, but it does make a difference in the amount of attention your letter will receive.

Figures 12.7 and 12.8 show the two most widely adopted letter formats: the **full block** and the **modified block.** Study the spacing and indentation. Side margins are 1 to 1½ in. The top margin is a minimum 1 in. but may be deeper if the letter is short. A letter usually has more white space at the bottom than at the top, and it is conventional to type the body, or text, of the letter over the centerfold of the page.

One thing seldom considered in format selection is the option of a justified or jagged right margin offered by word processors. Although many people feel that the lined-up right margin gives a letter a finished, booklike appearance, some professionals and business psychologists suggest that the jagged edge provides a psychological advantage in that the receiver tends to find it more personalized. Thus, in correspondence of a sensitive nature, it may be wise to use the jagged edge rather than the justified margin.

Parts of Letters

All letters have six major parts: the heading, the inside address, the salutation, the body, the complimentary closing, and the signature. In addition, many letters contain subject, reference, or attention lines; typist's initials; and enclosure and distribution notations.

HEADINGS

If you are using printed letterhead stationery, add only the date two spaces below the letterhead. If you are using plain, white stationery, type your street address, city, state, zip code, and the date. Decide, too, whether to use the full or modified block format.

Example

1414 Southwest Ninth Street
Florence, South Carolina 29501
12 November 200X

Heading

409 Northeast Twelfth Avenue
Irving, Texas   75060
12 February 200X

(3–6 lines)

Inside
address

Mr. James W. Nelson
General Manager
Ace Equipment Company, Inc.
1092 East Eleventh Street
Salutation
Seattle, Washington   98122
(2 lines)
Special
element
Dear Mr. Nelson:
(2 lines)
SUBJECT: Your letter of 15 January 199X
(2 lines)

_____
_____
_____ .

(2 lines)

Body

_____
_____
_____
_____
_____ .

(2 lines)

_____
_____
Compli-
mentary
closing
_____ .

(2 lines)
Very truly yours,

Signature

(4 lines)

Carol Hopper
(2 lines)
Notations
CH:cm
(2 lines)
c: David L. Meeks

FIGURE 12.7   *Full block format for business correspondence*

Heading
⎡ 2107 West New Boulevard
⎢ Sterling, VA   22170
⎣ 3 August 200X

                                          (3–6 lines)

Inside
address
⎡ Mr. Ray Adams, Associate
⎢ Sun Air Corporation
⎢ State Road 7
⎣ Auburn, WA   98002
    (2 lines)

Salutation ⎡ Dear Mr. Adams:
    (2 lines)

_____
_____
_____.

    (2 lines)

Body
⎡ _____
⎢ _____
⎢ _____
⎢ _____
⎣ _____.

    (2 lines)

_____
_____.

    (2 lines)

Compli-
mentary   ⎡   Sincerely,
closing
                      (4 lines)

Signature ⎡   Marcia Jordan
                      Personnel Specialist

    (2 lines)
MJ:ts
Notations ⎡ (2 lines)
             Enclosures (2)

FIGURE 12.8 *Modified block format for business correspondence*

Do not include your name. Each line begins at the same margin. Refrain from abbreviating. Write out *Street, Avenue, Boulevard, East, Northeast,* and so on. Notice also that in street addresses one-digit numbers are written out, but two or more digits are written in Arabic numerals.

Examples

One Landmark Plaza
Twenty-two West Third Avenue
123 Forty-second Street
10 October 200x

but

2134 West 114th Street

Leave two spaces between the state and the zip code. The heading is placed flush left in the block format or flush to the right margin in modified block format.

INSIDE ADDRESS

The inside address includes the full name, position, company, and address of the recipient of your letter. It is spaced three to six lines below the heading.

Examples

Dr. Mary Jones, President
New Community College
101 South Palm Avenue
Fairmont, West Virginia 26555

Mr. Horacio L. Fernandez
Director of Human Resources
Ace Manufacturing Company, Inc.
Davenport, Iowa 90521

Notice that short titles may be placed on the same line as the name, whereas long titles are placed on a separate, second line. If possible, address your letter to a specific person rather than just to a position within a company. You may abbreviate titles such as *Dr., Mrs., Mr.,* and *Ms,* but do not abbreviate *the Reverend* or *the Honorable* or titles denoting rank, such as *Lieutenant, Captain, Professor.*

    The inside address is always placed flush to the left margin.

    You may abbreviate the state using the Postal Service two-letter abbreviations, as shown in Figure 12.9.

SALUTATION

The salutation is your greeting to your reader. It is typed two lines below the inside address and is followed by a colon. Further, it must agree with the addressee of the inside address. We have all seen letters ad-

Alabama	**AL**	Montana	**MT**
Alaska	**AK**	Nebraska	**NE**
Arizona	**AZ**	Nevada	**NV**
Arkansas	**AR**	New Hampshire	**NH**
California	**CA**	New Jersey	**NJ**
Colorado	**CO**	New Mexico	**NM**
Connecticut	**CT**	New York	**NY**
Delaware	**DE**	North Carolina	**NC**
District of Columbia	**DC**	North Dakota	**ND**
Florida	**FL**	Ohio	**OH**
Georgia	**GA**	Oklahoma	**OK**
Guam	**GU**	Oregon	**OR**
Hawaii	**HI**	Pennsylvania	**PA**
Idaho	**ID**	Puerto Rico	**PR**
Illinois	**IL**	Rhode Island	**RI**
Indiana	**IN**	South Carolina	**SC**
Iowa	**IA**	South Dakota	**SD**
Kansas	**KS**	Tennessee	**TN**
Kentucky	**KY**	Texas	**TX**
Louisiana	**LA**	Utah	**UT**
Maine	**ME**	Vermont	**VT**
Maryland	**MD**	Virginia	**VA**
Massachusetts	**MA**	Virgin Islands	**VI**
Michigan	**MI**	Washington	**WA**
Minnesota	**MN**	West Virginia	**WV**
Mississippi	**MS**	Wisconsin	**WI**
Missouri	**MO**	Wyoming	**WY**

FIGURE 12.9   *U.S. Postal Service two-letter abbreviations*

dressed to "Dear Sir or Madam" or "To Whom It May Concern." These vague salutations are useful when the receiver of the letter can be anyone in a certain department or division, as in an order letter. However, in negative situations (such as collection letters or complaint letters), it is easy to have the letter misplaced or even ignored. In such situations people are likely to avoid responsibility by being neither "sirs nor madams,") and your letter will probably concern not a single "whom."

To avoid this situation, especially when money is involved, *always* find a name to place on the letter. A simple telephone call to the company to find out the name of the person in charge of complaints or the name of an executive officer will ensure that the letter arrives into the hand of a real person. This person will be more likely to respond in a timely fashion because you have established a personal relationship with him or her.

Further, the salutation must agree with the addressee of the inside address.

Examples

Dr. Susan Clark, Dean
New Community College
101 South Palm Avenue
Miami, FL 30212

Dear Dr. Clark: (agreement with person)

Director of Personnel
Ace Manufacturing Company, Inc.
4012 West Grand Street
Hilo, HI 96720

Dear Sir or Madam: (agreement with title)

Ace Manufacturing Company, Inc.
4012 West Grand Street
Augusta, GA 30906

Ladies and Gentlemen: (agreement with corporate body)

League of Women Voters
Ten Northeast Datepalm Drive
Phoenix, AZ 85062

Ladies: (agreement with gender)

Because many women are joining the corporate ranks of business and industry, it is not unusual to see salutations, such as *Ladies and Gentlemen:, Hello:,* or *Dear Director of Training:*. Unless a woman has expressed a desire for *Miss* or *Mrs.,* use *Ms* (optional period) whether she is married or unmarried.

Avoid *Sir:* (too formal), *My Dear Sir:* (too pretentious), *Dear Sirs:* (sexist), and *To Whom It May Concern:* (too impersonal).

The salutation is always typed flush to the left margin.

BODY

The body, the text of your letter, begins two lines below the salutation. Notice in Figures 12.7 and 12.8 that paragraphs are not indented in the block format but are indented five spaces in the modified block. Single space within paragraphs and double space between them. The body should fall over the center of the page.

COMPLIMENTARY CLOSING

The complimentary closing is two lines below the body and is followed by a comma. Only the first word is capitalized.

Examples

Very truly yours,	(formal)
Yours truly,	(less formal)
Sincerely,	(emotional)
Respectfully,	(if addressee outranks you)
Cordially,	(warm)

The complimentary closing is flush to the left margin in block format and at the horizontal midpoint or at the heading margin in modified block format.

SIGNATURE

Your full name is typed four lines under the complimentary closing. Sign your name in black ink between the two. If the addressee is known to you, you may sign less formally than the typed signature.

Example

Very truly yours,

Judy S. VanAlstyne

Judith S. VanAlstyne

SPECIAL ELEMENTS

Occasionally, a subject, reference, or attention line is used to alert the reader to the subject, file reference, previous correspondence, account number, or other emphasis.

Examples

Below salutation	Subject: Invoice #20947
Above salutation	ATTENTION: Mr. D. W. Clark
	RE: Your letter of 12 June 199X

Such special elements are typed flush to the left margin two lines below the inside address but above the salutation.

TYPIST'S INITIALS

If your letter is typed by someone other than you, place your initials in capital letters and the typist's initials in lowercase letters flush to the left margin two lines below the typed signature. Use a colon or a virgule between them. If you type your own letter, your initials after the signature are optional.

Examples

JSV:mt

JSV/mt

ENCLOSURE NOTATIONS

If you send materials or documents with your letter, add an enclosure notation two lines below the typist's initials.

Examples

Enclosure

Enclosures (2)

Enc: Copy of Check #1029

DISTRIBUTION NOTATION

If you are sending copies of your letter to other readers, add a distribution notation two lines below the last element. You may use *CC:, cc:,* or *Cc:* (for *carbon copy*) or *Copy:, pc:,* or *c:* for photocopies. Follow the notation with a colon and include the names of others who are receiving copies.

Examples

CC:	David Little, Chairman	Copy:	Charles Clooney
cc:	Carolyn Houser	pc:	office file
	Lawrence Clarkson		Treasurer Jay Blakely
Cc:	Marvin Greer	c:	Jane Howell
	Mary Greer		Samuel Murphy

Second Pages

If a letter requires a second page, type the recipient's name, the page number, and the date in a block flush to the left margin or across the page.

Examples

Ms Sally Queen
Page 2
15 July 200X

Ms Sally Queen   –2–   15 July 200X

ENVELOPES

Your envelope should be 9½″ x 4½″ and of the same quality as your stationery. The recipient's name, title, company, address, city, state, and zip code are centered horizontally and vertically. As a guideline, begin 12 lines down from the top. Single space between lines. Your own name, address, city, state, and zip code are placed in the upper left-hand corner. Use the Postal Service abbreviations for states. Figure 12.10 shows a sample envelope.

```
Dr. Richard Grande
1022 Northeast First Street
La Mesa, CA   92041




                        Ms Julie Maney
                        Director of Personnel
                        Ace Manufacturing Company, Inc.
                        4092 West Grant Street
                        Manchester, CT   06040
```

FIGURE 12.10   *Sample envelope*

CONTENT

Whether composing electronic or print mail, adhere to a tight organization with an introductory purpose, a "You" perspective body, and a purposeful conclusion.

Organization

Organize your message into three parts:

1. A brief introduction that states the purpose of the letter immediately, unless you are conveying "bad news"
2. One or more body paragraphs that contain specific detail
3. A conclusion that establishes goodwill or encourages your reader to act

Introductory Purpose

State your purpose immediately. Do not just fill space until you get around to your purpose.

Vague purpose	It became apparent about five years ago that computerized bookkeeping was to be the answer to the problems which were plaguing our bookkeeping department. Therefore, we would like to investigate your software program . . .
Clear purpose	I am seeking answers to three questions regarding your software program, Computerized Automotive Reporting Service.

Other clear introductory purpose statements follow:

Examples

Here are the instructions for assembling the Ace 1 Trampoline that you requested by phone on September 20.

Please consider my resume and application for a junior management position at your Pompano resort.

This is in answer to your inquiry about leasing our trucks.

Your vacuum cleaner is repaired and is ready to pick up.

Congratulations on your promotion to Director of Affirmative Action.

You are right. You paid your bill exactly when you said you did.

Once you have stated your purpose, you may clarify why you need this particular information.

Example

Because I edit our company's in-house newsletter, it is essential that I purchase software that has both WordPerfect and Microsoft Word capabilities.

"You" Perspective Body

Provide the details of your correspondence in the body, using a "you" perspective. Put yourself in your recipient's shoes and consider how that person will respond to your words. Be courteous, direct, and confident. Avoid a slangy, abrasive, pompous, or abrupt tone. Consider the abrasiveness of the following phrases:

Examples

Your department should shape up . . .

I demand that . . .

I am appalled at your slow response . . .

I beg to advise you of my intent to . . .

Rush me information on . . .

Much of your correspondence will be highly repetitive ("Thank you for your order of July 25." "Our records indicate that your payment is past due."), yet trite and cliché expressions should be avoided.

Cliché Expressions	Plain English
Having received your letter, we . . .	We received your letter . . .
Pursuant to your request . . .	As you requested . . .
Per your memorandum . . .	As you noted . . .
Enclosed please find my report . . .	Here is my report . . .
It is imperative that you write at once . . .	Please write at once . . .
I am cognizant of the fact that my report is tardy . . .	I know that my report is late . . .
At the earliest possible date . . .	As early as possible . . .
I beg to differ with your . . .	I disagree with your . . .
Please be advised that the new policy . . .	The new policy is . . .
I hereby request that . . .	Please consider . . .
I beg to acknowledge receipt of your check . . .	I received your check . . .
We are in hopes that you succeed . . .	Good luck . . .

Every letter you write should sound fresh and conversational.

Word Processing

A final word about the "you" perspective should be added. Word processing by computers has removed the drudgery from written correspondence by allowing organizations to create standard text for repetitive form letters. If you are already using a word processing computer program, you know that you can delete anything from a character to a paragraph or relocate or insert words, phrases, sentences, and entire paragraphs with ease.

The word processing capability, however, may tend to depersonalize your letters. Great care must be taken to maintain a friendly, "you"-oriented tone.

Purposeful Conclusions

Your conclusion provides an opportunity for you to urge action or establish goodwill. A brief closing should motivate the recipient to follow up on your letter or, at least, to feel favorable to you, as illustrated in these examples.

Examples

May I have your answer by March 12?

If you will call us within the next few days, we can send our sales representative to demonstrate our software capabilities.

I would like an interview and am available weekdays for the rest of this month.

Thank you for pointing out this problem.

I appreciate your services . . .

Do not, however, be obvious or presumptuous. Avoid "I want to thank you in advance for . . ." and "Please feel free to call me if you have further questions." An advance thank-you implies that you may be too lazy to write a proper thank-you when your request is fulfilled. The second closing is unnecessary; the recipient will call you if he or she has questions whether you invite a call or not.

POSITIVE OR GOOD NEWS LETTERS

Recipients of many types of letters are happy to receive them. Some letters offer services or sales, place orders, tender a congratulation or thank-you, or transmit desired information. You will not only receive such letters but will also be called upon to write them in your career field. Good news letters include

- Inquiry and request letters
- Invitations
- Order letters
- Congratulatory and commendation letters
- Thank-you letters
- Sales or service offer letters
- Employment application and cover letters
- Transmittal letters

Inquiry and Request Letters

An inquiry or request letter is a good news letter because the recipient will benefit from the writer's interest. Nevertheless, an inquiry or request solicits a response, which will ask of the recipient time and, perhaps, effort. Therefore, certain strategies will help to motivate the recipient to respond quickly and accurately.

First, your introductory purpose statement may include the suggestion that you need an immediate response.

Example

I am seeking additional information on your Canon BJC-3000 Bubble Jet printer because I plan to purchase one this month.

Second, you may clarify why you need this particular information.

Example

Because I edit our company's in-house newsletter, it is essential that I purchase a printer that performs proportional spacing.

Third, you may subtly compliment the person or the company.

Example

I am interested in obtaining some additional details about your VID-80 Model III. Your company was the first to advertise such a modification more than six months ago, so I believe that you have the most expertise in this field.

Fourth, if you are asking questions, simplify and separate them into numbered items, make each specific, arrange them in a numerical table, and allow sufficient space between each to allow jotted answers right on your letter.

Example

I have three questions that will affect whether or not I upgrade my system at this time:

1. What type of format does your CP/M use? Is it compatible with the system Radio Shack computers use?
2. Do schematic diagrams come with the unit? If not, are they available for an extra charge?
3. If the added memory is purchased, is there any way to use the additional memory when operating with TRSDOS or similar operating systems?

You may also consider enclosing a self-addressed, stamped envelope. This double strategy of leaving space to answer on your letter and sending a return envelope will allow the recipient to respond on your letter and immediately place your response in the outgoing mail.

Finally, if it is appropriate, you may offer to share the results of your inquiry.

Example

Because I am compiling information on the technical writing programs of all Florida community colleges, I will be happy to share with you my final report.

Figure 12.11 shows a poor request letter. It exemplifies a "me" rather than a "you" perspective and is poorly organized and rude. Figure 12.12, however, illustrates a well-written request letter that motivates a quick response.

Invitations

An invitation is a type of request letter, too. You may wish to invite a speaker to address your group or organization. Or you may wish to invite members or guests to attend special functions, such as meetings or luncheons. Invitations should be brief, but detailed. First, extend the invitation including the function, date, time, and place. If necessary, mention guest status, fee, honorarium, or the like to be expected. Second, elaborate on the purpose and offer any other particulars that the invitee will need, such as the number of expected persons, length of a requested speech, or other program components. Finally, urge a response by enclosing a response card, requesting a call, or suggesting you will be calling in a few days for an answer. Figure 12.13 shows a typical invitation letter.

3072 Southwest Sixth Avenue
Norwich, CT   06360
4 May 200X

Mr. Fred Mandel
Radio-Electronics
200 Park Avenue
New York, NY   70908

Dear Mr. Mandel:

Demanding and insulting [Please rush me information on how to convert my transistor output voltage to 6VDC usable voltage for my portable radio. I read your article in Radio-Electronics, but it confused me.

Unclear questions [What I need to know is can I use the same 300 Ohm pot or something else for my 9VAC. Can I use the same number rectifier for my radio? Can you recommend a zenior diode? What else do I need to know?

Rude [Thank you in advance for answering my questions.

Very truly yours,

Walt

Walter Matthews

WM:jsv

FIGURE 12.11   *A poor request letter*

10111 Kennedy Boulevard
Atlanta, GA   21441
21 February 200X

Mrs. Dawn Michele
Customer Service Representative
Atlanta Bank
100 Peachtree Drive
Atlanta, GA   21445

Dear Mrs. Michele:

I have heard about your excellent loan rates through a friend who banks with
Atlanta Bank. I am considering buying a home, so I wish to obtain some
information about your loan practices.

What are your current interest rates for mortgage loans?

How many points do you currently charge?

Please send me any brochures you have available on your equity loans.

I expect to buy the house within the next month; therefore, I would appreciate
this information as quickly as possible. I have enclosed a stamped, self-
addressed envelope for your convenience.

Sincerely,

Suzanne Kelley

Suzanne Kelley

SK:bc

Enclosure

FIGURE 12.12   *An effective request letter*   (Courtesy of student Suzanne Kelley)

Meet for Good Fellowship

PROVIDENCE JEWELERS CLUB

P.O. BOX 4350
EAST PROVIDENCE, RHODE ISLAND 02914

16 September 200X

Ms. Robin Revell
30 Ormsby Avenue
Warwick, Rhode Island   02886

Dear Ms. Revell:

The particulars — Please accept my invitation to attend as my guest the Providence Jewelers Club "Speakers' Luncheon" on 1 October 200X, 12:00 noon, at the Providence Marriott Hotel.

Elaboration — I would like to introduce you to our members and guests as the recipient of the Providence Jewelers Club 200X–200X scholarship. Your choice of meals should be indicated on the enclosed card and returned as soon as possible.

Urge to action — Should you not be able to attend, please call me at 738-8560 as soon as possible. I look forward to meeting you.

Very truly yours,

Paul J. Austin

Paul J. Austin
President

PJA/eeg

Enclosure

FIGURE 12.13 *An invitation letter* (Courtesy of Paul J. Austin)

Order Letters

Another good news letter is the order letter, one that informs a seller that you want to purchase a product or service. Three writing strategies will help you to be clear and accurate about your specific order, shipping instructions, and method of payment. First, accurately describe the product or service by specifying the name, brand, model, stock number, quantity, color, dimensions, unit price, and other details. Include an informal table for multiple product or service orders for easy reader reference. Second, include your shipping instructions, such as first-class or third-class mail, Federal Express, United Parcel, or special mailing address, department, or special attention notation. Third, mention the date needed, if this is an issue, and, finally, specify your method of payment: enclosed check or money order, credit card charge number, COD, installment, and so on.

Figure 12.14 illustrates a clear and accurate order letter.

Congratulatory, Commendation, and Thank-You Letters

Both of these good news letters are characterized by informality and friendliness. The salutation might address the recipient by a first name. The introduction should mention specifically the occasion for congratulations or appreciation. Add detail to underscore your sincerity and end with a warm complimentary close.

Figure 12.15 and Figure 12.16 illustrate typical congratulatory and thank-you letters.

Sales and Service Offer Letters

Even though a sales or service offer letter is written to persuade the recipient to purchase a product or service, it may be considered a good news letter in that it offers to enhance the recipient in some manner. Five writings strategies can aid you in obtaining a favorable response.

First, identify and limit your audience. Determine the needs of this group and bear in mind exactly what you want your audience to do after reading your letter. The "you" perspective is critical for your desired response. Keep in mind what you can do for your reader throughout the letter.

Second, begin your letter with an attention-getting statement. You may ask a question, offer a free gift, employ a *how-to* statement, or use flattery. In short, hook your reader into reading further.

World Mortgage Company
2740 Elm Street
Hartford, CT 10031
30 January 200X

J. C. Dine and Company
8190 West Presidential Boulevard
Fairfield, CT 10345

ATTN: Mail Order Department

Good Morning:

Please send me the following items from your 2000 Dine Office Furniture catalog:

Steel Mobile Utility Cabinet	060-333-230	2@$150.99	$ 301.98
O'Sullivan Drop-Leaf Table	060-824-557	1@$38.99	38.99
Hide Side 2-Drawer File	060-777-135	3@$29.99	89.97
			$430.94
		Tax 6%	25.87
		Shipping	52.00
			$508.81

Label the shipment "Attention: Jeffrey Johns" and use the above address.

I am enclosing a check for the full payment of $508.81.

Very truly yours,

Jeffrey Johns

Jeffrey Johns, Human Resources Associate

SK

Enclosure

FIGURE 12.14   *An effective order letter*

Senior Systems Engineer
Spherion Corporation
July 17, 2000

Donna Weston
Spherion Corporation — Consulting Group
2000 W Commercial Boulevard
Fort Lauderdale, FL 33309

Re: Award for Merrill Tritt

I want to recognize Merrill's outstanding performance working as the Data Conversion Analyst on the NIS payroll data conversion. He has done an excellent job of coordinating the conversion efforts with the Project Team in Oakbrook, the Spherion Payroll Department, the NIS businesses, and Ceridian. Merrill single-handedly learned the Ceridian data conversion layouts, wrote the conversion programs with his own mapping tool, and worked with the business and Ceridian to complete successfully a very difficult data conversion.

Merrill also took it upon himself to learn the Ceridian HR/Benefit data upload utility to mechanize completely an otherwise extremely time-consuming manual conversion process. This work has saved countless hours of manual data entry and, subsequently, shortened the data conversion process from days to hours. He also learned the export utilities to develop data extracts for use by the data validation team. These extracts have saved many hours of data validation.

Through his dedication and resourcefulness, Merrill has clearly demonstrated his value as one of the "Great" Employees.

Sincerely,

Dino Vigliano

Dino Vigliano

Cc: Bill Blog
        Tim Gorson
        Clark Kent

2000 W. Commercial Blvd., Suite 100-A
Ft. Lauderdale, FL 33309
TEL 954 351-3844 FAX 954 302-0735
www.spherion.com

FIGURE 12.15   *An effective congratulatory/commendation letter*   (Courtesy of Merrill D. Tritt)

3502 Southwest Palm Avenue
Athens, GA   30605
7 March 200X

Professor J. John Jenks
Community Service Division
Broward Community College
One East Las Olas Boulevard
Fort Lauderdale, Florida   33301

Dear Professor Jenks:

*To the
point,
informal*

    Your generous recommendation of me to Tech Laboratories, Inc. got me the job. Thank you!

*Specific
details of
appreciation*

    The word processing skills you taught me plus your pep talks on organization have opened new doors of self-confidence for me. I'm truly grateful for your interest.

*Friendly
closing*

    I'll keep you posted on my advancement.

                     Thanks again,

                     Ralph

                     Ralph Kennery

FIGURE 12.16   *An effective thank-you letter*

Examples

You will receive a free vacation for two, a new car, a dream cottage, or other valuable gift simply by making an appointment to inspect our resort.

Here's how to save $100.00 on your next automobile purchase.

You made a smart choice by enrolling at Broward Community College. Now let us help you make a smart choice in selecting your college wardrobe.

    Third, call attention to the product or service's appeal. Persuade the reader that your offer is so desirable that he or she cannot resist it.

Examples

Our time-share condominiums are caressed by gentle ocean breezes and within steps of your very own tennis courts, golf course, and spa. We offer the last word in glamorous vacations.

Ace Motors offers the world-recognized most economical car on the highways—the Ace 400ZT.

Designer jeans, polo shirts, a dazzling array of blouses, dresses, skirts, and accessories are waiting for you.

Fourth, present evidence of your product or service's application. Emphasize its convenience, usefulness, and economy. Endorsements, guarantees, and special features may be highlighted. Present the facts in a manner that emphasizes the attractiveness of your offer.

Examples

A member of Time-Share International, Driftwood Resort offers not only the most reasonable prices on the Gold Coast but also opportunity to vacation in 39 countries of the world. Spacious two and three bedroom plus studio accommodations are available to suit your precise requirements. Fully equipped kitchens and modern hot tubs allow you the casual lifestyle of a truly refreshing holiday.

Fifty-three motoring journalists from 15 European nations voted this newest Ace "Car of the Year." The 400ZT is aerodynamically designed to accelerate from zero to fifty in only 7 seconds. Disc brakes, rack and pinion steering, and a performance-tuned suspension system make this automobile a marvel to drive.

The world's most compact identification Friend or Foe Transponder for air superiority is now available from AlliedSignal, Inc. Representing the answer to the critical need to distinguish clearly platforms as friendly, enemy, or neutral, we offer you the smallest, lightest Mark XII Diversity Transponder in the world that provides complete military and commercial IFF capability.

Finally, urge your reader to action. Make it easy to return a postcard to order your product, suggest an appointment next week, invite readers in for a free gift, include a phone number to call, and so on.

Examples

Call 305-455-9000 to arrange a tour of our facilities and to find out what valuable gift is yours. You won't be disappointed.

Stop by this weekend to test drive your next car—the Ace 400ZT.

During Orientation Week we will be open until 9:00 P.M. for you to drop in and browse. Free textbook covers are yours with every purchase.

Figure 12.17 shows a persuasive service offer letter. The audience consists of busy top executives who may be frustrated by the poor qual-

Jane Hansen & Associates
409 East Seventy-second Street
Suite 9001
Los Angeles, CA   90028
12 May 200X

Mr. Frank Mahoney
Senior Vice-President
City National Bank
100 Northeast First Street
Los Angeles, CA   90066

Dear Mr. Mahoney:

Attention getter

DID YOU KNOW   —it costs $20.00 for each letter your
                                        staff types?
                              —your typical employee spends 25% of
                                        his or her time at work writing?
OF COURSE YOU KNOW —the ability to write well gets the
                                        results you want, and contracted
                                        training saves your bank money.

Appeal of offer

To upgrade the writing skills of your employees, you may now
contract WRITING SKILLS & STRATEGIES, a training seminar
tailored to your employees. Further, we will conduct the seminar
at your bank during the times most convenient for your busy
staff.

Application and appeal

Endorsement

WRITING SKILLS & STRATEGIES reviews troublesome mechanics,
grammar, and usage as well as offers indispensable tips on
correspondence, report writing, and much more. An experienced
writing instructor will tailor the materials to your specific needs.
Over 40 banks in south Florida will attest to the practicality of
this training program.

Urge to action

May I make an appointment, Mr. Mahoney, to discuss course
content, prices, and times? I will call your office within the next
ten days.

Very truly yours,

Jane Hansen

Jane Hansen
Writing Consultant

JH:jsv

Application

P.S. The enclosed brochure highlights features of WRITING
SKILLS & STRATEGIES.

FIGURE 12.17   *An effective service offer letter*

ity and time-intensive writing of their employees. The layout is catchy and appealing. The service's application and the desired action is effectively covered, yet the letter is brief.

Employment Application and Cover Letters

Because employment application letters often include a resume, a separate writing strategy, they will be discussed separately in Chapter 13.

Transmittal Letters

Letters of transmittal, sometimes called cover or face letters, announce the enclosure of attached material and reports. Content and samples are covered in Chapters 13 and 15.

NEGATIVE OR BAD NEWS LETTERS

Some letters must convey bad news. They may inform the recipients that they are not hired, cannot get a refund, are late with a payment, and so on. Again there are writing strategies to convey your bad news in a positive, result-producing manner. Bad news letters include

- Negative response letters
- Complaints
- Collection letters
- Solicitations

Legal Considerations

"Anything you say can and will be used against you in a court of law." These words from the Miranda Rights are familiar to anyone who has ever seen a film or television show depicting police officers arresting a suspect. They also apply to letter writing.

In the world of business, there is always concern for what is said and how it is said. Political correctness and the avoidance of sexual harassment are probably the most apparent concerns. However, too few people realize that accidentally insulting an applicant's character while refusing credit or a job can be grounds for legal action. Therefore, it is essential that your letters maintain a somewhat neutral tone.

Letters that sound threatening or abusive or appear to solicit bribes or favors can always come back to haunt you. Remember, you are signing your name to every letter you write and are, therefore, personally responsible for every word you write. This is especially true of bad news letters because the receiver is already apprehensive upon receiving the letter.

Negative Response Letters

A letter that must say *no* requires a buffer statement before the bad news. A buffer statement presents a valid reason before the negative response.

Examples

We lease our apartments only through registered real estate brokers; therefore, . . .

Because the criteria for the position of Personnel Specialist require a college degree in business administration, we can consider only those applicants with that credential.

I have referred your letter to Ms Jane Clifford, assistant to the director of computer services, because only that department is authorized to give you the information you request.

A second strategy is to avoid the word *no*.

Too harsh   I'm sorry to say no, I cannot address your engineers next week.

Buffered   Because I will be in Chicago all of July, I cannot speak before your group.

A third strategy in a negative response letter is to avoid an apology. Your reasons are valid, so eliminate *unfortunately, we regret, we are sorry, we wish we could,* and so on.

Fourth, do not leave the opportunity to reopen discussion or consideration. Avoid "If you wish to discuss this further . . ." and "We wish we could . . ."

Finally, conclude by establishing goodwill. Use "We hope we have an opportunity to serve you in the future," "We appreciate your interest in our organization," and the like. Sincerity is an important consideration. If your effort to establish goodwill appears contrived, forced, or formulistic, it will irritate your reader.

Figure 12.18 illustrates an effective negative response letter to a job applicant. It buffers the bad news, does not apologize, and establishes goodwill.

Complaints

All of us have found it necessary to complain about defective products, delayed orders, billing errors, or inadequate services. Although we usually write complaint letters when angry or frustrated, angry tones seldom elicit the action we desire. A complaint letter needs restraint, specificity, and a clear statement of desired action.

ACE COMPANY, INCORPORATED
2900 Northeast Seventy-ninth Street
Columbus, Ohio   43219

5 January 200X

Ms Jane Doe
1107 Northwest Twelfth Street
Dayton, Ohio   33331

Dear Ms Doe:

Buffer statement [We have received your letter and resume, exploring career opportunities at Ace Company. While your education and experience present an interesting background,

Negative response [your qualifications do not fit our particular requirements at this time.

No apology but positive action [We will place your materials in our potential file. Should a position for which you qualify open, we shall review your file and notify you.

Establishes goodwill [We appreciate your interest in Ace and wish to extend our wishes for success in the attainment of your career goals.

Sincerely,

Ralph Maran

Ralph Maran

RM:tac

FIGURE 12.18 *An effective negative response to a job applicant*

The first strategy is to provide a detailed description of the faulty product, service, or suspected error along with the specifics about your purchase or contract.

Example

I am returning for a full refund the U.D.S. Computer Telephone, model 333, for which I sent money order 40920 in the amount of $10.00 on 15 July 200X. I received the defective phone on 20 August 200X.

Second, state precisely what is wrong with the product or service.

Example

The phone malfunctions. The beeper activates on the third dialed digit so that dialing cannot be completed. A persistent, loud buzzing interferes with reception on incoming calls.

Third, consider describing the inconveniences you have experienced. This is not always necessary but may help to underscore the seriousness of your complaint.

Example

Because I conduct a great deal of business from my home telephone, I have lost sales and commissions by not having a properly functioning instrument.

Finally, state clearly what action you desire. You may want a refund, a replacement, copies of all records, or some other consideration.

Example

I am enclosing the warranty and am requesting a full refund for the purchase price. I will not consider a replacement because I have lost confidence in your merchandise. I shall appreciate a prompt refund check.

Figure 12.19 illustrates a complaint about a credit problem.

Collection Letters

Unfortunately, not all customers pay their bills on time. Therefore, companies must employ several correspondence strategies to urge payment. Frequently, companies send a series of collection letters, each employing a stronger tone than the former.

It is not wise to demand immediate payment in the first collection letter because there may be valid reasons for slow payment, such as misdelivered or misplaced bills or errors in the company's billing. Further, an early threat to begin legal action or collection services may cause the well-intentioned customer to avoid further profitable transactions with the company, or the company may suffer negative word-of-mouth publicity. A tactful "you" perspective is important in maintaining good relationships. Ultimately, you may need to threaten posting with credit agencies, legal action, a lawyer's intervention, or even serving a summons. Your intent is to avoid those actions.

The first letter should make the customer feel valuable, allow the customer to save face, urge prompt payment, but offer to establish a partial payment schedule if that course of action is appropriate.

1390 Southwest Twentieth Street
Davie, FL   33326
22 September 2000

The Doubleday Store
Customer Service Department
501 Franklin Avenue
Garden City, NJ   07769

Re: Account #96-299-38934

To Whom It May Concern:

Details of complaint

Please review my account for a credit. On 12 July 2000 I received the Pierre Cardin canvas luggage set from your company which I ordered on 15 June 2000. When I received the luggage from your company, it was on a trial basis for 60 days. After examining the luggage, I determined that it was not substantial enough for my needs, and I returned the entire set on 12 August 2000.

Expansion

The charge of $279.95 has continued to be shown on my last two monthly statements. I wrote a note on the statement each time indicating the date and return of luggage and sent the statement back to your company. Copies of these notes are attached to this letter. To date, I have not received an adjusted statement.

Desired action

Would you please credit my account for $279.95 and send me an adjusted statement?

I will appreciate your prompt attention to this matter.

Very truly yours,

Ruth Burrows

Ruth Burrows

RB

Enclosures: 2 statement photocopies

FIGURE 12.19 *An effective complaint letter*

After a reasonable length of time for turnaround mail, send a follow-up letter that refers to the first request for payment and asks for an immediate payment to avoid further action. In your final letter refer to the first requests for payment and alert the customer that the account will be turned over to a collection agency or attorney if partial or full payment is not made immediately. Figures 12.20 through 12.22 show three collection letters illustrating increasing pressure for payment.

Downtown Travel Centre

City Park Mall
140 Southeast First Street
Fort Lauderdale, Florida 33301
(305) 525-1303
Fax: (305) 525-1367

15 January 2000

Ms. Mary Houghton
301 Northeast 109 Street
Plantation, FL 33323

Re: Invoice #47091

Dear Ms. Houghton:

Shows customer value

Downtown Travel Centre values you as a trustworthy client. We appreciate your patronage and are here to serve your future travel needs. Our records, however, indicate that you have not paid the invoice for the balance due on your Travcoa tour to Vietnam and Cambodia issued to you on 15 November 1999. We trust that you enjoyed your holiday and that your nonpayment has been an oversight due to the press of travel and the holidays.

Polite request for payment

Please send your check in the amount of $1,985 as soon as possible. If this is impossible, please call me to establish a schedule of partial payments. If you have already mailed your check, please disregard this letter.

Sincerely,

Jeffrey Duckham

Jeffrey Duckham
Travel Associate

FIGURE 12.20 *An initial collection letter*

Downtown Travel Centre

City Park Mall
140 Southeast First Street
Fort Lauderdale, Florida 33301
(305) 525-1303
Fax: (305) 525-1367

10 February 2000

Ms. Mary Houghton
301 Northeast 109 Street
Plantation, FL 33323

Re: Invoice #47091

Dear Ms. Houghton:

Reminder

May we remind you that this account is overdue for payment and that we notified you of your lateness by our letter of 15 January.

Strong urge for payment

We remind you that our terms are 30 days, and it has now been two months since we issued our invoice for the balance of your Travcoa tour without receiving even a partial payment. In order to avoid further action, please send us a check for $1,985 by return mail.

Very truly yours,

Jeffrey Duckham

Jeffrey Duckham
Travel Agent Associate

Make us your Travel Headquarters . . . you'll be glad you did!

FIGURE 12.21   *A second, follow-up collection letter*

Downtown Travel Centre

City Park Mall
140 Southeast First Street
Fort Lauderdale, Florida 33301
(305) 525-1303
Fax: (305) 525-1367

21 February 2000

Ms. Mary Houghton
301 Northeast 109 Street
Plantation, FL 33323

Re: Invoice #47091

Dear Ms. Houghton:

Curt reminder — According to our records $1,985 is outstanding on your account and is now considerably overdue for payment.

Strongest urge for payment — Since you have not responded to our previous two reminders, I now have no option but to put this matter in the hands of our collection agency, which may affect your credit rating.

If you have any questions concerning your account, please call me at the above number. If we don't hear from you by March 30, we will take the necessary measures to recover this debt.

Very truly yours,

Jeffrey Duckham

Jeffrey Duckham
Travel Agent Associate

Make us your Travel Headquarters. . . . you'll be glad you did!

FIGURE 12.22   *A third, final collection letter*

Solicitations

A solicitation for money or volunteer activities may be classed as bad news because such solicitations offer nothing in return except, perhaps, the opportunity to further a cause that interests the reader. Alumni groups, political candidates, and supporters of various causes frequently solicit money and volunteers. Such letters require inventive, eye-catching, persuasive techniques, including those used in advertising: bandwagon appeal, snob appeal, humor, and endorsements.

The humorous approach is one effective technique, as illustrated in Figure 12.23, a letter soliciting a contribution from a college alumnus.

MIAMI UNIVERSITY
ANNUAL FUND

ARE YOU WILLING
TO TAKE A CHANCE

with a contribution?

The chance of tossing a coin three times and getting either three heads or three tails is 25%
. . .

The chance of rolling one pair of dice and getting greater than a seven is 42% . . .

The chance of two people in a group of 23 having the same birthday is 50% . . . in a group
of thirty people, 71% . . . in a group of fifty, 97% . . .

But the chance of helping Miami students with a contribution to the Miami University Fund
is 100%.

That's because thousands of Miami alumuni join together, every year, to make their contributions combine to do great things. They were doing it when you were in school. They're
doing it now. And, because so many share, all contributions help.

Contributors put their money where they like: for library software, and journals and for
faculty and undergraduate research; to help academic departments; for scholarships and
grants, for the athletic program; to enhance the arts at Miami; for any particular unit of
department; or for unrestricted use — wherever Miami's needs are greatest this year. Gifts
to every part of the University combine to make a better Miami.

Take a chance, won't you? Make a contribution to Miami. And, remember, your gift to
Miami is a sure-fire bet because it has a . . .

                                        100% chance of doing great things.

                                        Judy Schiller

                                        Judy Schiller '64
                                        Director, Alumni Relations

P.S. You can write a check and return it in the enclosed envelope . . . or you can go online
      and make a secure gift at www.muohio.edu/alumni/ However you choose to make your
      gift to Miami this year, I send my sincere thanks for your generosity.

JS:HO

FIGURE 12.23 *An effective solicitation letter*   (Original letter written by professional copywriter John
Yeck who authored solicitation letters for his Alma Mater, Miami University, Ohio, from 1953–1998.)

CHECKLIST

Correspondence

GENERAL

❑ **1.** Have I chosen the appropriate type of correspondence for the occasion?

  ❑ In-house memorandum?

  ❑ E-mail (usually by permission)?

  ❑ Formal letter?

  ❑ Faxed correspondence (if appropriate)?

❑ **2.** Have I used the International Standard for dating? If not, why?

MEMOS

❑ **1.** Have I used the necessary headings?

  ❑ Full name and title of the addressee?

  ❑ Full name and title of the sender?

  ❑ The date?

  ❑ The subject?

❑ **2.** Is my content to the point, timely, addressed to a single subject, and attractive?

E-MAIL

❑ **1**. Have I included a brief, clear subject line?

❑ **2.** Is my clock accurately set to send the correct date and time of posting?

❑ **3.** Have I included the full name and title of the addressee in the To: line?

❑ **4.** Have I included my full name and title in the From: line and/or a signature block?

❑ **5.** Have I included a polite salutation?

❑   **6.** Is my content to the point, timely, addressed to a single subject, and attractive?

    ❑ In a response, have I clearly indicated by quoting what the original queried?

    ❑ Have I limited the length of my message to one page?

❑   **7.** Have I adhered to all the particulars of Netiquette?

LETTERS

❑   **1.** Have I used a consistent block or modified block format?

❑   **2.** Are all parts of the letter included?

    ❑ A heading, if not using letterhead, which includes my full address and the date?

    ❑ An inside address to a particular person with his or her title and complete address?

    ❑ A salutation to a particular person, if possible?

    ❑ Engaging and clear body paragraphs?

    ❑ A complimentary close, typed name, and written signature?

❑   **3.** If there are two or more pages, are they identified by receiver, page number, and date?

❑   **4.** Is the envelope the same quality paper as the stationery, and its content complete and appropriately placed?

❑   **5.** Does the opening of the letter engage the reader's interest and reveal the purpose of my letter?

❑   **6.** Is my content written with a "you" perspective?

❑   **7.** Is my content free of cliché expressions?

❑   **8.** Is my tone appropriate for a good news letter (an inquiry or request, an invitation, an order, a congratulation, a thank-you, a sales or service letter, a transmittal letter)?

❑   **9.** Is my tone appropriate and does it include a buffer statement if a bad news letter (a negative response, a complaint, a series of collection letters, solicitations)?

❑ **10.** Does my closing urge the reader to act?

❑ **11.** Is my content letter perfect?

❑ **12.** Does my content enhance my image?

EXERCISES

1. **Subject Lines.** Rewrite the following memo or e-mail subject lines to make them more precise and descriptive:

Test Results	Minutes
Picnic	Training
Hours	Policy Change
Overtime	New Personnel

2. **Letter Formats.** Correct the errors in the letter elements of the following modified block letter format:

Ms. Julie Wilson
2 N. W. Park Ave.
Chicago, Ill.   33302

Mr. Daniel C. Taylor
four-one-two E. 72nd St.
Dayton, Ohio,   40727

Dear Sir,

Re: Account #407-201 E

_____
_____
_____

_____
_____
_____

                              Best,

                              Marcia Morris

                              Marcia Morris, Treasurer

                              mm:tc

3. **Letter Content.**   Rewrite the body of the following complaint letter by dividing it into an introduction, body, and closing, and by eliminating "letterese," a "me" perspective, and an inappropriate tone:

Dear Sir:

In reference to your lousy iron which I purchased recently, I want my hard-earned money back. If you don't refund me the price in full, I beg to inform you that I will take legal action. It spews water all over the clothes I iron, scorches things even on a low setting, will not stand up on its base, and the plug broke the last time I plugged it in. If you have any questions, do not hesitate to call me. I am appalled at the workmanship of this piece of junk. Get with it.
Cordially,

---

4. **Buffer Statements.**   Write buffers to the following blunt, negative response statements:
   a.   We cannot send you the items you ordered because you did not enclose return postage.
   b.   We're sorry to inform you that we cannot hire you at this time.
   c.   I must say no to your request for a writing seminar in June.

WRITING PROJECTS

1. **Memo or E-mail.**   Write a brief memo to a subordinate to urge that a previously assigned written report be turned in a week earlier than previously scheduled.

2. **Memo or E-mail.**   Write a memo to your employer (real or imaginary) to point out a minor problem at your place of work, such as a scheduling mix-up, inadequate lighting, the need for more storage shelves or files, your inability to perform an assigned task, or an error in your paycheck. Be brief and to the point.

3. **Good News Letters.**   Write a good news letter based on one of the following suggestions:

a. Write **an inquiry or request letter** to a company or other organization in response to an advertisement or article in a professional journal in your field. Ask at least four technical questions about a product or service. Review the writing strategies that will motivate a quick response.

b. Write **an invitation** to a professor or an authority in some field to make a 20-minute address at your annual kickoff meeting of a special-interest organization. Make up all of the details the invitee will need to know.

c. Write **an order letter** for merchandise from a particular company. Specify the stock or model number, quantity, size, color, unit cost, and total cost as appropriate. Include all details about payment and delivery.

d. Write **a congratulatory or thank-you letter.** Some suggested topics are congratulations to a friend who has been hired or promoted, won a professional award, completed a degree or other special training, been elected to public or organizational office, or opened up a business. Thank-you letter topics might be a response to a letter of recommendation, a letter of appreciation to a teacher or counselor whose advice you followed, a letter to a hotel or restaurant that hosted a group function or dinner, or to a person who directed some business opportunity your way.

e. Write **a sales or service offer letter.** Consider your audience's age, occupation, geographical location, needs, and interests. Some suggested topics are the merits of a particular automobile, college, restaurant, new store, personal computer, bank, travel bureau, or flower shop or maintenance service for lawn, pool, or snow removal.

4. **Bad News Letters.**   Write a **bad news letter** based on one of the following suggestions:

a. Write **a negative response letter.** Some suggested topics are turning down an offered job, declining an invitation to speak, declining to serve on a committee or join an organization, refusing to volunteer time to an organization, refusing a job applicant, inability to fulfill a reservation request at a hotel or travel group, declining to refund money for a specific piece of merchandise, or refusing a sales or service offer.

b. Write **a complaint letter.** Some suggested topics are an error in your credit card or telephone bill, rude service you received at a store or restaurant, late delivery of merchandise, poor quality of some product recently purchased, or damaged goods delivered by a carrier.

c. Write **a series of three collection letters,** each firmer than the one before. Some suggested topics are late dues payment to a club or organization, late payment for a service you rendered, or late payment for an automobile, credit account, or bank loan.

d. Write **a solicitation letter.** Select a cause that interests you: save the whales or manatees, abortion legislation, a political candidate, proposed zoning changes, or a halfway house for troubled youth. Solicit donations or an appearance at a civic forum to discuss the issue with top policymakers.

5. Write **an order letter** to be faxed or e-mailed to an overseas company. Be sure to use the International Standard for dating, check multicultural/international language considerations, and convert any currency difference.

NOTES

CHAPTER 13

Preparing Resumes, Cover Letters, and Interviews

SHOE   by Jeff MacNelly

S K I L L S

After studying this chapter, you should be able to

1. Define *resume* and *cover letter*.
2. Appreciate the importance of a detailed, letter-perfect resume and cover letter.
3. Compile a file of job descriptions with information on working conditions, necessary qualifications, job outlooks, earnings, potential employers, and employment services.
4. Compile a file of personal education, employment, skills, valuable activities, and accomplishments.
5. Develop a portfolio of outstanding writing assignments, special recognitions, and the like.
6. Compile a list of keyword nouns, adjectives, and action verbs defining your skills and accomplishments, which can be scanned into a company database for future search.
7. Obtain prospective references plus actual letters.
8. Understand the distinctions between traditional, functional, narrative, and online resume formats.
9. Develop and write a personal resume in the appropriate format for your situation.
10. Write a specific and general cover letter with appropriate inclusions.
11. Understand and practice the *do's* and *do not's* of an employment interview.
12. Prepare yourself for the interchange of questions between an interviewer and yourself.
13. Identify and employ strategies you may use if an interview is going sour.

INTRODUCTION

Résumé is a French word that means "summary." A personal resume is a concise summary of pertinent facts about yourself—your employment objectives, your employment and educational history, personal data, and reference lists. Although most dictionaries include the accent marks (résumé), it is common practice to omit them. A resume and its accompanying cover letter are indispensable job-hunting tools that are submitted to employers to "sell" yourself as a prospective employee and to obtain an interview.

Another term you might encounter is *curriculum vitae,* or *vita* for short. These Latin words, meaning "life's course of events," are used in place of the word *resume* by people holding higher degrees, such as the M.A., Ph.D., Ed.D., D.D.S., M.D., or J.D., and by high-level executives with many years of experience in the business or professional worlds. Although the terms *curriculum vitae* and *resume* are sometimes used interchangeably, you should use the title best suited to your career field and your qualifications, as the misuse of one may be perceived as pretentiousness and the misuse of the other as underselling oneself or a lack of professionalism.

Your resume allows you to organize and amplify your data in a manner that highlights your strengths in order to obtain an interview. A job application form (Figure 13.1) usually is restricted to mere listings of schools and previous employment names and addresses, but introduces the basic information concerning a job applicant. Traditionally, resumes and cover letters have been mailed to companies that have advertised for applicants or to those companies where you know an opening exists. You may, in some instances, present a resume during an interview that you have obtained by other means. Today, posting resumes online in response to Web page listings, job bank listings, and numerous career sites is a must for a wide job hunt.

Employers receive hundreds of resumes for their positions and must quickly scan them, giving only a few seconds of preliminary attention to each. In order for yours to be put on the review pile, it must be letter perfect, immediately convey your skills, and convince the reader that you are qualified, capable, and worthy of an interview.

RECORD KEEPING

Long before you actively seek initial employment in your chosen career or seek to change jobs, compile two files: one to contain all the information you can gather on future employment possibilities and the other to contain all of your educational and employment background information, materials for a portfolio of personal achievements, and prospective references.

Prospective Positions

In your first file include a list of all prospective employers and their addresses and phone numbers. Also include a listing of all interesting job descriptions with information on working conditions, qualifications, job

application for employment

We are an equal opportunity employer, dedicated to a policy of non-discrimination in employment on any basis including race, color, age, sex, religion or national origin.

PERSONAL INFORMATION

Date _____   Social Security Number _____

Name _____
Last   First   Middle

Present Address _____
Street   City   State   Zip

Permanent Address _____
Street   City   State   Zip

Phone No. _____

Referred By _____

EMPLOYMENT DESIRED

Position _____   Date You Can Start _____   Salary Desired _____

Are You Employed Now? _____   If So May We Inquire of Your Present Employer? _____

Ever Applied to this Company Before? _____   Where _____   When _____

(Side tab labels: Last, First, Middle)

EDUCATION	Name and Location of School	Circle Last Year Completed	Did You Graduate	Subjects Studied and Degree(s) Received
Grammar School			☐ Yes ☐ No	
High School		1 2 3 4 5	☐ Yes ☐ No	
College		1 2 3 4 5	☐ Yes ☐ No	
Trade, Business or Correspondence School		1 2 3 4 5	☐ Yes ☐ No	

Subject of Special Study or Research Work _____

Activities Other Than Religious (Civic, Athletic, etc.) _____

EXCLUDE ORGANIZATIONS, THE NAME OR CHARACTER OF WHICH INDICATES THE RACE, AGE, SEX, COLOR OR NATIONAL ORIGIN OF ITS MEMBERS.

Form M660-26NR   Printed in U.S.A.   (Continued on Other Side)   APPLICATION FOR EMPLOYMENT
© 1985 Wilson Jones Company

FIGURE 13.1   *Sample employment application*

FORMER EMPLOYERS   List Below Last Four Employers, Starting With Last One First

Date Month and Year	Name and Address of Employer	Salary	Position	Reason for Leaving
From				
To				
From				
To				
From				
To				
From				
To				

REFERENCES   Give Below the Names of Three Persons Not Related To You, Whom You Have Known At Least One Year

Name	Address	Business	Years Acquainted
1			
2			
3			

PHYSICAL RECORD   Do you have any physical condition which may limit your ability to perform the job applied for?   This question is voluntary, and any answers will be kept confidential.

In Case of Emergency Notify

Name   Address   Phone No.

I authorize investigation of all statements contained in this application. I understand that misrepresentation or omission of facts called for is cause for dismissal. Further, I understand and agree that my employment is for no definite period and may, regardless of the date of payment of my wages and salary, be terminated at any time without any previous notice.

Date   Signature

DO NOT WRITE BELOW THIS LINE

Interviewed By   Date

REMARKS:

Neatness   Ability

Hired   For Dept.   Position   Will Report   Salary Wages

Approved   1.   2.   3.

Employment Manager   Dept. Head   General Manager

FIGURE 13.1   *continued*

outlooks, earnings, and other related information that will help you to define your job search. Some sources for the job search file include:

1. ***Dictionary of Occupational Titles (DOT).*** A good place to start is to check your library, career center, or the Internet for this yearly publication of the U.S. Bureau of Labor, which lists in two volumes 12,000 types of jobs and their required job skills. You may measure yourself against those skills and take courses to obtain new skills. Here is an entry for a technical writer:

131.267-026 WRITER, TECHNICAL PUBLICATIONS (profess. & kin.)

Develops, writes, and edits material for reports, manuals, briefs, proposals, instruction books, catalogs, and related technical and administrative publications concerned with work methods and procedures, and installation, operation, and maintenance of machinery and other equipment. Receives assignment from supervisor. Observes production, developmental, and experimental activities to determine operating procedure and detail. Interviews production and engineering personnel and reads journals, reports, and other material to become familiar with product technologies and production methods. Reviews manufacturer's and trade catalogs, drawings, and other data relative to operation, maintenance, and service of equipment. Studies blueprints, sketches, drawings, parts lists, specifications, mock-ups, and product samples to integrate and delineate technology, operating procedure, and production sequence and detail. Organizes material and completes writing assignments according to set standards regarding order, clarity, conciseness, style, and terminology. Reviews published materials and recommends revisions or changes in scope, format, content, and methods of reproduction and binding. May maintain records and files of work and revisions. May select photographs, drawings, sketches, diagrams, and charts to illustrate material. May assist in laying out material for publication. May arrange for typing, duplication, and distribution of material. May write speeches, articles, and public or employee relations releases. May edit, standardize, or make changes to material prepared by other writers or plant personnel and be designated Standard-Practice Analyst (profess. & kin.). May specialize in writing material regarding work methods and procedures and be designated Process-Description Writer (profess. & kin.).

2. ***Occupational Outlook Handbook (OOH).*** Another helpful publication, also put out by the U.S. Bureau of Labor, is available both in print and online. It analyzes 250 jobs by discussing the nature of the work, working conditions, employment statistics in the field, training requirements, job outlook, and salaries. Each of these comprehensive analyses may be five or six pages long, so are not reproduced here.

3. ***Journal of Career Planning and Employment.*** Peruse this monthly journal for information on position requirements, resumes, cover letters, and all up-to-date articles on careers.

4. **Libraries.**   Ask the reference librarian to help you locate other occupational handbooks, government publications, and newsletters
   that contain information on openings and qualification requirements in your field.

5. **Internet sites.**   Search the Internet using keywords *Careers* or
   Resumes and explore the following Web sites where you will find
   more than a million resources, including occupational profiles,
   guides to job hunting, company profiles, resume and cover letter
   tips, sites to post your resumes, and available job listings all over
   the world. Many of these sites include bulletin boards where you
   may post questions about companies, positions, and salaries. Some
   specialize in helping college students. If your newspaper offers a
   digital edition, check out all the information on job listings and employment tips. Start with **<http://www.hotresume.com>,** which
   has hundreds of links to jobs and other career sites. Some other major sites are

<http://www.careermosiac.com>	<http://www.monster.com>
<http://www.headhunter.com>	<http://www.ajb.dni.us>
<http://www.jobtrak.com>	<http://www.jobdirect.com>
<http://www.jobweb.com>	<http://www.occ.com>
<http://www.careerxroads.com>	<http://www.fcc.gov/job>
<http://www.careerbuilder.com>	<http://www.careerpath.com>
<http://www.jobbankusa.com>	<http://www.careerexchange.com>

   Here is a copy of a position opening that was posted online within
   <http://www.careermosaic.com>:

   Racal-Datacom
   Engineering

   Title
     SOFTWARE ENGINEERS
   Location
     FT. LAUDERDALE, FL 33340, USA
   Description
     SOFTWARE ENGINEERS
     Opportunity In Fort Lauderdale, FL!

   Join the winning team at Racal-Datacom, a $550 million leader in managed
   networking systems! We currently seek MTS 3 Software Engineers with a
   Bachelor's Degree in Electrical or Computer Engineering or equivalent
   combination of education/training/experience.

   You will need at least 5 years related work experience within a high-tech
   manufacturing/research and development environment to include software
   design ("C" and assembly, embedded systems 68302/68360) experience
   within a data communications company.

In-depth knowledge of ISDN networking requirements, an understanding of overall project concepts, as well as project leadership experience are also necessary. If you have a focus toward the future, consider joining a corporation on the cutting edge! We offer highly competitive compensation and benefits including health/dental/life insurance, paid vacation, 401 (K) tuition reimbursement. For consideration, please forward a hard copy of your resume to: Racal-Datacom, Staffing Dept., PO Box 407044, Ft. Lauderdale, FL 33340-7044, FAX: 305/846-5025. Or you may e-mail your resume to us at:

racal_staffing@usa.racal.com

For more information on our products, benefits and vision, please visit our web site within CareerMosaic at:

http://www.careermosaic.com

Racal-Datacom is an Equal Opportunity/Affirmative Action Employer.

Please use our Online Response Form

Learn more about Racal-Datacom. Visit our home page.

6. **Civil service offices.**   Call federal, state, county, and city civil service offices. These offices list government positions in such fields as engineering, building and construction, computer technologies, general technology, law enforcement, and parks and recreation. Some offices offer prerecorded recruitment information and telephone tapes listing open positions.

7. **Help-wanted columns.**   Read your newspaper classified section to determine the openings in your field. Note the number of openings, the job descriptions, the qualifications, and salary ranges. If you are willing to relocate and are able to obtain various out-of-town newspapers, search them for openings. Many major newspapers are online. List the companies and details for future reference.

8. **Classified telephone directories.**   Thumb through the Yellow Pages of your city and other desirable cities. List the names of companies or agencies that might be prospective employers.

9. **Trade magazines and newspapers.**   Buy or subscribe to magazines or newspapers that specialize in your field. These publications offer help-wanted sections.

10. **College placement offices.**   Visit your college or university career service center. Most offices offer interest inventory tests, career counseling, career guidance software, occupational briefs, recruitment brochures, and job files of local, state, and national full- and part-time positions. Hundreds of companies conduct on-campus interviews for those students who sign up on their postings.

11. **Interviews and networking.**   Actively arrange to speak to faculty in your major field and people already employed by companies that interest you for practical advice and employment tips. Use the

names of those you contact to network effectively with even more people. Many top managers or directors in companies in your area may give you time to discuss career information.

12. **Employment agencies.**   Many companies do not advertise openings but put their employment search tasks in the hands of public employment agencies and private executive search companies. Some agencies specialize in certain types of employment, such as electronics, computers, money and banking, and allied health fields. Search your Yellow Pages and make an appointment to discuss your employment opportunities. Incidentally, employers who hire you often pay the fee to the agency at no cost to you. Some agencies will post your resume online for you. Seek those agencies that can help you.

In summary, be informed and realistic about your job prospects and the companies that might hire you. Learn how your qualifications fit the needs of prospective employers and what you need to learn to make yourself more marketable.

Personal Portfolio

Include in this file information concerning your education and employment records, skills and achievements, and potential references.

1. **Education.**   Compile a list of the names and addresses of all educational institutions you have attended: high schools (if you are still in college), community colleges, universities, trade and vocational schools, and military schools. Write down the inclusive dates (months and years) of your attendance. Include degrees, fields of study, major and minor, if any, grade point averages, skills obtained, honors, awards, outstanding activities and leadership positions in organizations, and other telling achievements. Do not forget scholarships and honor societies.

2. **Employment.**   Keep written records of your employment history, including the dates, company names, addresses, your position titles, and summaries of your duties, responsibilities, and skills obtained.

3. **Outside organizations.**   List other community or collegiate organizations with which you are associated, officer and chairperson positions, and other pertinent data.

4. **Special achievements.**   Each time you obtain a new skill, such as mastering a computer program, earning a certificate for achieving a new skill, earning a new license, or attending a management workshop or seminar, drop a note and the appropriate paperwork into your portfolio.

5. **Presentation portfolio of original work.** Include copies of your best professional writing class assignments or outstanding samples of writing done on the job. If you have published a brochure, newsletter, feature story, proposal, or special report, save it in your portfolio for a later presentation. Save also any stationery, envelopes, business cards, personal logos, and promotion materials for a project you have developed in school or on the job. Develop a presentation folder of these materials by placing them into a binder or album. The folder will demonstrate your ability to write, design, and organize your own materials. Plan to take the portfolio to interviews, and be alert to an appropriate opening to present the portfolio to demonstrate your abilities. Wait until you have been asked about your professional abilities. Do not just foist your portfolio on the interviewer.

6. **Prospective references.** List the names, addresses, and phone numbers of prospective references. These may include present and past employers, teachers and professors of courses in your chosen field, recognized leaders in your field with whom you have been in work-oriented contact, clergy, and other professional and experienced men and women who can attest to your skills, abilities, and personal attributes. Ask permission to use them as prospective references. If you are ready to search actively for a position, ask those whom you believe will write the most informed and positive references about you to actually write a letter, leaving the salutation blank. Keep these on hand to send them when requested, or give copies to your career center if you are going to have that center supply your references on request.

THE RESUME

Your resume is your ambassador to the work world. Resume format continues to change as more and more job searches are conducted online. Today employers are looking for single page resumes containing *keywords,* which can be scanned by software such as OCR (optical character recognition) to index your resume into a computer database, allowing management to then access by those keywords all related resumes for a particular position. Traditional resumes are organized either *chronologically* or *functionally.* Sometimes a *narrative* format is appropriate. All of these may be mailed or carried to prospective employers or posted online.

Overall Tips

For easy preparation and reading, try to contain your information on one page. You may use 10-, 11-, or 12-point type in easy-to-read fonts, such as Times New Roman and Arial. Use a high quality printer and white, heavy bond paper if you are printing the resume. If you are mailing your resume to a company that is likely to scan it into a database for a key-word search of applicants, mail it between cardboard in a large envelope to avoid folds, creases, and wrinkles, which will distort type. All resumes are divided into similar sections to provide a complete inventory of your objective, qualifications, and experience. The sections include

- Name, address, and phone number (plus fax number and e-mail address if appropriate)
- Career objective
- Summary of qualifications (optional, but desirable)
- Educational background
- Employment experience
- Special elements (honors, skills, and special activities); optional, may be covered in summary, education, and/or employment data
- References or reference statement

Employers skim resumes in less than 60 seconds, sorting out those that are uninteresting or messy. Therefore, you want to present a letter-perfect resume, which, as the result of good overall document design, em-phatic features, and clear headings, not only is readable but also highlights your qualifications through keyword nouns, adjectives, and ac-tion verbs. Use boldface type, bullets, and small caps to make elements stand out. Headings may be centered or placed flush left to the margin. Bracketed items are optional, depending on the position for which you are applying. Do not include unnecessary personal data. It is illegal for employers to consider your height, weight, age, religion, sex, sexual ori-entation, or political affiliations. Figures 13.2 and 13.3 show two sample resume formats; the first is more traditional.

Traditional Resumes

CHRONOLOGICAL LISTINGS

Resume information on education and employment is listed in reverse chronological order; that is, list your present employment and work back, and list your present or latest educational institution and work back. Figure 13.4 shows a sample traditional resume:

NAMES AND ADDRESSES

Decide whether to use a centered, flush left, or right and left page design. Type your full legal name. If fellow workers, teachers, or supervisors ad-dress you by a different name, include that in the middle in quotation

Jane C. Doe

Street Address
City, State Zip
Phone: (000)-000-0000
E-mail: JDoe@aol.com

OBJECTIVE

Name of position or broader statement indicating long-term objective. Include statement if willing to relocate.

SUMMARY (Optional)

Number of years experience, overall strengths, outstanding qualifications

EDUCATION

Month, year to Present	**Name of present institution** Street address City, State Zip ■ Phrase on degree sought, major, and expected date of graduation, high GPA ■ [More bulleted listing of honors, activities, organizations, certificates earned, skills obtained, if not included in Summary or Special Skills]
Month, year– Month, year	**Previous schools** Street address City, State Zip ■ Bulleted types of degrees, diplomas, major courses, GPA ■ [Expansion of honors, activities, etc. if not included in Summary of Special Skills]
Dates	Miscellaneous educational experiences, such as company courses, correspondence courses, seminars, home study, computer training

EMPLOYMENT

Month, year to Present	*Name of Company* Street address City, State Zip ■ Position title ■ Bulleted phrases that amplify the duties performed, skills, promotions, awards, and, possibly, the reason for leaving. Stress action verbs.
Month, year– Month, year	*Name of Company* Street Address City, State Zip ■ Position Title ■ Bulleted phrases of amplification

SPECIAL SKILLS, AWARDS, AND ABILITIES (optional)

Dates	In reverse chronological order include scholarship honors and awards, scholarships and grants, languages, professional memberships, skills not otherwise listed, etc.

REFERENCES (include three or four if listing)

Name, Title
Company or Institution
Street address
City, State Zip
Phone: (000)-000-0000

Name, Title
Company or Institution
Street address
City, State Zip
Phone: (000)-000-0000

or

Available upon request from Career Center, Complete Address and Phone (or from self)

FIGURE 13.2 *Sample resume format—traditional*

John C. Doe

Street Address, City, State Zip, Phone, E-mail address

OBJECTIVE
Name of position or broader statement including long term objective and highlighting summary of qualifications and skills. Include statement if willing to relocate.

EDUCATION

Name of present institution   **Month, year to Present**
- Phrase on degree sought, major, expected date of graduation, high GPA
- (More bulleted listings of honors, activities, organizations, certificates earned, skills obtained)

Previous school   **Month, year–Month, year**
- Bulleted types of degrees, diplomas, major courses, high GPA
- (Expansion of honors, activities, skills obtained, etc.)

Previous school   **Month, year–Month, year**
- Bulleted types of degrees, diplomas, major courses, high GPA

EXPERIENCE

Name of company, Street Address, City, State Zip   **Month, year to Present**
- Position title
- Bulleted phrases that amplify the duties performed, skills, promotions, awards, and, possibly, the reason for leaving. Stress keywords and action verbs.

Name of company, Street Address, City, State Zip   **Month, year–Month, year**
- Position title
- Bulleted phrases of amplification

Name of company, Street Address, City, State Zip   **Month, year–Month, year**
- Position title
- Bulleted phrases of amplification

SKILLS

Management skills	**Computer skills**	**Other skills**
(Bulleted phrases of experience)	(Bulleted specific list of software, programs)	(Bulleted foreign languages and other achievements)

REFERENCES

References names and addresses or statement "References available upon request."

FIGURE 13.3   *Sample resume format—modern*

Therese V. Lazzari-Lindo

4100 Southwest Tenth Street
Boca Raton, FL 37001
Phone: (561)-322-3166
E-mail: Lazzari-Lindo@aol.com

OBJECTIVE

Associate position in psychological counseling with opportunity to utilize my skills with children, youth, and families. Willing to relocate.

SUMMARY

Eight years experience in children, youth, family counseling plus vocational training including behavioral modification, counseling, crisis intervention, group management, in-service training and supervision.

EDUCATION

August 2000– May 2001	**Florida Atlantic University** 900 Glades Road Boca Raton, FL 37002 • B.A. degree in Social Psychology • Magna Cum Laude, GPA 3.75
May 1998– July 2000	**Broward Community College (North Campus)** 1000 Coconut Creek Boulevard Coconut Creek, FL 33066 • A.A. degree in Liberal/Arts, 2000 • Courses in Psychology, Abnormal Psychology, Sociology, Speech, Technical Writing, Advanced Computers

EXPERIENCE

August 2002 to Present	**Florida Department of Rehabilitative Services** Adolescent Unit 1403 Northwest 40th Street Lauderhill, FL 33313 • Children, Youth, and family Counselor • Counseling and care management of adolescents; monitoring progress in foster homes; advising foster parents as needed, referrals to community-based services • Maintenance of records and liaising with the judicial system
January 1997– June 2002	**Covenant House** 733 Breakers Avenue Fort Lauderdale, FL 33301 • Case Manager, Interim Overnight Shift Manager • Counseled residents and families, acting resident supervision, implemented crisis intervention, and maintained records
June 1995– January 1997	**Head Start Program** 905 West Atlantic Boca Raton, FL 37001 • Part-time Counselor • Social work and counseling parents and children

REFERENCES AVAILABLE UPON REQUEST. CONTACT ME.

FIGURE 13.4   *Sample traditional reverse chronological resume*   (Courtesy of student Therese V. Lazzari-Lindo)

marks (e.g., Francis "Frank" Jones). Include your street address, city, state, zip code, and phone number with area code. You may include both your home and business phone, or your school phone and home phone if you wish to be contacted during vacation or summer break. Include your fax number and e-mail address if you have them. Review the sample resumes in this chapter to examine a variety of layouts.

EMPLOYMENT OBJECTIVE

Either center or place flush left the heading "Employment Objective," "Career Objective," or just "Objective." Use boldface type. The heading may be in all capitals or upper- and lowercase. Keep all of your headings consistent. Beneath or beside it write the exact position you are seeking in a brief phrase that spells out your short- and long-term objectives. If you are willing to relocate, include that statement along with your objective. Avoid full sentences.

Examples

Legal Secretary. Willing to relocate.

Position in Customer Service or Sales Field.

Position in drafting leading to design responsibilities.

Sales position with emphasis on e-commerce.

Human Resources Generalist.

Searching for an opportunity to work for a leader in innovative network and telecommunication solutions.

SUMMARY OF QUALIFICATIONS

You may include this summary along with your employment objective listing or separately under the heading "Summary." Again, use just phrases and keywords plus action verbs. You may bullet each separate qualification if you have room.

Examples

- Offer extensive hands-on supervisory experience, skilled at building strong team environments and fostering open communications.
- Experience with technical manual writing and design, speech writing, article publications, interviewing techniques, and computer skills including C++, Java, and Oracle.
- Expertise in Microsoft Systems Management Server 2.0, NT 4.0, NT Workstation, and NT servers. Experience in designing, implementing, and maintaining Web sites.

- Experience in implementing enterprise-wide database-driven applications. Background in multidimensional data modeling based on relational databases. Exposure to data warehousing techniques and OLAP. Strong technical leadership abilities.

EDUCATIONAL BACKGROUND

If your educational preparation is stronger than your employment experience, develop this section first. Type the heading "Education." Beneath it begin with your most recent school and list all other schools in reverse chronological order. List the inclusive dates, including the months, in one column. In another column include the names of the institutions, addresses, cities, states, and zip codes. Cite your major field of study, degrees earned or expected, and graduation dates. Highlight your academic record (if high), awards, special activities, organizations, special skills, and job-related courses. Include only those achievements, however, which are slanted toward the job. A prospective employer in banking might be impressed if you were treasurer of your student government association but would not care if you were on the tennis team.

Grade point average (GPA) may be a concern of employers and may be presented in either of two ways. If you have a better than average GPA, then indicate both your overall average and your average in your major. It is most likely that you did better in courses within your major area of study than in most of your other coursework. Thus, if you believe your overall GPA might hinder your chances at getting an interview, put only your GPA in your major on the resume.

Examples

- Communications major, concentration in broadcast production and news
- Graduated Cum Laude
- Study Abroad Program (City University, London)—broadcast studies courses
- Dean's Award for Excellence in Leadership and Service, 2000
- Alumni Leadership Scholarship, 1998–2001

EMPLOYMENT EXPERIENCE

If your previous or present employment is more indicative of your qualifications than your educational background, place this section before your educational data. Type the heading *Employment, Work Experience,* or *Experience.* Present your information in reverse chronological order. List the inclusive dates, including months. Employers are often looking for gaps in both educational and employment histories. A continuous record of employment or education suggests that you are a responsible individual with work-oriented goals. List the names of the companies, street addresses, cities, states, and zip codes. List your position title and whether the work was part-time or full-time. Use brief keyword phrases with strong action verbs to highlight your duties, achievements, awards, and affiliations. Check out job listings in the newspaper and online to determine the desired skills in demand for the job you seek. Some suggested keyword verbs, nouns, and adjectives follow:

Verbs

accomplished	coordinated	participated
achieved	created	performed
adapted	devised	presented
analyzed	directed	proposed
collaborated	established	published
communicated	implemented	reorganized
compiled	increased	specialized
completed	initiated	standardized
conducted	instructed	supervised
consulted	organized	trained

Nouns And Adjectives

administrator	e-commerce experience	morale booster
analytical	effective	motivator
capable	efficient	multilingual
collaborator	exceptional	negotiator
communicator	experienced	reliable
competent	flexible	responsible
consistent	global	team worker
creative	imaginative	troubleshooter
dedicated	innovator	professional
diversified	interpersonal skills	proficient

You may wish to list your reason for leaving each position (i.e., "Offered higher-paying position," "Returned to college full time").

SKILLS AND SPECIAL ELEMENTS

Employers are seeking applicants with a multitude of skills. Whether or not you have already summarized your qualifications, use an appropriate heading, such as *Awards, Skills, and Activities,* or just *Skills.* Then list your honors and awards; include specific skills not already mentioned, such as multilingualism and specific computer skills. Add accomplishments, such as publishing a book or article, mastering braille, teaching, or tutoring. If you have not already included organizations (academic, professional, or community) in which you hold office or in which your are active, indicate them here. Remember to include volunteer work. Such participation shows professional involvement and concern. Likewise, professional licenses and certification pertinent to the specific position should not be overlooked. Some sample listings follow:

- Computer skills: Microsoft Excel, Microsoft Access, Visual Basic 6.0, SQL Server 7.0, Oracle, MCSE (Microsoft Certified Systems Engineer), C++, Java, Web page design, HTML.
- Communications skills: Published three articles on global marketing in Company A newsletter, constructed graphics for textbook inclusion for professor, fluent in Spanish and French.
- Accounting skills: Reconcile accounts, develop spreadsheets, tutor high school students in basic accounting activities.
- Volunteer work: Active member of Big Brothers for Abused Boys; door-to-door canvas work for American Cancer Institute; Treasurer of Renaissance 2000, an art museum fund-raising organization.

REFERENCES

Under or next to the heading "References," consider listing three or four personal references. These should not be close personal friends, but professionals with whom you have worked or with whom you are acquainted. Past employers, professors, and leading people in your community are best. In the modern job market, the need for quick hiring created by faster and faster job turnover has made immediate referral a necessity. You should have in your files or in a career center file letters of reference written before you distribute your resume. If you do not include the reference names and addresses in your resume, use a statement, such as "Immediate references available; contact me" or "References available upon request from the Career Planning and Placement Center" (or from your college or university), and give the complete address. Take along a file of references to interviews and present them if asked.

Functional Resume

If you are short on employment history but strong on skills and other accomplishments learned at school or elsewhere, use a functional resume format to highlight your skills in brief paragraphs of phrases and bulleted listings. Essentially, use the same headings as you would in the chronological resume, but move your skills, awards, achievements, and other activities to the top. Figure 13.5 shows a functional resume with strong skills and achievements sections.

PATRICIA D. TYLOR

1701 Southwest Tenth Place, Miami, FL 33325 - E-mail: TylorP@icanect.com
Phone: (305) 475-9237, Fax: (305) 475-7233

OBJECTIVE

Medical laboratory technician at a modern, expanding hospital laboratory

SUMMARY

- Experienced laboratory technician
- Three years training in advanced instrumentation, advanced medical laboratory techniques, advanced mycology, and other procedures
- Advanced computer skills including Microsoft Word, Microsoft Excel, Quicken, Corel Presentations 8

SKILLS

Laboratory Procedures: Extensive practice in hematology with Coulter Model S-Plus, Coulter Diff 3 System, and Coag-A-Mate

Perform urinalyses and serology procedures, whole blood calibration procedures, patient blood-drawing procedures, manual chemistry tests for Glucose, BUN, SGOT, Alkaline Phosphatase, Uric Acid, and Cholesterol

Communication. Revised laboratory procedures manual for Sheridan Vocation Center. Presented numerous oral training sessions for first-term students. Excel in computer skills (document design, desktop publishing, word processing, spreadsheets, Oracle, and Java). Top grade in Technical Writing class. Speak fluent Spanish and French.

SPECIAL AWARDS, ACHIEVEMENTS, VOLUNTEER WORK

August 2000 – Full-time Brewster Scholarship Recipient, Broward Community College
Dean's List every term
December 1999 – Participant in National Medical Laboratory Technician Seminar, Baltimore, MD
June 1998 to Present – Volunteer in County Hospice, 872 Hollywood Boulevard, Hollywood, FL

EMPLOYMENT

Florida Medical Center, 5000 West Oakland, Pompano, FL 33305   **August 2000 to Present**
- Laboratory Technician in Hematology
- Responsible for early morning start-up procedures
- Perform routine urinalyses and serology procedures
- Experienced in whole blood calibration procedures

EDUCATION

Broward Community College (Central)   **August 1999 to Present**
3501 Southwest Davie Road, Fort Lauderdale, FL 33314
- Majoring in Medical Laboratory Technology; will receive Associate in Science degree in May 2001
- Grade point average: 3.6 of 4.0

Sheridan Vocational Center   **January 1998–July 1999**
5400 Sheridan Street, Hollywood, FL 33021
- Completed the NAACLS accredited, 12-month Certified Laboratory Assistant Program
- Grade point average: 4.0 of 4.0

REFERENCES

Available upon request from the Career Center, Broward Community College, 3501 Southwest Davie Road, Fort Lauderdale, FL 33314, Phone: (954) 524-1111

FIGURE 13.5 *Sample functional resume stressing skills and activities*

Narrative Resumes

The very mature and exceptionally experienced person may choose the narrative format to cover fully past employment achievements. Using the same headings as in other resumes, you may present your resume in a first-person narrative, much like an expanded cover letter. Use complete sentences to tell the story if your professional life is extensive. Again, cover the material in only one page and use keywords.

People in show business, art, photography, or music will present resumes that are quite different from those in most professions. Such persons will submit a narrative resume and lists of works in which they have performed, a list of professionals with whom they have studied, or a list of their artwork, publications, and the like. This list will often be submitted in a personal portfolio or a video providing actual samples of their work. Persons in these specialized fields will learn more about special inclusions from their peers and from experience.

Figure 13.6 shows an example of a narrative resume.

Online Resumes

You may use any of the above formats if you plan to post your resume on the Internet at job banks, at specific companies which accept online resumes, on your own Web page, or to a Web board. However, if you are using one of the job sites, determine if that site has its own resume format preference. You must pay particular attention to simplicity because some materials may not post well on the Internet due to transfer problems of text and document formatting. Avoid columns, fancy fonts, underlining, horizontal or vertical lines, italics, or excessive bulleting. Figure 13.7 shows a sample online and easily scannable resume.

COVER LETTER

A cover letter, also called a *letter of application* or a *face letter,* accompanies your resume and serves both as an introduction of yourself and as a strategy to interest employers sufficiently to read your resume and to grant you an interview. The letter should reflect your personality and emphasize your potential. While a model letter may be developed, each letter should be tailored to a particular prospective employer.

Each letter should be individually typed on bond paper of the same quality as your resume. Address your letter to a specific name, if possible.

Your opening paragraph must attract the reader's attention without being overly aggressive or "cute." You must indicate the position or type of employment you are seeking and refer to your enclosed resume. You

GREGORY WELLBORN
334 South Bonita Avenue
Pasadena, CA 91107
Phone: (202)-596-7081

OBJECTIVE/WORK SUMMARY

Officer in a technology firm. Am willing to relocate. Thirteen years experience in all phases of top management.

EXPERIENCE

NATURAL HOME PRODUCTS COMPANY   January 1997 to Present

As **General Manager,** I developed this subsidiary of Prepared Products Company into a full line marketer of all natural consumer packaged goods. My responsibilities included all market and product development, materials sourcing, and establishing independent operating and management control systems. The result was the successful development of an all-natural consumer products line and introduction in targeted metropolitan markets.

ALLIANCE PRODUCTS   June 1992–December 1996

General Manager   June 1994–May 1996

I undertook the turn-around of the company's unprofitable performance, developed and implemented a strategy to transform the company from a single product distributor to a multiline marketer of synergistic products in key market segments. My accomplishments included successful implementation of the above strategy by broadening the product base to 11 lines, improving average gross margins from 20% to 40%, reducing dependence on distributor sales from 80% to 10%, and maintaining sales volume during the transition period.

Vice-President—Finance and Operations   January 1993–May 1994

I established accounting and management reporting systems, negotiated credit lines and equity issues, oversaw production, and managed vendor relations for this new venture. My accomplishments included raising $300,000 in venture capital and developing a flex-time production system which reduced average production time from 2 days to ½ day and eliminated overtime pay averaging 3% of sales.

BANK OF AMERICA   December 1987–December 1992

Marketing Director—Eastern L.A. Region   January 1991–December 1992

My responsibilities included incorporating 11 autonomous branches into one new region, developing and administering the region's first marketing plan. My accomplishments included obtaining $22 million in new loan commitments, and successful integrating of the branches' management reporting and credit authorization systems.

Assistant Vice-President—Corporate Banking   March 1989–December 1990

Commercial Loan Officer   March 1988–March 1989

My responsibilities included managing a $15 million portfolio and supervising a team of commercial loan officers.

THE SIGNAL COMPANIES   June 1986–February 1988

Acquisition Analyst—Entertainment Division

My responsibilities included all financial and marketing analyses of prospective med acquisitions.

EDUCATION

University of Southern California, 1986, M.B.A. with Finance and Marketing emphasis

SPECIAL ACTIVITIES

I am active in Door of Hope, Harambee Center, and Friendship Corner charities and have served on numerous community boards such as the Chamber of Commerce, Downtown Development Association, and County Industrial Group. Have served as Chairman of the Opera Society and the Philharmonic Symphony.

   REFERENCES AVAILABLE UPON REQUEST

FIGURE 13.6   *Sample narrative resume*   (By permission)

Merrill D. Tritt, PHR
7821 78 Terrace
Sunrise, FL 33321

Home (954) 720-4381
Pager (954) 983-4147
E-mail mtritt@media.net

OBJECTIVE

To contribute my knowledge and skills from past and present work experience to a quality, driven organization in a technically related or management position.

PROFESSIONAL EXPERIENCE

Spherion Corporation FKA Interim Service, Inc., Fort Lauderdale, FL, March 2000 to Present
Technology Consultant

Specialize in human resource and payroll data systems. Currently performing analysis of corporate-wide human resource and payroll data conversion from third party to in-house system. Previously oversaw conversions for recently acquired companies and completed same with custom programming. Ensured data integrity and assisted data testing team by developing custom reports for validation.

Precision Response Corporation, Miami, FL. June 1999 to March 2000
Manager of Human Resource Administration

Assisted Human Resource personnel with policy and procedural exceptions and inquiries. Participated with bringing company to COPC-2000 compliance and created necessary documentation. Developed Human Resource policies and procedures. Developed and administered comprehensive audits to ensure adherence to corporate policies. Assisted with benefits administration, management, and audits. Guided further growth and development of Human Resource information systems and ensured data integrity.

Precision Response Corporation, Miami, FL, March 1997 to May 1999
HR Generalist/MIS Business Analyst/Programmer-Analyst

Assisted in the planning, architecture, development, and setup of Oracle NCA HRMS and Oracle NCA Payroll to fit Precision Response Corporation's HR/Payroll policies, practices, and procedures. Participated in design workshops. Cleansed proprietary data and assisted with data conversion. Developed SQL scripts for HR and Payroll reporting. Created a comprehensive end-user manual and ensured quality training. Acted as trainer and end-user/system support for both Human Resource and Payroll departments.

Incredible Universe, Hollywood, FL, August 1996 to April 1997
Human Resource Manager / Payroll Administrator

Implemented, coordinated, and executed Human Resource programs and functions. Maintained an effective employment and recruitment process. Directed all training activities. Administered and processed weekly payroll and monitored payroll budget. Administered and coordinated corporate benefit programs. Supervised other Human Resource personnel. Participated in resolving employee relations issues. Attended Worker Compensation hearings. Performed a key role in the closing of the Hollywood, FL location.

EDUCATION

Florida International University, Miami, FL 1996
Bachelor of Arts in Business Administration. Double Major: Personnel Management and Business Management. Graduated Magna Cum Laude. Overall GPA: 3.758.

HONORS AND CERTIFICATES

Professional of Human Resources – 1999

SHRM Learning System – 1999
Certificate of Completion

Florida International University
Graduated #1 in FIU's School of Business Management. Member of Golden Key National Honor Society, The Honor Society of Phi Kappa Phi, and The Honor Society of Beta Gamma Sigma.

The National Dean's List
1995-1996

SKILLS

Strong organizational, leadership, and interpersonal skills. Technical knowledge of IBM-based microcomputer hardware. In-depth knowledge of DOS, Windows 3.x, Windows 95/8, Quick adaptation to software applications with proficiency in Word, Excel, Word Perfect, Kronos Time and Attendance V8 and C/S, ADP HR Partner/Ad-Hoc Reporting, Ceridian Ensemble, Data Junction, Oracle Browser, Visual Basic 6.0, and SQL. Typing speed: 65 wpm.

FIGURE 13.7 *Typical, clear, online resume for easy scanning* (By permission)

may want to mention how you learned of the position and subtly praise the company. Consider these opening paragraphs:

> I am an ideal candidate for the electronics technician position that you advertised in the July 2 edition of *The Miami Herald.* My two years of experience in the field and an Associate of Science degree in Electronic Technology qualify me for employment with your progressive corporation. My enclosed resume amplifies my background.
>
> Mr. Jack Smith, Chair of the Data Processing Department at Port Washington Community College, has alerted me that there will be a computer technician opening in your computer maintenance group. As my resume details, my specific qualifications prove that I would be an asset to your organization.

Your body paragraph(s) should summarize and emphasize your specific qualifications and personal qualities as they relate to the position. Use an enthusiastic tone and project self-confidence. Mention your outstanding personal attributes. Consider this sample:

> I can offer you four years of experience in employment and education. I will earn my Associate of Science degree in radiology technology this May and have been employed as a medical assistant to Dr. James E. Perry, radiologist, here in Fort Lauderdale for two years. My personal attributes include determination, courtesy, and attention to detail.

The body material should be short, direct, and persuasive.

Your closing paragraphs should urge action on the part of the reader. Ask for an interview or suggest that you will phone shortly to arrange for an interview. If it is appropriate, seek additional information, applications, and so on. Mention your flexibility and availability. Consider these closings:

> May I have a personal interview to discuss your position and my qualifications? I will call your office on Monday to arrange a convenient date.
>
> I would welcome your consideration for an entry-level position and can start work any time next month. Any information you may have regarding my prospects with your company will be greatly appreciated. You may reach me after 3:00 P.M. weekdays at the number listed on my resume to arrange an interview.

Ultimately, remember that your cover letter is an overall view of your personality, your attention to detail, your communication skills, your enthusiasm, your intellect, and your qualifications. Your cover letter and resume are usually all a prospective employer has to decide whether or not you will reach the next phase in your employment search process—the interview.

Figures 13.8 through 13.10 show the complete texts for three sample cover letters. Figures 13.8 is for an entry level position; Figure 13.9

7661 Hood Street
Troy, NY 12181
6 July 200X

Mr. Jerry Diener
Human Resources Director
American Express Company
777 American Expressway
Plantation, FL 33324

Dear Mr. Diener:

I am an applicant for the entry-level position in your information resources department.
Dr. Ted Smith, head of the Computer Sciences Department at Queens Community College,
South Campus, who developed your system, has told me about your progressive practices,
and I believe I can offer a substantial contribution to American Express Company. My resume highlights my qualifications.

My main experience is in the accounting field. To further my understanding of the complete
accounting cycle, I enrolled in 2000 at Hudson Valley Community College in Troy, New
York. At that time I studied advanced computer courses and found that I have a natural
ability for the logical thinking that is required for computer programming. Because a career
in Information Technology seems the ideal goal for me, I am now working towards my A.S.
degree in data processing and will be graduated this August, when I plan to relocate to
South Florida.

I will call you on Wednesday, 13 July to seek an appointment for an interview in mid-
August.

Yours truly

Sandra J. Burnette

Sandra J. Burnette

SJB

Enc: 1

FIGURE 13.8   *Sample cover letter for entry-level position*   (Courtesy of student Sandra
J. Burnette)

7821 NW 78 Terrace
Sunrise, FL 33321

February 11, 2000

Jack Hawkins,
Director of Human Resources
Olash Enterprises, Inc.
1514 East Washburn Way
Chicago, IL 61606

Dear Mr. Hawkins:

Please accept this letter and my resume as an expressed interest in joining your corporation, where my education and experience can be immediately and profitably utilized to our mutual benefit.

After five years of balancing college courses and employment, I secured positions both as a Human Resource Generalist and Human Resource Manager. When my former employer, Precision Response Corporation, decided to upgrade their aging payroll and HRIS systems with Oracle database products, they requested that I assist them with the implementation, as I have not only a broad base of Human Resource knowledge but also a considerable technical background.

In the past, I have also taught college level courses at Broward Community College. As such, I have excellent communication and interpersonal skills. Additionally, my day-to-day output, both presently and in the past, reflects a high level of motivation, efficiency, and ability to meet any objective, alone or as part of a team effort. I have a proven ability to troubleshoot, perform under a minimal amount of supervision, and demonstrate a high degree of initiative and good judgment. I am confident in my abilities and have garnered many compliments from coworkers and supervisors. I am confident I could provide many valuable contributions to your organization and shall welcome the opportunity to discuss this with you in person.

I look forward to hearing from you to schedule an interview at your earliest convenience, and I greatly appreciate your time and courtesy in reviewing this material.

Sincerely,

Merrill D. Tritt

Merrill D. Tritt, PHR
HM(954) 720-4382
WK(954) 351-4209
email: mtritt@media.net

FIGURE 13.9   *Sample cover letter for a position requiring some experience*   (Courtesy of Merrill D. Tritt)

Robert W. Johns   2200 Southwest 16th Terrace   (954)-555-7266 Home
                             Fort Lauderdale, FL   (954)-555-2255 Fax
                             33316   (954)-556-6899 Pager

7 August 2000

Mr. Mark Lauer
AlliedSignal Marketing, Sales & Services
1001 Pennsylvania Avenue, Suite 7005
Washington, D.C. (Northwest) 20004

Dear Mark:

I would like to express my strong interest regarding the position opening for Marketing
Sales & Services in Florida, which we discussed in July in your Washington, D.C., office. I
am pursuing opportunities that will leverage my extensive expereince within the aeorspace
industry.

Over the last sixteen years in the industry, I have become acquainted with most aspects of
design, manufacturing, operation, maintenance, and financing of aerospace equipment.
Furthermore, I especially enjoy prospecting and developing new business arrangements.

My primary responsibilities over the last five years have been to develop business relationships and technical strategies. As you will note from the attached resume, I have been successful in creating AlliedSignal alliances on both a technical and executive level. Further, I
have outstanding ability to sell products and services by quickly assessing a customer's
needs, mutually consolidating those needs into a set of workable requirements, and creating
a solution with contingencies for unforseen circumstances. I believe that this "big picture"
approach to selling would carry over quite effectively to promoting and selling within the
Marketing Sales & Services division of AlliedSignal.

I understand that the majority of my experience relates to avionics and many of the opportunities at P & W are more applicable to AES products, but I feel my ability and commitment to learning would easily overcome challenges such as this one.

Finally, I believe my contacts, experience, education, and overall breadth of knowledge of
the aerospace industry would be better utilized if I were operating from within a broader
sector of the organization like Marketing, Sales & Services. Combining our expertise would
provide opportunities which would mutually benefit both of us. Therefore, I look forward to
having the opportunity to discuss this with you further and urge you to contact me at your
earliest convenience.

Best Regards,

Robert Johns

CC: Reed Childers

FIGURE 13.10   *Sample cover letter for a position requiring extensive experience*
(Courtesy of Robert W. Johns)

is for a position requiring some experience; and Figure 13.10 is for a position requiring extensive experience.

THE EMPLOYMENT INTERVIEW

The final step in the employment search is the interview. You may be interviewed by one or more people, individually or in a group. Your first interview may be considered preliminary and is often called a **screening interview.** In this interview, numerous applicants are invited to speak with a number of interviewers, perhaps a search committee, as well as the officers and directors of the organization. The objective is to narrow the pool of candidates for more intensive follow-up interviews. If you are asked back for a second interview, the indication is that you are a good candidate for the job. This will most probably entail a **line interview,** during which you will be more intensively interviewed by fewer personnel. Here you may meet other supervisors and the people with whom you will work if you are hired so that they can offer the interviewer opinions. It is essential that you are prepared, informed, gracious, and self-contained.

The "Do's"

It is at the interview that your prospective employer evaluates you—your appearance, your personality, and your abilities. Chapter 17 discusses a number of verbal strategies to employ in an information-gathering interview, other verbal transactions, and nonverbal messages that you may want to review quickly. To prepare yourself for the employment interview, here are a few tips:

1. **Be on time.**   Know the exact time and location of your interview. Arrive ten minutes early and check yourself out in a restroom mirror. State clearly your name and purpose as well as the name of the person who has established the interview to the secretary, receptionist, or security desk guard. You have some control over an appointment. Although the time, place, and type of interview will be set by the interviewer, you have alternatives to accepting those details, such as suggesting alternate times for the interview that may alleviate rushing, unpreparedness, or a tight schedule.

2. **Dress appropriately.**   If you are a male, dress conservative and neatly in a dark, but not black, suit, white or light solid shirt and tie, knee-length dark socks, and polished leather shoes. If you are a female, wear a gray, black, navy, or tan suit, closed heels with a moderate heel, and natural-colored hose. Either sex might wear just one article that is eye-catching and memorable, such as a striped silk tie, a lapel rose, or an unusual, but not showy, piece of jewelry. Since 55 percent of communication is nonverbal, your appearance is critical.

3. **Be prepared.** In advance, learn as much as you can about the company so that you can project your interest and ask informed questions. Take several copies of your resume along to refresh your memory in response to direct questions about your background, to hand to the interviewer in case he or she has misplaced your resume, and to submit as an extra to be given to someone else at the company at the interviewer's discretion. Be prepared that you may be asked to stay and take tests associated with the job. Be ready to answer simple and tough questions (see the section "Questioning Skills" in this chapter).

4. **Modulate your voice.** The acoustics of the room, the distance from the interviewer, and your own anxiety should be noted in order to modulate your voice to create a friendly, yet assertive, impression. If you talk too loudly or your voice is too high pitched, you will be offensive and boorish. If you talk too softly, you may appear "wishy washy" or uninterested. Accept or ask for water or coffee if your voice becomes raspy. A few sips will relax your vocal chords.

5. **Show your personality.** Be enthusiastic and excited about the job. Use your interviewer's name more than once. Smile and shake hands firmly. If you are a woman, offer your hand first, for many males are still uncertain if it is proper etiquette to initiate a handshake with a woman. A handshake establishes confidence and warmth. Wait to be asked to be seated. Lean towards your interviewer to show interest and enthusiasm. Listen well. Know at least five things about yourself that make you unique in the marketplace, but do not brag. Ask open-ended questions about company policies, working conditions, advancement possibilities, and salary range. Let the interviewer know you want the job. Ask if your prospects of being hired are good.

6. **Be courteous.** At the end of the interview, thank the interviewer for the time given you. Do not dawdle; your interviewer is a busy person. Be prepared to shake hands again. Even if you are disappointed with the outcome, use your best manners. Leave decisively.

8. **Follow-up letter.** Immediately, write a one-page follow-up letter to your interviewer expressing your appreciation for the time given to you and reiterating your interest and availability. Include a paragraph reminding your interviewer of your strengths and refer to your conversations about key skills and issues. This puts you back into the mind of the interviewer.

The "Do Nots"

Conversely, guard against making the wrong impression. Consider the following items to avoid:

1. **"Winging" it.**   Remember your competition is coming to the interview prepared.

2. **Familiarity.**   Avoid being too familiar with the interviewer. Ask how he or she would like to be addressed, and do not ask personal questions. Do not drop names; the name you drop may be a political or personal enemy of the interviewer.

3. **Nervousness.**   Do not fiddle with your hands or rings. Do not smoke, chew gum, or suck on a mint. Keep your hands and pencils away from your face. Sit still instead of fidgeting. Breathe normally rather than in gasps or sighs. Sit in a comfortable, but not slouchy, position, and keep both feet on the floor with one foot slightly more forward than the other.

4. **Language.**   Do not tell off-color jokes or use even mild profanity or overused slang. Avoid "ums" and "you know." Avoid pretentious words. Be as articulate as you are able.

5. **Questions.** Do not fail to answer questions, or at least acknowledge that you do not know the answer but would be interested in finding out.

6. **Endings.**   Let your interviewer indicate when the interview is over. Do not mention next appointments, nor look at your watch.

Questioning Skills

Your resume and application will have presented basic information about you, but both your interviewer and you will ask questions. The interviewer will expect you to summarize your qualifications. In addition, the employer wants to learn about your potential, your expectations, your attitude toward the organization, and your personality and character. By thinking about the following questions that you may be asked, you can prepare oral responses that are complete and confident.

Interviewer's Questions

1. Tell me about yourself. Expand on your resume. Can you give me a summary of your employment, education, and personal background?

2. Why do you believe you are qualified for this position? (Stress your skills.)

3. What do you see as your strongest (weakest) personal qualities? (Name two or three each; negatives can be balanced against positives or recognized with a goal to improvement through your experience and training.)

4. How would your friends describe you? (Are you prepared, smart, a people person, capable, patient, eager to learn, a diplomat, self-directed, other? Avoid negatives.)

5. Why did you select us in your job search? What are your reasons for wanting to work here? (Demonstrate that you have done your homework about the organization.)

6. What are your long-term career goals? What do you want to achieve in five years? Ten years? (Demonstrate that you have goals and expectations that are realistic.)

7. What have been your most satisfying and most disappointing school or work experiences? (Recite your accomplishments and stress how you overcame disappointments.)

8. How do you handle stress? How would you handle conflict between yourself and a subordinate? Supervisor? Peer? (Summarize how you cope with personal stress, such as by physical exercise, reading a book, gardening, fishing, frank discussion, and the like.)

9. Describe a few situations in which your work was criticized. How did you handle the criticism? (Admit you were hurt but after working on the project, it worked out well.)

10. Suppose your boss at a weekly team meeting unexpectedly began to criticize aggressively your performance on a current project. What would you do? (Be either stoic and accepting or politely confrontational, but seek a follow-up meeting in order to seek advice and remedial steps.)

11. Tell me about a time when you had to do several things at once. How did you handle the situation? (Be honest, but don't admit going bonkers. Did you multitask, reorganize, delegate authority, check with your boss about priorities?)

12. Suppose you are in a situation where you have two very important responsibilities that both have a deadline that is impossible to meet, and you can't accomplish both. What do you do? (You should probably decide which task you are most suited for and then go to your boss to ask that the other one be reassigned or rescheduled for accomplishment, if possible.)

13. What are your attitudes towards unions (absenteeism, punctuality, geographical transfers, shift work, weekend assignments, travel)? (Know in advance what the organization expects and evaluate your attitudes accordingly.)

14. What salary do you have in mind? Would you accept x amount? (Be prepared to give a minimum or a range, emphasizing needs, goals, and past earnings. Ask for clarification about work hours, promotion potential, and raise and bonus potentials.)

Throughout the interview, you should feel free to ask questions yourself. Seek clarification of any vague questions asked of you. Let your own line of questioning reflect that you are interested and serious about the organization. Decide which of these typical questions you want answered.

Interviewee's Questions

1. What are the organization's long- and short-term plans? Is the organization expanding or shrinking?

2. What is the potential for advancement in this position?

3. How does your organization support employee advancement? Are there educational opportunities? Grants? Training programs? Performance reviews?

4. What is the leadership structure? Will I have any decision-making authority?

5. What are the organization's relations with the community? Consumers? Related organizations?

6. What are the grievance procedures? Layoff policies? Leave policies?

7. What are your hours of operation? How much travel is expected? Do employees normally work many hours of overtime?

8. Are there any benefit provisions? (Insurance, pension plans, stock options, cars or housing provided, travel per diem, employee discounts, and the like.)

9. How is the housing market in this area? What cultural, recreational, and social opportunities are available in this city? How would you rate public transportation, schools, municipal services?

10. What are the wages and salaries? Are increases based on performance or indexes? What are the means of advancement?

Salvaging the Sour Interview

Sometimes, despite all of your preparation, the interviewer may seem uninterested in what you have to say, and you detect that you are not coming across well. Don't give up. It *is* possible to salvage a job interview that is not going well. Robert Half of Robert Half International, a corporate recruiting company, offers this advice:

1. Analyze the problem. Are you stressing the wrong information? Try to change the focus of discussion.

2. If the interviewer has missed your strongest points, don't be subtle; tell about them confidently.

3. If the interviewer seems rushed or preoccupied, offer to return when things are calmer.

4. Return to a previous question that you feel you may have fumbled and add any more necessary information.

5. Find a trigger of common interest in the office, such as a book, art piece, or other item that you can discuss as an ice-breaker.

6. Be honest and ask the interviewer what you're doing wrong. Example: "I'm not sure I've impressed you enough today. I know I can do this job. Would you tell me where I'm going wrong so I can address your fears?"

7. Ask for the job. The most successful salespeople know the best way to get an order is to come out and ask for it. Example: "I like this company and I would like to work for you. And if you hire me, I won't let you down."

A realistic examination of your employment prospects, a letter-perfect resume, a self-confident cover letter, and a thoughtful interview are going to get you the position you seek.

CHECKLIST

Resumes, Cover Letters, and Interviews

RESUME

❏ 1. Have I compiled a record of my prospective positions by examining sources?

❏ The *Dictionary of Occupational Titles?*

❏ The *Occupational Outlook Handbook?*

❏ The *Journal of Career Planning and Employment?*

❏ Career handbooks, government publications, and newsletters in my field?

❏ Internet sites?

❏ Civil service offices?

❏ Help-wanted columns?

❏ The Yellow Pages?

❏ Trade magazines and newsletters?

❏ College placement office materials?

❏ Interviews with professionals?

❏ Employment agencies?

❏ 2. Have I compiled a personal portfolio demonstrating my background?

  ❏ Education details?

  ❏ Employment details?

  ❏ Community involvement?

  ❏ Special achievements?

  ❏ Bound presentation portfolio of my best and original work?

  ❏ Prospective references?

❏ 3. Have I decided on the appropriate format for my needs?

  ❏ Traditional?

  ❏ Functional?

  ❏ Narrative?

  ❏ Online?

❏ 4. Does each resume include scannable keywords: nouns, adjectives, action verbs?

❏ 5. Have I included all the proper sections with the necessary details?

  ❏ Name, address, and phone/fax/e-mail address?

  ❏ Career objective?

  ❏ Summary of qualifications?

  ❏ Educational background?

  ❏ Special elements?

  ❏ References or reference statement?

❏ **6.** Have I used a pleasing document design on white bond paper with proper placement of sections, consistent headings and indentations, and appropriate boldface/italics?

COVER LETTER

❏ **1.** Have I prepared a letter with a consistent block or modified block format?

    ❏ Heading with date?

    ❏ Inside address?

    ❏ Proper salutation to a specific person?

    ❏ Body paragraphs?

    ❏ Complimentary closing?

    ❏ Signature?

    ❏ Enclosure notation?

❏ **2.** Does my first paragraph attract favorable attention by indicating the position, my source of learning about the position, subtle company praise, and reference to my resume?

❏ **3.** Do my body paragraphs summarize and emphasize my specific personal and professional qualifications?

❏ **4.** Does my closing urge action on the part of the reader?

❏ **5.** Have I reviewed my letter for an enthusiastic tone and projection of self-confidence?

INTERVIEW

❏ **1.** Is there a specific time and place scheduled; do I know exactly how to get there and how much time is required?

❏ **2.** Do I have the suitable clothing for a professional interview?

❏ **3.** Have I reviewed what I know about the company so that I can engage in intelligent discussion?

❏ **4.** Am I familiar with my resume and my strengths and weaknesses?

❏ **5.** Do I have a resume on hand as well as my personal portfolio?

❏ **6.** Am I prepared to ask significant questions?

❏ **7.** Am I prepared to be enthusiastic, mannerly, courteous, and calm?

❏ **8.** Do I know how to behave at the end of an interview?

EXERCISES

1. **Prospective Positions.**   Compile an employment opportunities file for yourself. Include photocopies from the DOT, OOH, and other position requirements from journals, government publications, newsletters, Internet sites, civil service offices, help-wanted columns, Yellow Pages, trade magazines and newspapers, your college placement office, interview notes, and employment agency telephone or in-person interviews. Submit this file to your professor for analysis.

2. **Employment and Educational History.**   Compile a file of your employment and educational background including all of the institutional or company names, complete addresses, degrees and achievements, position titles, responsibilities, extracurricular activities, honors and awards, GPAs, and three to five greatest character and skill strengths. Include military records, community organization data, and other pertinent records. Be prepared to submit this file in a folder to your professor for analysis.

3. **Presentation Portfolio.**   Place in a binder or album your outstanding class or on-the-job writing projects plus any brochures, newsletters, feature stories, proposals, and special reports. Also include professional items, such as letterheads, fax cover sheets, business cards, banners, or other items that you have designed. Submit the portfolio to your professor for a personal critique.

4. **References.**   Obtain permission and actual letters of reference from outstanding past employers, professors, and other professionals to submit to your teacher for analysis.

5. **Development of an Employment Section of a Resume.** Using the following information, develop an employment section of a chronological resume. Use headings.

    > The applicant is seeking a position as an electronics technician with the opportunity for advancement. His first job was as an auto mechanic with Bird Ford Agency, 101 North Garfield Avenue, Davenport, IA, where he serviced cars and repaired radios for two years, January to January. (Add zip codes to addresses. Provide months and years.) Next he worked from February 199X to May 199X as an automotive electronics supervisor at Borg Corporation, 6025 South 50 Street, Philadelphia, PA. Third, the applicant is currently employed as a service manager for Ace Auto Company at 1102 Broad Way, Delray Beach, Florida, where he began work in June of 199X.

6. **Cover Letter.**   Revise the following cover letter content to make it more assertive, specific, and persuasive. The applicant is seeking a position as a draftsperson with a large architectural firm.

    > Pursuant to your recent ad, I am applying for a job with your company. I don't have much experience but am willing to learn. I will be graduated from college this June, and although I changed my major three times, I will earn a degree in drafting technology.
    >
    > I have studied some very pertinent courses and held a number of part-time jobs with local architects, learning about materials, office procedures, and on-site supervision. I think I have a pretty good design ability and am a careful draftsperson.
    >
    > I realize you want a more experienced person, but I would like to talk to you about myself.

7. Using the chart on keyword action verbs, nouns, and adjectives, compile a list of keywords that reflect your personal accomplishments and skills to include in all resumes.

WRITING PROJECTS

1. **Personal Traditional Resume—Chronological or Functional.** Write either a chronological or functional resume—whichever would be most suitable for you if you were looking for a particular position that you could fill immediately while continuing your college education. If you have solid employment experience in the field in which you seek employment, the traditional chronological format will probably be your choice. If you are short on experience but long on the skills that this job will require, select the functional. If you are devising your resume in response to a real job listing, copy the listing and attach it to your resume.

2. **Collaborative Project—Narrative or Online Resumes.**   Find three or four classmates who are pursuing the same field of study as you are. Assume that all of you all have extensive experience in your field. Review the duties of a collaborative team, as discussed in Chapter 2.

   Step 1.   Parcel out the following tasks: one member of the team should obtain copies of the DOT and OOH job descriptions and any other library materials that define the training requirements, job outlook, and salaries in your field. Another team member should obtain job postings from the Internet, local or out-of-town classified ads, and trade magazines or newspapers. The third member should check the college Career Center and call one or two employment agencies to obtain additional position postings.

   Step 2.   Collaboratively, devise realistic details for a narrative resume and design it carefully. If you are responding to an actual position, attach it to the resume.

   Step 3.   Using that same information, convert the data into an online resume suitable for posting on the Internet at a specific job bank or your own Web page.

3. **Collaborative Project—Cover Letter.**   Each member of your team is to use his or her traditional or functional resume format and write a cover letter to a specific person at a specific company, agency, or institution. Be brief, persuasive, and self-confident. Exchange letters with each member of the group and critique each other's work. Be specific about possible additions, deletions, rewording, and other possible improvements. Peer evaluation is one of the most valuable tools there is in writing an effective cover letter.

VERBAL PROJECTS

1. **Group Project—Interviews.**   Divide the class into two groups. Each member of each group will be prepared to submit his or her resume and cover letter to a member of the other group for a mock interview. Allow the interviewer to review the resume and cover letter. Arrange the desk and chairs so that class members can see clearly the interviewer and the interviewee. Conduct the mock interview. Then change groups and have the interviewers be the interviewees. After each interview, critique the work. Remember, these will not be perfect but an exercise toward honing your skills.

2. **Mock Interviews.**   Invite a professional interviewer to your classroom or to your television recording studio to conduct mock interviews. If the studio is not available, arrange for someone to videotape the proceedings in the classroom. Usually, someone from

the Human Resources (Personnel) department of your college or someone from your Career Center, or any other experienced interviewer (a member of a faculty selection committee, perhaps) will be willing to give the time for the interviews. Have your professor present three or four resume/cover letter combinations to this volunteer so that he or she may become familiar with the material. Following the interviews, play back the videotape of each mock interview in order to critique the pros and cons of the interviewee. Be honest but tactful in your criticism.

NOTES

CHAPTER 14

Writing Brief Reports

S K I L L S

After studying this chapter, you should be able to

1. Appreciate the diversity and purpose of a wide variety of standard brief reports.

2. Explain the differences between analytical, evaluation, and recommendation reports.

3. List and define five common types of analytical reports.

4. List and define three common types of evaluation reports.

5. Incorporate the appropriate language, organization, and content into brief reports.

6. Devise appropriate formats including headings, visuals and graphics, and emphatic features for clarifying the content of brief reports.

7. Write a periodic or progress report.

8. Write a laboratory or test report.

9. Write an incident report.

10. Write a field report.

11. Write a feasibility report.

INTRODUCTION

Engineers, scientists, technicians—in short, all employees in an organization—must assume the everyday task of informal and formal report writing. Reports are the written record of the events and progress of work. They assist professionals and management in making decisions, or they may become the working documents to help all employees carry out a program.

Informal reports are usually short, tightly focused on one subject, and addressed to a single receiver or small audience. Formal reports, on the other hand, cover more complicated topics; contain title pages, tables of contents, extensive body content, appendixes, and the like; and are frequently addressed to overlapping audiences. This chapter addresses the briefer types of typical reports, and the following chapter covers the longer, formal report particulars.

RECEIVER/AUDIENCE

Professional and technical reports are characterized as intensely factual and objective and are written in an authoritative and serious tone. You should avoid opinionated and judgmental language even though your goal may be to persuade your audience to certain conclusions or actions. Consider this biased paragraph from an accident report:

> A minor accident occurred in the Data Storage Department recently. One of the departmental technicians carelessly overturned a cup of coffee on a microfilm machine, which resulted in a massive short circuit when the machine was turned on. Luckily, there were no personal injuries.

The emphasis here is on blame rather than on the facts of the incident itself. An improved version would emphasize the facts in a detailed and objective account:

> An explosion occurred on the second floor, east wing of Building C, Avtec Corporation, on 13 June 199X, at 3:42 P.M.
>
> The cause of the explosion was an overturned cup of coffee spilled by Technician Seth Deutsch onto an Acme 710 microfilm projector (Inventory #703-11201-8). The liquid penetrated the felt light seals and dripped onto the main power supply, resulting in an explosive short circuit when the machine was turned on by Technician Brenda Bailey.
>
> Although there were no personal injuries, damages to the machine and surrounding area total $1,480.00.

The revised version contains essential facts and presents them in an unbiased manner.

FORMAT/GRAPHICS

Brief reports are often presented on preprinted forms, in a memo, or in a letter format. If you are devising your own format, use emphatic features such as topical headings, capital letters, underlining, boldface print, variable spacing, numbers, and bullets to enhance and emphasize the data of even the briefest report.

Consider visuals and graphics to organize your data and to convey a clear message. Study your message to determine if informal tables, formal tables, pie diagrams, maps, charts, and other graphics will present your information more clearly than lengthy paragraphs of dry material.

An incident report detailing a work-related accident might include a brief location diagram and an informal table of repair or replacement costs. A progress report may call for a formal table of materials, costs, or scheduling dates. A lab or test report will best display testing results in tables or performance curves. A feasibility report may include the results of surveys presented in circle graphs, bar charts, or tables. In addition, tables comparing features of optional equipment may be appropriate. Be alert to graphic possibilities that quickly convey vital information and are appropriate to technical writing.

TYPES OF REPORTS

You may classify typical reports in several manners. One approach is to group them as analytical, evaluation, or recommendation reports. The first two types are generally brief, while recommendation reports may entail much more detail, persuasive tactics, and supporting evidence.

ANALYTICAL REPORTS

Analytical reports, in contrast to evaluation or recommendation reports, emphasize findings without expressing strong opinion. They are confined to facts and a few logical inferences. Further, they are objective and informative rather than persuasive.

Here is a list of typical analytical reports required in business or industry:

- **Work logs:** Control cards or sheets indicating work assignments, locations, and work performed accounts
- **Expense reports:** An itemized accounting of all expenses incurred while performing duties, such as promoting sales, attending conferences, or inspecting field progress
- **Request for leave reports:** Requests for vacation, work-related travel, sick leave, or other absences
- **Periodic or progress reports:** The written information on the status of a project
- **Laboratory and test reports:** The written results of laboratory or testing experimentation

Work Logs/Expense Reports/Leave Requests

Many companies preprint forms for the types of reports that are used most often, such as work logs, travel expense reports, and leave requests. Figures 14.1 through 14.4 show four typical preprinted report forms. Determine which forms your company uses and maintain the required reports as directed.

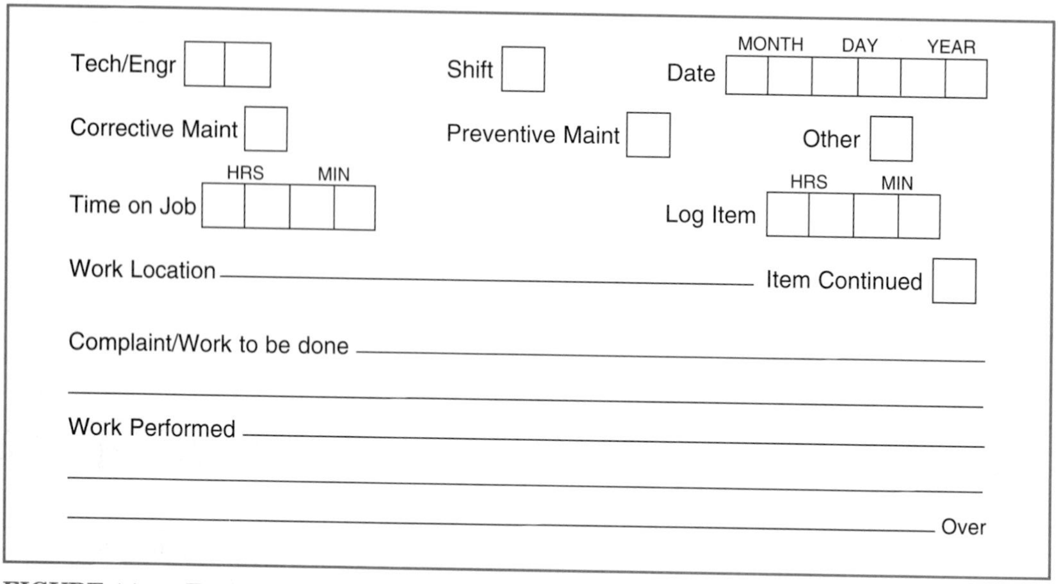
Tech/Engr
Shift
Date MONTH DAY YEAR
Corrective Maint
Preventive Maint
Other
Time on Job HRS MIN
Log Item HRS MIN
Work Location
Item Continued
Complaint/Work to be done
Work Performed
Over

FIGURE 14.1 ***Typical preprinted work log report***

Periodic Progress Reports

PURPOSE

A periodic or progress report provides a record of the project status over a specific period of time. The report or reports are issued at regular intervals throughout the life of a project to allow management and workers to keep projects running smoothly. A periodic or progress report states what has been done, what is being done, and what remains to be done. The report usually reviews the expenditures of time, money, and materials; therefore, decisions can be made to adjust schedules, allocate budget, or schedule supplies and equipment. Often a company employs preprinted forms to report routine progress. Employees simply fill in the blanks at the completion of tasks.

Similar periodic reports may be required daily, weekly, quarterly, semiannually, or annually. They ensure uniformity and completeness of data.

ORGANIZATION

All narrative reports in a series of progress reports should be uniform in organization and format. A progress report covers

TRAVEL EXPENSE REPORT

RYDER TRUCK RENTAL, INC.

NAME _____

TITLE _____

WEEK ENDING _____

VENDOR NUMBER	REFERENCE NUMBER	LOC CODE	MO REC.

VEHICLE NUMBER	ACCOUNT NUMBER	AMOUNT
W/E DATE	TOTAL AMOUNT	

AIR TRAVEL CHARGED TO RYDER
(FOR H.Q. AND REGION MANAGER USE ONLY)

TICKET NUMBER (LAST 3 DIGITS)	TICKET DATE	TICKET AMOUNT
		$

BUSINESS PURPOSE OF EACH TRIP

DATE OF		EXPLANATION (IF NOT CHECKED BELOW)	
DEPART.	RETURN		
			☐ VISIT TO COMPANY LOCATION ☐ SALES CALL
			☐ VISIT TO COMPANY LOCATION ☐ SALES CALL
			☐ VISIT TO COMPANY LOCATION ☐ SALES CALL

	DATE	SUN	MON	TUES	WED	THURS	FRI	SAT	TOTAL
DAILY ITINERARY	FROM								
	TO								
	TO								
AIR TRAVEL PAID BY EMPLOYEE									
CAR RENTAL*									
PERSONAL CAR EXPENSE (DETAIL ON REVERSE SIDE)									
ROOM*									
MEALS PLUS TIPS**									
COMPANY CAR EXPENSE (DETAIL ON REVERSE SIDE)*									
MISCELLANEOUS (DETAIL ON REVERSE SIDE)*									
ENTERTAINMENT (DETAIL ON REVERSE SIDE)*									
TOTAL									

ACCOUNTING FOR ADVANCES

* ATTACH RECEIPTS
** INCLUDE ENTERTAINMENT MEALS ON REVERSE

IN ADDITION TO THE REQUIRED RECEIPTS AS SPECIFIED ABOVE, RECEIPTS FOR EACH EXPENDITURE OF $25.00 OR MORE MUST BE ATTACHED.

ADVANCE RECEIVED		
EXPENSES THIS VOUCHER		
BALANCE DUE COMPANY		
OR BALANCE DUE TRAVELER		

TRAVELER'S SIGNATURE

APPROVED BY (SIGNATURE)

6-22 (11/80)   SIDE ONE   10395   Litho By 1R In U.S.A.

FIGURE 14.2 *Typical preprinted travel expense report* (Courtesy of Ryder Truck Rental, Inc.)

PERSONAL CAR EXPENSE									
DATE	SUN	MON	TUES	WED	THURS	FRI	SAT	TOTAL	
ODOMETER READING-ENDING									
ODOMETER READING-BEGINNING									
MILEAGE									
LESS PERSONAL MILEAGE									
NET COMPANY MILEAGE									

AMOUNT AT _____ ¢ PER MILE

DATE	AMOUNT	VEHICLE NO.	COMPANY CAR EXPENSE (DETAIL BELOW)

DATE	AMOUNT	MISCELLANEOUS (DETAIL BELOW)

ENTERTAINMENT EXPENSE					
DATE	AMOUNT	TYPE OF ENTERTAIN-MENT	PLACE NAME, ADDRESS, OR LOCATION	BUSINESS RELATIONSHIP OF INDIVIDUALS OR GROUP ENTERTAINED (Give Name, Title, Etc., Include Names of Co. Employees.)	BUSINESS PURPOSE Date, Duration, Place, and Nature of Association Business Discussion or how otherwise related to active conduct of the business.

FIGURE 14.2 *continued*

REQUEST FOR LEAVE OF ABSENCE*
FIRE DEPARTMENT

NAME _____
　　　　　　　　　　　　　Last　　　　　　First

DATE _____

*1　Leave requests for vacation and anticipated leaves must be submitted two weeks prior to requested starting date.

*2　Leave requests for sick leave and emergency leave must be submitted by 9:00 A.M. the day before your shift works.

_____
　　　　　　　　　　　　Signature

LEAVE CODES	
V	—Vacation
S	—Sick with pay
FS	—Family Sick with pay
I	—Job Injury with pay
Z	—Sick without pay
W	—Personal Absence w/o pay
A	—Absence w/o leave
PG	—Maternity Leave w/o pay
FL	—Funeral Leave with pay (next of kin)
M	—Military Leave (calendar days)
CL	—Conference Leave
JD	—Jury Duty
SV	—Vacation from Sick
MV	—Management Vacation
CT	—Comp Time

Department _____

_____ Days _____ from _____ to
　　　　Code

_____ (inclusive dates).

_____ Days pay in advance requested on

_____ (last shift worked).
　　　　　　　Date

Regular pay checks of: _____

_____

Explanation _____

_____

_____

STATION OFFICER'S SIGNATURE NECESSARY FOR SICK LEAVE

DISAPPROVED
APPROVED　———————————————
　　　　　　　Station Officer

DISAPPROVED　　　　　　　　DISAPPROVED
APPROVED _____　　APPROVED _____
　　　　Commander　　　　　　　　Chief Officer

FORM AA-108 Rev. 4/82

FIGURE 14.3 *Typical preprinted leave request form* (Courtesy of Fort Lauderdale, Florida, Fire Department)

APPAR. NO. ___ RADIO NO. ___ MO. ___ YEAR ___ CO. ___ MAKE ___

WEEKLY APPARATUS REPORT

	1st Monday	2nd Monday	3rd Monday	4th Monday	5th Monday
Check pumps—Record vacuum test					
Check pumps—Record pressure test					
Were all drains flushed?					
Did connections or packing glands leak?					
Does relief valve or governor operate satisfactorily?					
Change Hurst Tool Fuel					
Check and flush foam pick-up					

	1st Tuesday	2nd Tuesday	3rd Tuesday	4th Tuesday	5th Tuesday
Resuscitator—Blood pressure of lowest cylinder					
Air Chisel—Record pressure of lowest oxygen cylinder					
Demand Regulator Mask—Record pressure					
Portable Spotlight—Running test					
Wench & Cord—Operating check					
First Aid Kits—Inventory check					
Hurst Tool & Compressor—Running test					
Tires—Record pressure					

Form AA-122 Rev. 8/77

FIGURE 14.4 *Sample periodic report* (Courtesy of Fort Lauderdale, Florida, Fire Department)

- A review of the aims of the project, highlighting accomplishments or problems
- A summary or explanation of the work completed
- A summary or explanation of the work in progress
- A summary or explanation of future work
- An assessment of the progress

The accounting of work in progress may be chronological, topological, or even a combination of the two. A chronological organization covers the work done by time (first A was done, then B, then C; work in progress includes first A, then B, then C; and remaining work includes first A, then B, then C). A topological organization covers the work done by task (Task 1, Task 2, Task 3, etc.). A combined organization covers the tasks within each time period (first time period: Tasks 1, 2, and 3; second time period: Tasks 1, 2, and 3, and so on).

Headings for the sections, however brief, are usually used. Consider these topical headings:

Introduction	Overall Goal
Work Completed	Work Completed
Work in Progress *or*	Expenses Incurred
Work Remaining	Present Work
Appraisal	Future Work
	Conclusions

The major factor of a progress report is time. Other features, such as costs, materials, and personnel, may be incorporated into the appropriate sections or attached as support materials. If the reports are being prepared frequently during the course of a project, they are characterized by brevity—phrases rather than sentences and paragraphs. Figures 14.5 and 14.6 show two progress reports.

Laboratory and Test Reports

PURPOSE

Students and employees in the fields of chemistry, data processing, fire science, electronics, nursing, and other allied health areas must frequently write laboratory or test reports. The reports present the results of research or testing.

MEMORANDUM

TO: Susan Niles, Law Department

FROM: Catherine Robertson, Executive Secretary *C.R.*

DATE: 12 May 200X

SUBJECT: Ace Company, Inc. Acquisition—Progress Report

INTRODUCTION

Here is the requested progress report on the Ace Company, Inc. acquisition.

WORK COMPLETED

- The stock purchase and Noncompete Agreements have been typed in final draft.
- Copies of the above have been sent to the Seller and acknowledged as received.
- Hotel and plane reservations have been confirmed by Downtown Travel Centre for arrival in Detroit on 25 June at 3:30 P.M. and a suite at the Downtown Hilton.

WORK IN PROGRESS

- Limousine has been requested from Detroit Limousine Service for transfer of all parties to the Hilton Hotel; awaiting confirmation.

WORK REMAINING

- Meeting with the Seller and John Galt Associates (law firm) to take place 26 June.
- Stock Purchase and Noncompete Agreements to be executed by all parties.
- Upon return, file on acquisition to be completed, and all documents to be filed as previously discussed.

APPRAISAL

Although this acquisition was originally scheduled for a 1 May closing, all work is up to date, all parties have been notified of the delays, and all parties are scheduled for the 25–26 June closing. With the final execution of all documents, this acquisition will be concluded.

CR:js

FIGURE 14.5   *Sample progress report in memo format*

Guardian-American
1901 Southeast 16th Terrace
Hollywood, Florida 33318
August 1, 200X

Ms. Marilyn Thompkins
1699 South Ocean Drive
Jacksonville, FL 30902

Dear Ms. Thompkins:

As you requested, the following is a progress report on the Hidden Court Townhomes, a
project of Guardian-American Company, L.L.C., in Hollywood, Florida.

WORK COMPLETED

We have completed the steps and obtained the permits, and we are preparing the site.
- All engineering drawings and architectural drawings are completed and submitted
  to the City of Hollywood for permits, which have been granted.
- Permits obtained for a site sales trailer.
- Completed design of a $4' \times 6'$, two-sided roadside sign, which is being fabricated.
- Established a site phone number (954-961-1919) to be used for sales efforts.
- Appointed John Ligori as primary phone operator to control core buyer
  demographics at 2% sales commission.
- Obtained City of Hollywood approval to remove southwest and southeast parking
  lots to define our lot boundaries more effectively and to provide more visual
  evidence of project progress.

EXPENSES INCURRED

Impact fees for the permits have been paid to the City of Hollywood. They include

Description	Total	Per Unit	Payment
Educational Impact Fees			
Prorated portion payable at issuance of building permit	$68,796	$1,323	$68,796
Recreational Impact Fees			
Prorated portion payable at Issuance of building permit	$18,200	$350	$18,200
Transit Impact Fees			
Prorated portion payable at issuance of building permit	$12,584	$242	$12,584

FIGURE 14.6   *Sample progress report with incurred expenses*   (Courtesy of Robert W. Johns,
Guardian-American)

City of Hollywood Park
Impact Fees
Payable in four payments
Starts when 1st building
permit is pulled $21,537 $5,384

                          Total $121,117 $104,964

WORK IN PROGRESS

Site preparation, sales campaign activities, and financing qualifying are ongoing.
- Roadside sign fabrication to be complete August 10.
- Fabrication, printing, and distribution of 1000 5.5″ × 8.5″, double-sided postcards for advertising.
- Designing sales campaign, including newspaper ads, radio, and television, to be "unveiled" in September.
- Qualifying the site for FHA financing.

WORK REMAINING

Final site preparation, construction, and sales campaign to be done.
- Erect roadside sign by August 15.
- Position sales trailer by August 20.
- Begin construction on September 17.
- Launch sales campaign on September 20.

APPRAISAL

Progress meets our projected time frame. All of the development monies are in hand. Attached is a sales promotion postcard for your information. The next progress report for investors will be issued September 20, 2000.

Very truly yours,

Robert Taylor

Robert Taylor, President

FIGURE 14.6   *continued*

ORGANIZATION

Typically, the report includes

- Statement of purpose
- Review of method or procedure of testing
- Results
- Conclusions and recommendations

Purpose Statement.　Here you clarify what is being tested and for what purpose (durability, colorfastness, safety factors, customer preference, and so forth).

Method/Procedure.　Explain the particulars of your testing method.

Results.　Present the test results. Use graphics for quick comprehension or comparison to preset standards.

Conclusions.　Explain the implications of the test results and make recommendations.

A test report is usually less formal than a true laboratory report. The test report may be transmitted in a memorandum or business letter. Figure 14.7 shows a test report from an avionics engineering firm to determine the efficiency and cost effectiveness of a dimmer design for a light behind a liquid crystal color display on an airplane flight instrument panel. Although a layperson could understand it, the test report is written to an informed audience and includes many technical terms. It is clear that the *purpose* of the tests was to determine the better of two designs by comparison testing. The *procedures* used to conduct the tests are explained only to the extent that the reader needs to know. The *results* of the tests are reviewed and presented in text and graphics for clear comprehension, and the *conclusion* is presented in the final paragraph of the report.

Laboratory reports generally cover the same four considerations but are more formally structured. Figure 14.8 shows a report on the preparation of aspirin. It includes flowcharts of the main chemical reactions, potential side reaction, and the separation scheme. The procedure section of the report details the method used to prepare aspirin with attention to the equipment used. The conclusions are covered under separate headings.

EVALUATION REPORTS

Evaluation reports are investigative in nature and emphasize strictly logical conclusions. They include an overview of a problem, an evaluation of findings, and a few conclusive evaluations or reasonable recommendations tightly based on the evidence. Evaluation reports may be spatial, chronological, or topical in organization. Some typical evaluation reports include:

- **Incident reports:** The written record of unforeseen occurrences, such as accidents, machine breakdowns, delivery delays, cost overruns, production slowdowns, or personnel problems
- **Field reports:** The written evaluation of data to determine appropriate action, such as estimating real estate value, determining services costs, or establishing claims for damage

Memorandum

Alltec Aerospace Company

Aircraft Avionics Division
3100 S.W. 22nd Avenue
P.O. Box 4582
Melbourne, CA 35821
Telephone (217) 338-5900

Allied
Signal

DATE: 9 June 200X

TO: Distribution

FROM: Robert Johns *R.J.*

SUBJECT: Evaluation of GAAD Dimmer Driven Circuitry

A parallel effort to devise a cost-effective and efficient dimmer design was undertaken by GAAD. The data which they presented to us was not relevant to the Boeing specifications. Moreover, the data could not be directly compared with previous experimental data relating to our design. Consequently, an experiment was designed to provide a direct comparison of the two designs so that we could determine the better design for our needs.

The following aspects of the design were compared:

- Efficiency
- Circuit performance
  —Operation over temperature
  —Dimming ratio
  —Light level controllability
- Lamp life
- Reliability of the circuit itself
  —Breakdown mechanisms
  —Part count
- Manufacturing costs
  —Repeatability
  —Part cost

Efficiency

Initially, it appeared that the efficiency of the resonant circuit was considerably less than the efficiency of the flyback design. This was due to the difficulty of measuring the voltage and current waveforms at the lamp itself. The determining factor, however, is the amount of light out for a given value of power in. The power dissipated in the circuit is not that important since most of the power in the lamp is reflected back into the LRU anyway. Figure 1 shows the light out versus power in for each circuit.

FIGURE 14.7   *Sample test report*   (Courtesy of Robert W. Johns, Guardian-American)

Circuit Performance

Figure 2 shows the power needed to light the lamp at full power over a wide temperature range. Clearly, the resonant circuit produces the light more efficiently over most of the temperature range.

A comparison of the dimming ratio between the two circuits is illustrated in Figure 3. It should be noted that the quality of the light output associated with the flyback design was superior to that of the resonant design. That is, the resonant design shows some perturbations (barber-polling) at low light levels. Some of these perturbations may be tolerable in that they cannot be noticed after the light travels through the LCD. Both circuits require some modification of the reflector backplate to even out the capacitance between the lamp and the backplate. This is required to maximize the dimming ratio of the lamp from a 100 foot-lambert reference. Figure 3 indicates that the resonant supply is superior in dimming capability. However, this data is somewhat misleading because the flyback design has a discrete dimming control mechanism. (At low duty cycles, the duty cycle control does not have the resolution necessary.)

Lamp Life

It is our opinion, based on experimental data, that the lamp life/reliability is clearly a function of the peak to R.M.S. current through the lamp. Figure 4 presents the current peak to R.M.S. value for each light level at room temperature. In an effort to reduce the ratio, the resonant circuit has a dual method of dimming the lamp controls. The bursting is common to both circuits, but the resonant method operates at 100 percent duty cycle over much of its high brightness range. The light is controlled by varying the high voltage DC into the bridge circuit. This results in a current ratio near unity. At the lower duty cycles, it should be noted that the peak currents in resonant circuit are only one cycle in duration and much smaller in absolute magnitude (13 to 15 mA peak), whereas in the flyback design the high peak currents (200 + mA peak) last the entire duration of each burst on time. It is believed that this is more detrimental to the cathodes.

Circuit Reliability

No real time tests using the resonant circuit have been started. However, nine lamps are undergoing life test using the flyback design. Of these nine, one failed completely in less than 120 hours, and a second lamp is indicating deterioration of the cathodes.

FIGURE 14.7   *continued*

The greatest virtue of the flyback design is its simplicity with a total of fifteen parts. Although it presently does not meet the dimming requirements (poor resolution in dimming control), it could probably be modified to do so with relatively little added cost.

Manufacturing Costs

The primary cost in the circuit is the magnetics associated with the flyback transformer. I speculate that the cost of the parts for the flyback circuit will be about $60.00 per circuit. However, the circuit requires some improvements in the design. That is, it is very sensitive to any fluctuations in the 16-volt power input: the transformer fails catastrophically. Also, there is no protection circuitry in the event that the lamp fails.

The resonant circuit is more complex with approximately 100 parts costing about $110.00 per unit. The corresponding reliability is, therefore, reduced. However, the reliability of the lamp is the critical factor. Consequently, the increased complexity of the circuit was a tradeoff for increased lamp life. The circuit includes protection circuitry for all modes of lamp failures. Planer magnetics were used to lower costs through increased reliability and repeatability. Additionally, a phase locked loop was incorporated to compensate for component tolerances—both part to part and over time. The circuit requires no tuning or setting during manufacturing.

Conclusion

The overall conclusion regarding the preferred circuit is very much dependent on the final application of the unit being designed. In most applications, however, the life of the lamp is critical to the overall MTBF of the equipment. I, therefore, recommend that the resonant driver approach be used. While the cost and simplicity of the flyback design are beneficial, those factors are outweighed by the expected disadvantage of the limited lamp life.

Robert Johns

RJ/rj

cc: W. Antle, O. Tezucar, V. Frazier, G. Catsimpiris, L. Evans, S. Hammack, T. Mickie—GAAD, W. Tombs—BGCS

FIGURE 14.7 *continued*

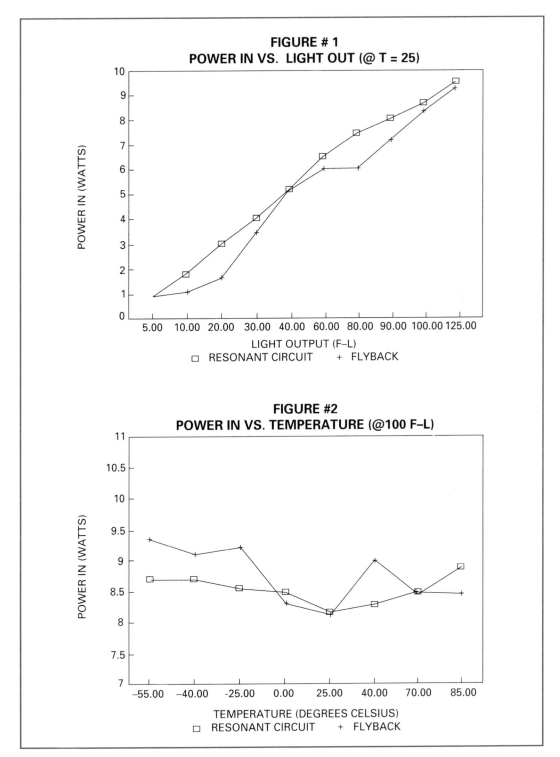

FIGURE 14.7 *continued*

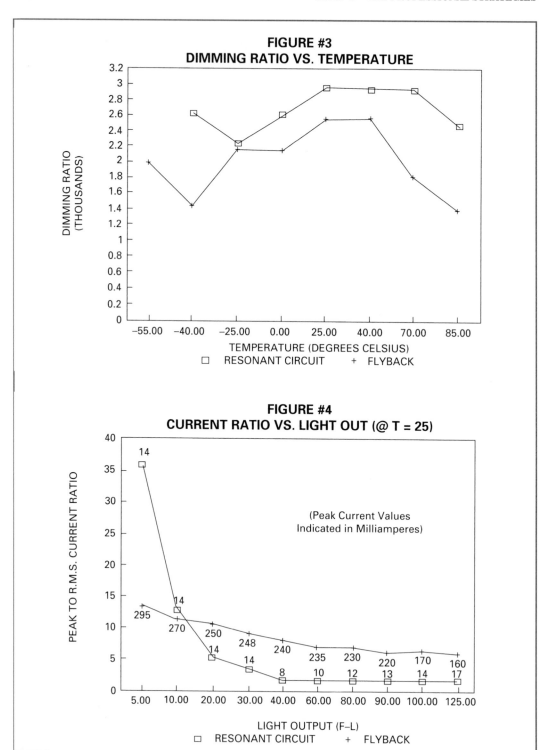

FIGURE 14.7   *continued*

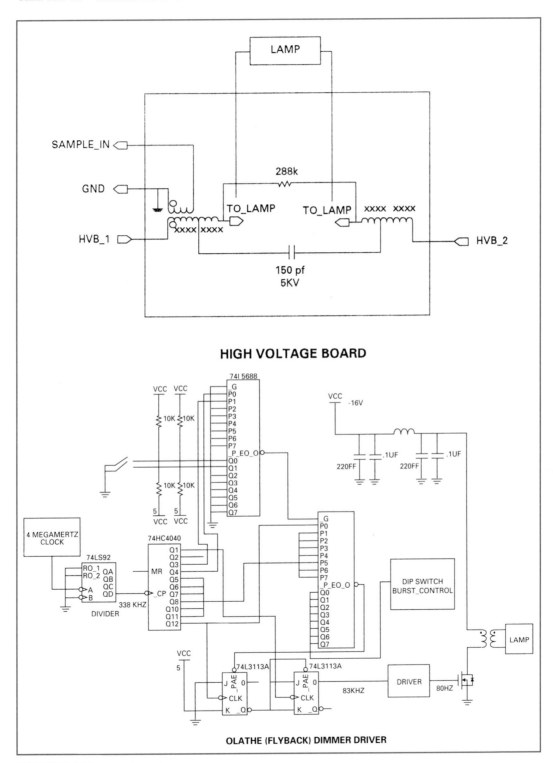

FIGURE 14.7 *continued*

THE PREPARATION OF ASPIRIN

Main Reaction

$$\text{(salicylic acid)} - \text{OH} + CH_3C-O-C-CH_3 \xrightarrow{H^+} \text{(acetylsalicylic acid)} -OH + CH_3C-OH$$

From Literature:

Salicylic Acid	M.P.‖157@159®
Aspirin (Acetylsalicylic Acid)	M.P.‖135@136®
Aspirin‖$C_9 H_8 O_4$	M.W.‖180

Calculations

$C_7H_6O_3$ M.W. 138

$$\left(\frac{1 \text{ mole Salicylic Acid}}{138\text{g. Salicylic Acid}}\right) = \frac{.0014 \text{ Moles}}{\text{Salicylic Acid}}$$

(2og. Salicylic Acid)

$(CH_3CO)_2O$ $C+H_6O_3$ MW 102 Density = 1.08 g./ML.

The limiting reagant is salicylic acid.

Side Reaction

Separation Scheme

COOH / OCOCH₃	Crystallize in H_2O	COOH / OCOCH₃	NaHCO₃
COOH / OH		COOH / OH (Some)	Insol. Polymer
$(CH_3CO)_2O$ H_2SO_4 H_2O CH_3COOH Polymer	SOL'N	Polmer	

COOH / OH

$(CH_3CO)_2O$
H_2SO_4
H_2O
CH_3COOH

COO Na / OCOCH₃

COO NA / OH (Some)

NAHCO₃ H_2O

CO_2

HCl

Recrystallize from Ethyl. Acetate

COOH / OCOCH₃ Pure

SOL'N

COOH / OCOCH₃

COOH / OH (Trace)

H_2O (Trace)

COOH / OH

H_2O

SOL'N
NaCl
H_2O

COOH / OH

FIGURE 14.8 *Sample laboratory report*   (Figures and excerpt adapted from *Experiments in General Chemistry* by Kenneth W. Whitten and Kenneth D. Gailey. Copyright © 1981 by Saunders College Publishing. Reproduced by permission of the publisher.)

Procedure

Salicylic acid: Sample + Paper 3.12g.   2.00g. = 0.014 moles

Paper 1.12g.   138g./mole

Sample 2.00g.   Salicylic acid

The salicylic acid was placed in a 125 ml Erlenmeyer flask. Acetic anhydride (5 ml) was added along with 5 drops of conc. H_2SO_4. The flask was swirled until the salicylic acid was dissolved. The solution was heated on the steambath for 10 minutes. The flask was allowed to cool to room temperature, and some crystals appeared. Water (50 ml) was added, and the mixture was cooled in an ice bath. The crystals were collected by suction filtration, rinsed three times with cold H_2O, and dried.

Crude yield: Product + paper 3.07g.

Paper 1.15g.

1.92g. aspirin

Theoretical yield = (0.014 moles) (180g. aspirin/mole) = 2.52g.

Actual yield   = 1.92g.

Percentage yield = 76%

The crude product gave a faint color with $FeCl_3$. Phenol and salicylic acid gave strong positive tests.

The crude product was placed in a 150 ml beaker, and 25 ml of saturated NaHCO was added. When the reaction had ceased, the solution was filtered by suction. The beaker and funnel were washed with CA. Two ml of H_2O dilute HCl was prepared by mixing 3.5 ml conc. HCl and 10 ml H_2O in a 150 ml beaker. The filtrate was poured into the dilute acid, and a precipitate formed immediately. The mixture was cooled in an ice bath. The solid was collected by suction filtration, washed three times with cold H_2O, and placed on a watch glass to dry overnight.

Yield: Paper and product: 2.78g.

paper: 1.08g.

product: 1.70g.

Theoretical yield = 2.52g.

Actual yield = 1.70g.

Percentage yield = 67% mp. 133-135°C.

The solid did not give a positive $FeCl_3$ test. The final product (CA. 075g.) was dissolved in a minimum amount of hot ethyl acetate. Crystals appeared. The crystals were collected by suction and dried.

MP. 135-136°C

FIGURE 14.8 *continued*

- **Feasibility reports:** The written evaluation of data to determine the practicality of future products, land development and environmental impact, expansion programs, new equipment, or services

Incident Reports

PURPOSE

No matter where you work, the unexpected frequently occurs. Such digression from normal operating procedure generally requires an incident report to supervisors or others to prevent the incident from recurring. The incident report is a written investigation of accidents, machine breakdowns, delivery delays, cost overruns, production slowdowns, or personnel problems.

The incident report may be reviewed when the next budget is planned if your evaluations involve finances. The report may constitute the basis of a longer proposal to improve procedures. It may even be used as legal evidence in a follow-up investigation. A carefully detailed report becomes part of the written record of what goes on in your place of work.

ORGANIZATION

The incident report adheres to fairly conventional organization. Its parts cover

- What happened (factual, not opinionated)
- What caused it (detailed and chronological)
- What were the results (injuries, losses, delays, costs)
- What can be done to prevent recurrences (evaluations)

In a lengthy incident report, it is a good idea to include topical headings, such as

Incident		Accident Description
Cause		Analysis of Causes
Results	*or*	Corrective Action
Evaluations		Evaluations

Your reader will be able to cull the appropriate information quickly by glancing at the headings.

Incident Description. In your introductory material, write a concrete statement detailing what happened. Include the exact date, time, and location. Personnel details, such as employees' names, titles, and departments, should be included. If personal injuries occurred, include the name(s) of victim(s), titles, and departments, or in the case of victims who are not employees, include home addresses, phone numbers, and places of employment. Describe the actual injury. If equipment is involved, identify it by including brand names, serial numbers, inventory numbers, or other pertinent descriptive detail.

Analysis of Causes. In this section write a chronological review of what caused the incident. Include what was happening prior to the incident and each step that caused the incident.

Results. In this section, explain what happened as a result of the incident, such as the action that was taken immediately. If anyone was injured, describe the extent of the injury and how, when, and where the person was treated. Explain who was immediately involved. This may include paramedics, police, repair experts, or extra workers. In the case of equipment failure, explain how it was required or replaced, how late deliveries were speeded, or how high costs were curtailed. Detail what was done to settle a personnel problem or to satisfy a customer demand.

The results section may require an actual or estimated expense review. Include a precise breakdown of medical expenses, equipment replacement, repair costs, profit loss, or other applicable costs.

Evaluations. This section should include concrete suggestions to prevent the incident from recurring. Consider what should be done, who should do it, and when it should be done. Include as much detail as your position authorizes.

Figure 14.9 shows a brief accident report with headings, a visual, and an easy-to-read continuation table.

Field Reports

PURPOSE

A field report presents an analysis of a location, site, or situation to record and determine appropriate action. Realtors prepare field reports on undeveloped, commercial, and industrial properties to determine the value and prospects of such properties. Service people from every field inspect property to determine costs and plans for building on, improving, or repairing the property. Firefighters, health inspectors, and others report their work in the field. Insurance adjustors inspect sites to establish claims for damages.

MEMORANDUM

TO: James Friedman, Supervisor
FROM: Cathy Victor, Night Manager *CV*
DATE: 6 October 200X
SUBJECT: Busboy Fall Near Kitchen Entrance, 6/10/0X

INCIDENT

On Monday, 6 October 200X, at 2:15 P.M. Mike Sullivan, busboy, fell at the kitchen entrance located in the rear of the dining room on floor 20. Mr. Sullivan was not injured, but there was considerable breakage of dishes and glassware. Figure 1 shows the exact location of the fall.

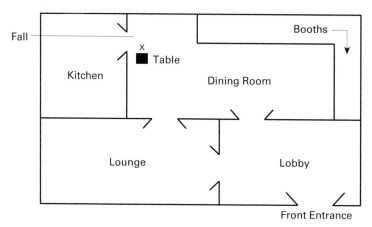

Figure 1   Exact location of incident

CAUSE

Prior to the fall, Mr. Sullivan had cleared dishes from three tables, as is customary. He was returning to the kitchen when his foot caught on the leg of the table next to the kitchen. The weight of the dishes he was carrying caused him to lose his balance and to fall forward. As he fell, the tray dropped, and the dishes broke.

RESULTS

As a result, the replacement of the dishes and glassware will cost $199.55, which includes:

FIGURE 14.9   *Sample brief incident report*

$60.00 for Anchor Hocking stemware (6 @ $10.00 ea.)
37.55 for Whitehall water glasses (5 @ $7.50 ea.)
72.00 for Sango dinner plates (6 @ $12.00 ea.)
30.00 for Sango coffee cups and saucers (6 @ $5.00 ea.)

EVALUATION

In order to prevent such a fall from recurring, I recommend that the table next to the kitchen entrance be removed. This will give the busboys and waiters more clearance when entering and exiting the kitchen area.

CV:jv

FIGURE 14.9   *continued*

ORGANIZATION

Often, preprinted forms with space for narrative reporting are used. Many companies devise organization formats in outline form to be followed by their reporting personnel. These regulating outlines list specific topical headings along with brief instructions about the data to be included under each.

Despite the diversity of headings required for any one specific field report, all such reports include

- Essential background data
- Account of the field inspection
- Analysis of findings
- Conclusions and recommendations

Background.   Explain what is being investigated and for what purpose (safety or health factors, cost determination, need for improvement, validity of claims, etc.).

Account of Inspection.   Clarify what the field investigation found.

Findings Analysis.   Interpret the significance of the findings as related to the purpose of the investigation.

Evaluations/Conclusions.   Detail what action should be taken.

Figure 14.10 shows a typical field report concerning a site inspection to determine a potential violation.

MEMORANDUM

TO: Enforcement Administration

FROM: Gerald Mills, Marine Resources *G.M.*

DATE: 28 April 200X

SUBJECT: NOV Request to Redress Complaint

BACKGROUND

On 10 February 200X, Herbert Johnson, a resident of Beautiful
Waters Condominium, Longboat Key, Florida, called to complain that
his condominium's board of directors had ordered a contractor to
bulldoze several large sand dunes and their associated vegetation on
the beach fronting the property. Mr. Johnson indicated that the
dunes had been flattened and the vegetation destroyed on
6 February 200X.

On 20 February 200X, I received a visit from Mr. Johnson to
present documentation that included the accompanying photographs
and a copy of the contract between the board president of Beautiful
Waters Condominium and Coastal Construction.

INSPECTION/FINDINGS

A visit to the site by Marine Resources staff on 13 April 200X confirmed that no dunes or vegetation were present. Following consultations with Board President Lee Santini and the zoning board, it was
determined that the dune destruction was a violation of Section
Z145773 and that Coastal Construction operated without a permit.

CONCLUSION

It is respectfully requested that NOV procedures be initiated in this
matter. I will attend to this personally with your approval. Please
contact me at (561) 542-3455 if you have questions or need additional information.

GM:acd

FIGURE 14.10   *Sample field report*

Feasibility Reports

PURPOSE

You may be assigned to look into a new project—a new product, the development of a new program, a relocation, the purchase of new equipment—to determine the practicality of the project. The feasibility report presents the evidence of your investigation and analysis plus your conclusions and evaluations.

Typically, feasibility reports analyze data to answer specific questions:

- Will a given product, program, service, procedure, or policy work for a specific purpose?
- Is one option better than another option for a specific purpose?
- How can a problem be solved?
- Is an option practical in a given situation?

ORGANIZATION

The feasibility report usually includes

- Explanation of the problem
- Preset standards or criteria
- Description of the item(s) or subject(s) to be analyzed
- An examination of the scope of the analysis
- Presentation of the data
- Interpretation of the data
- Conclusions and evaluations

A brief feasibility report does not require headings; however, for a longer report consider these headings:

Background		Introduction
Standards		Problem
Options	*or*	Criteria
Method		Options
Data		Limitations
Conclusions		Evaluations

Background. This section includes all introductory material, such as the purpose of the report, a description or definition of the question, issue, problem, or item(s). You may discuss the scope or extent of the report.

Standards.   Here you present a detailed explanation of the established criteria, aims, or goals of the question being investigated.

Options.   If applicable, present each alternative according to your established criteria. Consider costs, capabilities, procedures, personnel involved, required training, or other appropriate features of each option.

Method.   In this section, explain how each option was analyzed to determine its practicality. This may include description of testing methods, survey instruments, research source material, qualifications of consultants, and discussion of limitations to your investigation.

Data.   In this section, present the test results, survey results, or research findings.

Conclusions.   Finally, the feasibility report summarizes the investigation, drawing logical conclusions. Discuss the limitations, if any, of your study. Build in a time schedule for action and a review of the results. In short, interpret the data and offer your recommendations.

Figure 14.11 shows a brief feasibility report.

TO: J. D. Big, Director, Accounts Department

FROM: Jane White, Secretary, Accounts Department *JW*

DATE: 9 May 200X

SUBJECT: Leasing New Copy Machine

Inasmuch as we have been experiencing difficulties with our leased Atlas XC23-AS 076-364622 copier, I have investigated the feasibility of leasing a new computer. Our copier costs $125.00 per month for 30 months. It makes only 8 copies per minute, copies only 50 copies per setting, requires 15 seconds for the first copy, can hold only 150 sheets in the paper tray, does not have two-sided printing capability, and costs $0.04 per copy. In the past 30 days we have needed three service calls costing $105.00 each, for a total of $315.00 repair cost. At this rate, our annual repair cost could total $4,200.00, approximately $1,000 more than the cost of a new copier.

Due to our tenfold increase in contracts, our criteria for a new copier include

- low lease cost ($125.00 per month or less)
- minimum 10 copies per minute
- 90 to 100 copies per setting
- minimum 10-second first copy speed
- minimum 250 sheets in the paper tray
- two-sided print capability
- low cost per copy

Only two copiers are available within our leasing price range. Table 1 compares cost, feature capabilities, and limitations of Brand X and Brand Y:

FIGURE 14.11   *Sample brief feasibility report*

TABLE 1 *Comparative Features of Brand X and Brand Y Copiers*

Brand	Rental Cost Per month ($)	Capabilities	Limitations
X	$120.00 for 36 mos.	• 12 copies per minute • 100 copies per setting • 8-second first-copy time • $0.03 per copy • two-sided print	• 150 sheets in paper tray
Y	$120.00 for 36 mos.	• 14 copies per minute • 100 copies per setting • 10-second first copy speed • 500 sheets in paper tray • two-sided print	• $0.04 per copy

Both copiers meet our criteria. The Brand X tray capacity, although 50 short of desirable, is acceptable. The cost per copy of Brand Y is the same as our present cost. Brand Y offers increased copies per minute and the 500-sheet paper tray, which will require less time and fewer refills in any one print, making it more cost effective. Moreover, both leases cost less for a longer period of time than our present contract.

In order to assess the disadvantages, I surveyed five departments that already use the X or Y copiers to determine the number of service calls required in a one-year period. Table 2 shows the departments, copier brand used, number of service calls, and departmental costs per year:

TABLE 2 *Service Calls and Costs for Brand X and Y Copy Machines in One-Year Period*

Department	Brand	Number of Service Calls (#)	Average Cost Per Call ($)	Annual Service Cost ($)
Personnel	X	11	100.00	1,100.00
Payroll	Y	2	105.00	210.00
Info Technology	X	15	100.00	1,500.00
Records	X	7	100.00	700.00
Purchasing	Y	0	105.00	0.00

Both Brand X and Brand Y appear to be more reliable than our Atlas copier. Although Brand Y charges more ($5.00 more per call) for a base service fee than does Brand X, the survey data reveals that Brand Y is clearly the more reliable machine.

Therefore, I recommend that we lease a Brand Y copy machine, which will cost $60 less per year, give us faster copies per minute, require fewer paper tray refills, provide two-sided printing capability—speeding up production by 50%—and cut our maintenance/repair bills by approximately $4,000 per year.

May I have your authorization by Friday to negotiate a Brand Y lease?

JW:eg

FIGURE 14.11 *continued*

RECOMMENDATION REPORTS

Recommendation reports, in contrast to analytical reports (which emphasize findings) and evaluation reports (which emphasize conclusions), emphasize recommendations. Recommendation reports are strongly persuasive and urge action. Because they usually require extensive evaluation of the present situation or problem, plus complex supportive data on the included recommendations, we consider them in Chapter 15, "Devising Longer Reports, Proposals."

CHECKLIST

Brief Reports

ANALYTICAL REPORTS

Periodic or Progress Reports

❏ **1.** Have I selected a subject that requires a record of the status of a project over a specific period of time?

❏ **2.** Have I reviewed the aims of the project highlighting accomplishments or problems?

❏ **3.** Have I provided a summary or explanation of the work completed?

❏ **4.** Have I provided a summary or explanation of the work in progress?

❏ **5.** Have I provided a summary or explanation of future work on the project?

❏ **6.** Have I provided an assessment of the progress?

❏ **7.** Have I designed a document with headings, emphatic features, and graphics, where applicable?

Laboratory or Test Reports

❏ **1.** Have I selected a subject that presents the results of research or testing?

❏ **2.** Have I provided a statement of purpose?

❏ **3.** Have I reviewed the method or procedure of research or testing?

❏ **4.** Have I presented the results?

❏ **5.** Have I presented conclusions and recommendations?

❏ **6.** Have I designed a document with headings, emphatic features, and graphics, where applicable?

EVALUATION REPORTS

Incident Report

❏ **1.** Have I selected a subject, such as an accident, machine breakdown, delivery delay, cost overrun, production slowdown, or personnel problem, suitable for a written report?

❏ **2.** Have I described completely what happened in a factual, not opinionated, tone?

❏ **3.** Have I explained what caused the incident with sufficient, chronological detail?

❏ **4.** Have I explained the results, such as injuries, losses, delays, costs, repairs, paramedic or other auxiliary personnel involvement, problem solving, and satisfaction of customer demands?

❏ **5.** Have I suggested exact measures to be taken to prevent recurrences of the same nature?

❏ **6.** Have I designed a document with headings, emphatic features, and graphics, where applicable?

Field Report

❏ **1.** Have I selected a subject that involves an analysis of a location or situation requiring action, costs, plans, improvements, repairs, and so on?

❏ **2.** Have I presented all of the essential background data, including what is being investigated and for what purpose?

❏ **3.** Have I clarified exactly what was discovered by the field investigation?

❏ **4.** Have I interpreted the significance of the findings as related to the purpose of the investigation?

❏ **5.** Have I presented a logical and complete evaluation and other conclusions?

❏ **6.** Have I designed a document with headings, emphatic features, and graphics, where applicable?

Feasibility Report

❏ **1.** Have I selected a subject that requires an analysis of data to determine if a given new approach is attainable, sound, and practical?

  ❏ Am I investigating a new product, program, service, procedure, or policy to suggest improvement?

  ❏ Am I evaluating one option over another for a specific purpose?

  ❏ Am I recommending how a problem may be solved?

  ❏ Am I investigating whether an option is practical in a given situation?

❏ **2.** Have I provided a complete explanation of the problem?

❏ **3.** Have I established preset standards or criteria?

❏ **4.** Have I described the item or subject to be analyzed?

❏ **5.** Have I presented test results, survey results, or research findings to support my recommendations?

❏ **6.** Have I adequately interpreted the supporting data?

❏ **7.** Have I presented conclusions and evaluations for implementation, including limitations, if any, a time schedule for action, and related costs?

❏ **8.** Have I designed a document with headings, emphatic features, and graphics, where applicable?

❏ **9.** If a test, survey, or research was involved, have I presented the instrument(s)?

EXERCISES

1. **Progress Report.** Use the following information to write a progress report in letter format. Arrange the data in chronological, topographical, or combined order. Use headings and any other emphatic features and graphics that make your report clearer. Elaborate on the tasks as you see fit. Make up names, company name, and so on.

   Position: You are a training specialist who works for New World Communications. You are reporting to writing consultant Paula Kline, whose services have been contracted to conduct on-premises technical writing instruction at your company.

   Project: Technical Writing seminar to upgrade employee skills—scheduled 5, 6, and 7 March 2000.

   Data: **a.** Consultant evaluation forms to be devised.

   b. Negotiating luncheon menus with George Kelly, cafeteria manager; all three dates.

   c. Twenty registrants from 15 departments confirmed.

   d. Writing samples from each registrant being solicited to forward to consultant for evaluation before the instruction begins.

   e. Training Room C scheduled from 9:00 to 5:00, 5, 6, and 7 March.

   f. Certificates of completion to be designed and printed.

   g. Seminar announcements sent to 54 departments on 1 February.

   h. Printing and binding of 20 sets of instructional materials in progress.

   i. Overhead projector to be ordered for 6 March, 9:00–5:00.

   j. Wayne Salsbury to be notified to prepare introduction of Consultant for 5 March.

   k. Memos to supervisors explaining cost center billing sent on 15 February.

   l. Tables to be arranged in U for registrants on 5, 6, 7 March.

   m. Lectern for consultant to be placed in front of blackboard at the opening of the U arrangement.

2. Incident Report. Use the following information to write an incident report in memorandum format. You are a work supervisor reporting to the head of the department. Use the appropriate format, emphatic features, and graphics to make your analytical report clear and readable.

Incident:	Water pipe in ceiling burst; water damage to office equipment, carpet, and ceiling panels.
Where:	Work Station A, Engineering Department
When:	12:40 P.M., Friday, 18 April 2001
Damage:	IBM laptop computer soaked and short-circuited; requires overhaul maintenance and troubleshooting test.
	IBM laser printer (serial no. 809-71-2654) exploded due to short circuit; requires replacement.
	Carpet ($20' \times 16'$) needs replacement.
	Three ceiling panels soaked; need replacement.
	Water pipe repairs required by a plumber.
Costs:	Computer overhaul: $1,000
	Printer replacement: $2,500
	Carpet replacement: $900
	Ceiling panels replacement: $45
	Plumber: $150
Further results:	Estimated 1-week delay in preparation of proposal project in progress by engineer Thomas Waters.
Analytical analysis:	Position engineer Thomas Waters into Workstation 5 for one week. The workstation does not include a printer, but he can reprocess information that was not on back-up or zip disks at the time of the incident.

3. Test Report. Write a three or four paragraph test report based on the following information. Indicate the test results in a table as well as in words. Include your conclusions and recommendations. Use other document design features as appropriate.

You are an engineer at Universal Testing Laboratories. You have been given four electronic pool alarms to compare for costs, attachment method and ease, alarm length, battery information, remote receiver features, connectors to home security system, and warranty information. It is known that some 500 people drown in backyard swimming pools each year. More than half are children under five. These pool alarms are triggered by waves from a fall into the pool. Each begins beeping within 30 seconds. Your testing reveals the following:

a. **Unit #1**—Cost: $160. Warranty: One year. Attaches to pool edge with Velcro-style pads. Can be turned off and left in place when using the pool. Alarm sounds for eight minutes and resets automatically. Can be set off by gusty wind and rain. Requires a nine-volt battery. Has a low-battery indicator. Has a remote receiver to sound inside the house. Can be connected to home security system. Not very prone to false alarms due to gusty winds and rain.

b. **Unit #2**—Cost: $260. Warranty: one year. Rests on the edge of the pool. Must be removed before people can use the pool. Alarm sounds for seven minutes and resets automatically. Requires nine-volt battery. Does not have a low battery indicator, but a weak or irregular alarm signals time for a new battery. Has remote receiver to sound alarm indoors. Can be kicked into the water accidentally. Not prone to false alarms due to gusty wind and rain.

c. **Unit #3**—Cost: $150. Warranty: one year. Floats on the water with a 16-foot tether to attach it the side of the pool. Must be removed before pool use. Alarm sounds for three minutes and resets automatically. Requires a nine-volt battery. Has a low-battery indicator. Has remote receiver to sound alarm indoors. Prone to false alarms due to gusty wind and rain.

d. **Unit #4**—Cost: $150. Warranty: one year. Floats on the water. Has a 26-inch lanyard to attach to a pool ladder. Must be removed before using the pool. Alarm sounds for two minutes and resets automatically. Requires a nine-volt battery. Sounds a weak alarm when time for a new battery. Has remote receiver for indoor alarm. Only one that requires a square-head (Robertson) screwdriver to install battery. Prone to false alarm in gusty winds and rain.

4. **Feasibility Report.** Use the following information to write a feasibility report in memorandum format. You are the Grounds Committee Chairperson reporting to the Golden Lakes Condominium Association. Rearrange the data into a logical organization. Present the data in appropriate graphics. Based upon the information, draw logical recommendations in your conclusion.

488 PART 4: THE PROFESSIONAL STRATEGIES

PART 4: THE PROFESSIONAL STRATEGIES

Problem: Remodeling of Golden Lakes Condominium recreation building has resulted in grass damage in common areas.

Fact: Rainy season begins 15 June.

Data: Luxury Landscape will require 3 days to resod at bid of $4,839. Guarantee includes six inspections in 4-month period with necessary sod replacement at no extra charge.

Landscaping Professionals will require 2 days to resod at a bid of $3,984. Guarantee includes six inspections in 6-month period with necessary sod replacement at no extra charge.

Green Company will require 3 days to resod at a bid of $3,707. No guarantee offered.

K-Mart Professional Crew will require 3 days to resod at a bid of $4,000. Guarantee includes six inspections in 12-month period with necessary sod replacement at no extra charge.

Criteria: Twenty-square-foot area needs resodding. Budget allows $4,500 expenditure. Guarantee required.

WRITING PROJECTS

Any of the following five projects may be handled individually or collaboratively. Include a bibliography for each report in the style appropriate to the subject or in the style requested by your professor.

1. **Periodic or Progress Reports.** Select one option.

   a. In letter form, write a periodic report on your monthly expenses to your parents or spouse. Include a circle graph or bar chart on the percentage and the actual dollar expense of each category. Include, or comment on the lack of, the following categories:

Housing	School Supplies
Utilities	Insurance
Food	Charge Accounts and/or Car Payments
Transportation	Leisure
Clothes	Miscellaneous

   Include other categories as they are appropriate to your expenses. Conclude with an appraisal of your expenses.

   b. In memo form, write a progress report to your academic advisor showing your progress toward completing a degree or receiving a certificate or license. Begin with a statement of your overall goal and its requirements. Include tables of your courses, credits, grades, and grade point averages for courses completed, courses in progress, and courses remaining. Conclude with a discussion of your career plans.

c.   Write a progress report on a project in which you are involved either in school or on the job. Include an introduction and sections on completed work, current work, future work, and an appraisal of progress. Be alert to graphic possibilities.

2.   **Lab or Test Reports.**   Select two simple products (ballpoint pens, glues, paints, stepladders, brooms, car waxes, toothbrushes, garlic presses, etc.). With the purpose of determining which is the better product, devise a method or procedure to test each. Carry out your testing and then write a test report that states the object or purpose of the experiment, the explanation of the test method, a step-by-step analysis of the test and the results, and your analysis of which is the better product. The results section should offer the opportunity to present data in a table or performance curve.

3.   **Incident Reports.**   Write an incident report on a real or imaginary business/industrial accident for your "employer." Suitable subjects are damaged equipment, brief fire, broken merchandise, minor burns or sprains, collapsed shelving, broken windows or doors, and so forth. Write a concise description of the incident or accident. Next, write a sequential analysis of the cause. Follow this with a review of the results, and, finally, present your recommendations to prevent the incident from recurring. Include graphics of the location and cost breakdowns.

4.   **Field Reports.**   On your campus or at your place of employment, select a location to conduct a field examination that will include conclusions and recommendations based on your evaluation of the efficiency or safety of the location. Suggested fields to investigate are

On the Campus	*At Work*
parking lot layout	room furnishings
cafeteria food arrangements	office layout
classroom layout	restroom facilities
registration procedures	lounge or coffee room
recreational areas	locker space
library study carrel layout	emergency exit doors/stairwells
campus bookstore displays	storeroom arrangements
a piece of equipment	a piece of equipment
a small structure	security arrangements

Write a field report structured to include the purpose of your inspection, the methods of gathering data, the facts and results of your investigation, and your analysis, which will make the chosen field more efficient or safer. Use appropriate headings.

5. **Feasibility Reports.**   Select one option.

   a. Write a feasibility report that analyzes a new purchase. State your purpose and the requirements. Then describe two or more probable alternatives (cars, office equipment, water beds, appliances, etc.). Next explain a method for evaluating the products (survey, testing, research). Present your findings. Evaluate the data, and conclude with logical recommendations for purchasing one of the optional items.

   b. Write a feasibility report that analyzes a procedure or problem in your place of employment. Identify the problem (the present means of advertising a product, scheduling personnel, awarding salary increases, providing in-service training, promoting personnel, handling tasks, or other similar procedures). State the standards that should be set to remedy the problem. Devise two solutions to the problem and analyze the merits and limitations of each. Interpret the feasible solutions to draw logical recommendations for implementing one or the other.

NOTES

CHAPTER 15

Devising Longer Reports, Proposals

HI & LOIS   by Chance Brown

Reprinted with special permission of King Features Syndicate.

S K I L L S

After studying this chapter, you should be able to

1.   Identify the usual types of longer reports.
2.   Identify the special features of longer reports.
3.   Design title pages, tables of contents, and lists of illustrations.
4.   Write transmittal correspondence.
5.   Distinguish between and write several types of abstracts and summaries.
6.   Attach appropriate back matter—supplements, appendixes, exhibits, and the like.
7.   Understand the terms *internal, external, solicited,* and *unsolicited* as they refer to proposals.
8.   Design traditional and streamlined formats for proposals.
9.   Write a lengthy proposal with all appropriate features.
10. Design long reports and proposals using all document design features.

INTRODUCTION

Although the need for brevity is always present, a lengthy report is occasionally necessary. Quarterly and annual reports, long-range planning programs, systems evaluations, and proposals are a few typical long reports that most organizations produce. Frequently, these reports fall into the recommendation report category because they emphasize recommendations, are strongly persuasive, and include considerable supportive data. This chapter covers the special features of longer reports and reviews the particulars for a proposal, a widely used recommendation report.

PRESENTING THE LONGER REPORT

To avoid "gray material"—pages of dull, gray type—a longer report is distinguished by special features to make the information contained in it more accessible. These features may include

- Title page
- Transmittal correspondence
- Table of contents
- List of illustrations (tables and figures)
- Abstract or summary

- Body of report with topical headings
- Back matter

Title Page

Include an attractive and clarifying title page. This page should include

- A precise title
- The name, title, and company of the person(s) to whom the report is directed
- The name, title, and company of the writer(s)
- The date

A precise title, such as

**PROPOSAL FOR PURCHASE
AND INSTALLATION OF IONIZATION
AND PHOTOELECTRIC FIRE ALARM SYSTEMS
IN OCEANVIEW CONDOMINIUM UNITS**

is more effective than a vague title, such as

PROPOSAL TO DECREASE FIRE HAZARD

You have an opportunity here to use landscaped placement, multi-size type, and a variety of fonts. Occasionally, a company logo or art work is included on the title page. The first page of Figure 15.6 (page 510) illustrates a basic title page to longer reports.

Transmittal Correspondence

A letter or memorandum of transmittal accompanies most longer reports. The purpose of the transmittal correspondence is to orient the receiver to the long report or proposal in a suitable explanatory manner. It is usually very brief—three or four short paragraphs. It contains the following information:

1. The title and purpose of the report.
2. A statement of when and by whom it was requested or why it is being submitted.
3. Comments on any problems encountered (limited scope, unavailable data, deliberate omissions).

4. Acknowledgment of other people who assisted in assembling the report.

5. A statement eliciting feedback.

Do not include repetitions of the data from the actual report. These details will be covered in the **abstract** or **summary.** Figure 15.6 (p. 511) illustrates typical transmittal correspondence.

Table of Contents

Your reader(s) will want to be able to refer to sections quickly. A table of contents not only helps the reader(s) to turn rapidly to a particular section of the report, but also gives an initial indication of the organization, content, and emphasis of the report. A table of contents should accompany every written report that exceeds eight or ten pages and may be helpful in some shorter reports. Title the page *Table of Contents.* All headings used in the report are included in the table, and the subordination of sections is indicated by indentation. The starting page of each section is included as shown in the sample proposal in Figure 15.6 (p. 512).

List Of Illustrations

If graphics (tables and figures) are used throughout your report, include their table and figure numbers, titles, and page references on the same or a separate page from the table of contents. Center the title *List of Illustrations.* List tables separately from figures.

Back Matter

The back matter, or supplements to the body of the document, contains glossaries, tabulated survey results, financial projections, job descriptions and resumes of new personnel, equipment brochures, lists of references, and bibliographies. These supportive materials are attached as exhibits or appendixes and are grouped in the back of the binder of the parent document. Should you include back matter, follow these guidelines:

- Number and title each supplement.

  Examples: Exhibit 1   Interview Topic Outline

  Exhibit 2   Sample Policy Matrix

  Appendix A   Employee Safety Survey

  Appendix B   Resume of Donald C. French

- Refer to the exhibits or appendixes in the body of your text.

  Examples: Although I have interpreted the survey data here, the survey instrument and numerical tabulations are in Appendix C.

- Include the exhibit or appendix titles in your table of contents.

Figure 15.6 includes supplemental back matter following the recommendations sections of the sample proposal. Note their appendix numbers and titles, their listing in the appropriate table of contents, and references to them in the text.

ABSTRACTS AND SUMMARIES

Abstracts and summaries, short and concise reviews of a parent document, describe basic information contained within the document and are vital parts of longer reports. There are two kinds of abstracts and two kinds of summaries: descriptive and informative abstracts and executive and concluding summaries. In practice, the terms *abstract* and *summary* are often used interchangeably but should not be.

Abstracts and executive summaries are written to orient the reader to the material to follow, yet they are written after the main document has been written. All are reduced representations of the parent document and reflect the purpose and content, scope, methodology, and sometimes the conclusions, of the parent document. Both kinds of abstracts and executive summaries are located in the prefatory section of a long report; concluding summaries are placed at the end of sections within the document.

Abstracts accompany scientific and technical reports and articles, whereas executive summaries accompany business and industrial reports written for executives concerned with resource management. In short, abstracts and summaries share a number of similarities but serve different purposes and require different writing strategies. The potential audience of a long report expects an abstract or executive summary to accompany it. After reading the abstract or summary, the busy professional or executive can decide whether to read the entire report or article. Because the reader may not possess the technical knowledge and language that your full report embraces, be careful to avoid technical terminology in your abstract.

Abstracts

An abstract for scientific and technical reports and articles is a brief *descriptive* or *informative* preview of a longer report. Abstracts, written in an objective, detached tone, present an overview of the complete, longer report by identifying the organization of the paper to come, the scope, the methodology, and sometimes the conclusions and evaluations contained in the parent document. The information is usually presented chronologically.

Libraries subscribe to journals that provide brief descriptive or informative abstracts of articles published in a number of journals in a specific field. Some of these are written by the author of the article, although others are written by professional abstract writers. In a research environment these abstracts allow the researcher to obtain an overview of content, without a lengthy search for the article, and to read it thoroughly to determine its usefulness. The reader can review the abstract and then reject or pursue the original article.

DESCRIPTIVE ABSTRACTS

The *descriptive abstract* (also called the *indicative abstract)* represents the complete document and can be thought of as functioning separately from it. The reader of a descriptive or indicative abstract is usually an engineer, scientist, technician, or other expert. The abstract serves as an extended statement of the purpose, scope, and methodology used to arrive at the findings. A descriptive or indicative abstract is useful for a very extensive report because it indicates the organization of the report, although usually not the conclusions, results, or recommendations, which are covered in the parent document. The descriptive or indicative abstract is usually short (perhaps one to five sentences totaling 150 to 300 words). Figure 15.1 shows a descriptive abstract for an extended technical report.

INFORMATIVE ABSTRACTS

The informative abstract distills the essential information from the parent document. An informative abstract is frequently confused with an executive summary, but careful consideration elicits the differences. Like the descriptive abstract, the informative abstract is intended for the technical reader or expert who is interested in the methodology and validity of the results covered in the longer document. In contrast, the executive summary is intended for management personnel interested in the definition of a problem or a proposal and the best solutions recommended to solve the problem or realize the proposal.

An informative abstract will accompany research-related documents, such as laboratory reports or systems evaluations serving not as an introduction but as a complete description of the information in the parent article or report, the scope of the research or experimentation, the methods of procedures used as well as the results, conclusions, and rec-

Abstract

The AlliedSignal RDB-4B, a state-of-the-art digital airborne weather radar capable of forward-looking predictive windshear detection, is presented. The weather phenomena known as microbursts and windshears are overviewed. In reviewing the "physics" of windshear, the need for a forward-looking sensor is established, and the selection of digital radar as a means to the end is justified. The RDR-4B system capabilities are then summarized, followed by a top-level breakdown of the radar and associated windshear detection processing. Key discussions include formation of Doppler frequency domain data, adaptive rejection of clutter spectra, and statistical characterization of retained weather/windshear spectra. An explanation of how these statistics come to indicate a windshear's existence and severity, in terms of hazard factors, follows. The paper concludes by documenting an actual windshear's detection prior-to-encounter and subsequent penetration by an RDR-4B-equipped test aircraft.

FIGURE 15.1   *Descriptive abstract for a 12-page system evaluation paper on a new predictive/weather radar for aircraft*   (Courtesy of Robert Johns, AlliedSignal, Inc.)

ommendations. The informative abstract is longer and more detailed than the descriptive abstract. It is typically 300 to 500 words in length. Figure 15.2 shows an informative abstract that includes the results and conclusions of a 10-page report.

Summaries

Summaries are also of two principal kinds: *concluding summaries* and *executive summaries*. They differ from abstracts in that they are designed for management rather than for scientists and technicians. In addition to summaries to reports, busy executives often request assistants to present summaries of newspaper and other articles so that the executive may have access to the content of pertinent reading material in brief form.

Abstract

As part of a blind longitudinal study, 5,465 job applicants were tested for use of illicit drugs, and the relationships between these drug-test results and absenteeism, turnover, injuries, and accidents on the job were evaluated. After an average 1.3 years of employment, employees who had tested positive for illicit drugs had an absenteeism rate 59.3% higher than employees who had tested negative (6.6% vs 4.16% of scheduled work hours, respectively). Employees who had tested positive also had a 47% higher rate of involuntary turnover than employees who had tested negative (15.41% vs 10.51%, respectively). No significant associations were detected between drug-test results and measure of injury and accident occurrence. The practical implications of these results, in terms of economic utility and prediction errors, are discussed.

FIGURE 15.2   *Informative abstract containing essential information from the parent 10-page report*   (From *Journal of Applied Psychology,* December 1990 [629])

CONCLUDING SUMMARIES

A concluding summary is most often found following a section or chapter of a textbook or report. One may also be used in instruction manuals or system evaluations. Figures 15.3 and 15.4 show concluding summaries for a textbook chapter and a system evaluation.

EXECUTIVE SUMMARIES

An executive summary is similar to an abstract in that both are prefatory sections of a document and discuss what is to follow in the full text. Executive summaries often deal with resource allocation and preface long feasibility reports, proposals, and the like. Although they can be distributed separately, they usually precede the text. They serve to highlight the problem or proposal described, discuss its present implications, and offer solutions or recommendations to implement a program of action.

Engineers, scientists, technicians, and other employees of an organization will often write formal reports. Brief reports are the written record of events and the progress of the company's activities and assist all personnel in making decisions. They usually focus on just one subject and are addressed to a single receiver or small audience. Their tone is factual, objective, authoritative, and serious. Some types of reports are written in specific formats, whereas others may be presented as memorandums or titled documents. Graphics and visuals often characterize the content.

Brief reports tend to be analytical or evaluative. Analytical reports include work logs, expense reports, requests for leave or transfer, periodic or progress reports, and laboratory and test reports. Evaluation reports include incident reports, field reports, and feasibility reports. Each type of report, analytical or evaluative, requires a statement of purpose, background information, specific organization of material, and, frequently, results, conclusions, and calls to action. An analytical or evaluation report differs from the recommendation report in that the latter are strongly persuasive in tone, tend to be considerably lengthier, and call for implementation of the final recommendations.

FIGURE 15.3   *Sample concluding summary*

The **tone** of an executive summary is frequently persuasive; that is, it may be more personal and emotional than stringently objective. Consequences resulting from inaction may be stressed. Appeals to forward thinking and fair decision-making may be incorporated. Word choice may be more connotative.

The **organization** of material may be more causal than the chronological or procedural organization in an abstract. That is, the summary may consider first the most persuasive details and then the lesser ones.

The executive summary is typically from one to five pages in **length,** never exceeding more than 10 percent of the length of the original document. Figure 15.5 shows an executive summary from a 54-page home products business plan.

This summary includes background information on the company, a statement of focus, persuasive elements, the actual proposal to introduce three new products, costs and projected net income, management information, and projected consequences.

5.0 SUMMARY

Microburst and windshears impact aircraft performance in a substantial and sometimes fatal loss of aerodynamic lift. A solution to the problem requires a modified weather radar system to measure wind velocity and direction changes within an approaching air mass, to reject high clutter signals, and to achieve accurate wind characterization. A Doppler frequency-based digital signal process is required. AlliedSignal's RDR-4B Weather/Predictive windshear radar design, with its cockpit-mounted visual display, offers crucial information during all flight phases and especially addresses the challenges of windshear detection/avoidance during take-off and landing. The RDR-4B includes

- highspeed/throughput coherent processing spread over seventeen 32-bit floating point TTMS320-C31 digital signal processors (DSP) with a combined capacity of 270 million single to 540 million parallel operations per second;
- interference reduction via pulse-repetition-rate jittering and transmit frequency hopping capabilities;
- automated antenna tilt control for optimum detection in windshear mode;
- patented stabilization error correction algorithms utilizing ground clutter radar returns to dynamically compute and minimize antenna pointing erros;
- adaptive clutter discrimination/elimination processes that preserve weather returns while rejecting contributions from the ground, moving discretes, and harmonic interference; and
- automatic compensation for loss/lack of aircraft telemetry/system inputs.

A series of windshear penetration flights were conducted with the RDR-4B radar installed fully proved the efficacy of the system.

FIGURE 15.4   *Sample concluding summary from a 12-page system evaluation paper on a new predictive/weather radar for aircraft*   (Courtesy of Robert Johns, AlliedSignal, Inc.)

EXECUTIVE SUMMARY

National Home Products Corporation (NHP) was started in July 2000 by Mr. Thomas W. Lehmer and Mr. Gregory J. Welborn to develop unique nonfood or general merchandise products for distribution through Prepared Products Company's broker/distributor network in the United States.

Our specific focus is to take advantage of two predominant trends in consumer purchasing patterns: the demonstrated demand for greater convenience and better performance in consumer products, and the predominant interest in healthier and environmentally safer products.

It has been taken previously as an article of faith that product attributes that provide greater consumer appeal are incompatible with those features or components that are healthier or environmentally sound. We believe these attributes can be very compatible and that products can both provide greater convenience or performance as well as be health-conscious and environmentally safe.

Over the last year, NHP has researched and identified three market segments in which these demands overlap, the growth rate is substantial, and the potential exists for greater-than-category-average profitability. The market segments are grilling products, home fragrance, and home hearth products. Furthermore, we have developed two products for immediate introduction for the grilling market and three additional products for the home fragrance and home hearth markets. At the end of 2001, our projections indicate net income of $3,515.586 (11.8%) on revenues of $29,675,071.

NHP has assembled a strong and experienced management team and an affiliate relationship with Prepared Products Company (PREPCO), allowing us to draw upon PREPCO's successful record in innovative packaging and its distribution strength in the grocery, convenience, and club store trade channels, as well as in food service. NHP is operated as an autonomous company because of its unique and specific focus and its need to cultivate favorable media attention. We are seeking a $750,000 investment for promotion, advertising, and working capital needs.

We believe we offer a unique opportunity. NHP is at the forefront of two converging trends in consumer markets, has already developed five strong products, and has ready access to established distribution channels. NHP's conservative financial projects demonstrate substantial sales growth and profitability. The investment risk is minimized since utilization of the $750,000 investment will be dedicated to controllable expenditures. Finally, NHP will be well positioned in three to five years for sale or an initial public offering.

FIGURE 15.5   *An executive summary from a 54-page report*

Writing Abstracts And Summaries

To write an abstract or executive summary, follow these steps:

1. Read the entire report to grasp its full content.
2. Estimate the number of words and plan an abstract or summary that does not exceed 10 percent of the original length.
3. If you are writing an abstract, decide whether it should be descriptive or informative. In either case identify the organization, scope, and methodology.
4. If you are writing an informative abstract, include also the key facts, statistics, and conclusions.
5. If you are writing a concluding summary, reread the material and summarize the content, omitting specific detail, examples, graphics, and the like.
6. If you are writing an executive summary, condense the detail of the parent document and consider persuasive words and phrases that would predispose the reader to think favorably toward your report or proposal. Appeal to the public's desire for convenient, high performance, and environmentally safe products.
7. If you are writing an informative abstract or an executive summary, explain your key findings in a very condensed form. Omit or condense lengthy explanations, tabulated material, and other supporting explanation.
8. Edit for completeness and accuracy.

PROPOSALS

A proposal is an action-oriented report. While most reports include recommendations for ongoing accomplishments, a proposal suggests a future task and includes a complete plan of how to accomplish this task. That is, a proposal contains procedure or equipment analysis, cost analysis, the capabilities of existing facilities, information on involved personnel, and usually a timetable for accomplishing the work. A brief report may recommend that a new policy be devised. A proposal provides exactly what the policy must cover, a schedule for adoption, a procedure for implementing the policy, and the personnel who should be in charge. A proposal's purpose is to persuade the reader.

To be persuasive, a proposal must emphasize the advantages to the organization. You need to convince the decision makers to take action. As you develop your data and organize your material, stress one or more of the following advantages that your proposal will effect:

- Money savings or increased profits in the short and/or long term
- More efficient time applications
- Improved employee and client safety and comfort
- Compliance with laws or ethics
- Enriched employee morale

Proposals persuade people to act by appealing to their sense of responsibility and, perhaps, even playing to their fear of failure as leaders.

Types of Proposals

Proposals are classified as *internal* and *external, solicited* and *unsolicited.* Government and industry often solicit external proposals from independent agencies to solve problems or to develop services prior to awarding contracts or grants. A county commission may advertise for competing firms to submit proposals to develop a county-wide transportation system. A national airline may solicit proposals to develop a larger and faster jet. The federal government solicits grant proposals for services in education, environment, energy, science, medicine, rural development, and other areas. A university may hire a consulting firm to develop a long-range expansion program. The agency that solicits external proposals spells out general requirements, cost ceilings, deadlines, and criteria for evaluation.

In business and industry the internal proposal, one written by a member of the organization, may be solicited or unsolicited. Management may appoint an individual or a committee to devise a program to change or improve some existing procedure or practice. A cafeteria manager may ask employees to submit proposals to increase sales in the fast-food line. An office manager may solicit proposals for policy and procedure to avoid charges of sexual harassment. Or any employee or group of employees may initiate an unsolicited proposal to management to purchase new equipment, improve working schedules, or alter procedures.

Writing a good unsolicited proposal for your employer is an excellent way not only to improve working conditions but also to demonstrate your interest in and commitment to your company.

Organization and Format

A proposal explains an existing problem and proposes the concrete measures, procedures, or steps for its rectification, along with an explanation of costs, equipment, personnel needs, and a time schedule. A proposal usually involves

- A clear statement of what is being proposed and why
- An explanation of the background or problem
- A presentation of the actual proposal, including methods, costs, personnel, and action schedules
- A discussion of the advantages and disadvantages
- The conclusions, recommendations, and/or an action schedule

All formal reports are more readable if they contain headings. The following topical headings should be considered for organizational purposes for a proposal, although each actual proposal will suggest additional major and minor headings.

Traditional Format	*Streamlined Format*
Introduction	Subject
Purpose	Objective
Scope	Problem
Background	Proposal
Investigative procedure	Advantages
Findings	Disadvantages
Proposal	Action
Equipment	
Capabilities	
Costs	
Personnel	
Timetable	
Consequences	
Advantages	
Disadvantages	
Conclusion	

The streamlined format (SOPPADA) is usually used only for *brief* proposals.

Writing the Proposal

For purposes of instruction we concentrate on an unsolicited proposal, the kind you may originate in an entry-level career position or devise for consideration at your college or university. It is important to remember your audience. You essentially are writing a persuasive report; therefore, you must justify your proposal by presenting compelling reasons for its adoption. Although you may address your proposal to your immediate supervisor, a proposal is often reviewed by superiors further up the chain of authority. Your explanations, data, and language must be clear to those people who may not have any familiarity with the situation you describe. You must be objective and diplomatic.

INTRODUCTION

The introduction usually includes a *statement of purpose* or an *objective* and comments on the *scope* of the report. State briefly exactly what you propose, along with a general statement of why the proposal should be given serious consideration. The scope statement will orient your reader(s) to the material to follow. Examples follow:

Example 1—Purpose and Scope

Purpose. This proposal, to purchase and install bicycle supports and gate locks in the ABC Elementary School bicycle compound, is designed to eliminate prevalent vandalism and to decrease personal injuries.

Scope. In this report an examination of the problem, a tabulation of survey results, and a description of the proposed concrete bicycle supports and gate locks are presented, followed by the suggested layout, costs, product availability, and conclusions.

Example 2—Objective Heading Only

This report proposes the purchase of a table saw to increase production in our store fixture manufacturing plant and to increase profits. This proposal will document the problem, examine the capabilities of the proposed equipment, detail the costs, recommend the location, discuss the advantages, and present conclusions.

PROBLEM/BACKGROUND

The problem/background section details the existing problem, such as high costs, inefficiency, dangers or abuses, or low morale among employees. The solution you intend to propose will probably cost money or involve personnel in new responsibilities, so you must spell out that a very real and perhaps costly problem presently does exist. If it is not obvious how you researched the problem, you may need to include an explanation of your investigation techniques. If appropriate, research the operational costs of the present system. Project these costs over a week, a month, a year, or other appropriate time frames. Present data in tabular form.

Your problem may be that hazards or inconveniences exist under the present system. Document accidents, work slowdowns, late production schedules, or other related evidence. If the problem is causing low morale, research the turnover rate of personnel or incidents of friction.

The **deductive** organization pattern is easier to read than the **inductive** (see Chapter 2). In a proposal to purchase a computer for two floral shops, the problem was documented as follows:

Problem:

Bookkeeper transportation and telephone costs are excessively high. Because both of our floral shops maintain separate inventory control of customer account information, bookkeeper Mary Jacobs must frequently travel between the two shops and telephone for customer account information. In addition, this system requires twice the amount of time necessary to review the total daily sales. Table 1 shows the transportation and communication expenses of the bookkeeper under the present system:

TABLE 1   *Current Bookkeeper Transportation and Telephone Expenses*

Time Frame	Travel Time (hr)	Gas ($)	Telephone Time (hr)	Cost @ $10.00 per hr wage ($)
Day	1	2	1	20
Week	5	10	5	100
Month	20	80	20	400
6 Mos.	120	240	120	1,200
		TOTAL COST $1,200		

The $1,200 wage and gas expenditure can be put to more productive use, such as toward the purchase of computers for each shop.

The emphasis above is on inefficient costs. The following is from a proposal to establish a regulating committee to end sex discrimination at a community college. The problem does not entail costs, but details concrete evidence of unethical discrimination.

Problem:

Some sex discrimination facts present themselves:

1. Of the 215 faculty members, 79, or 36.7 percent, are women, a percentage that is not reflected in either administrative positions or standing committee membership.

2. Of the 36 administrators, only three (3) are women; of the ten (10) division chairpersons, none are women; of the twelve (12) department heads, only two (2) are women; of the twenty (20) area leaders, only five (5) are

women; of the total 78 positions, only ten (10), or 12.8 percent, are women.

3. Of the 207 faculty and administrators serving on standing committees, only 52, or 25 percent, are women. One committee has no women members.

4. Women staff have voiced concern that the inequitable number of women in administrative and standing committee positions is a "negative incentive" for innovative teaching, volunteer assignments, and requests for advancement.

5. Women students have voiced concern through the agency of the student government association that the college does not actively counsel and provide for women students to excel nor to set goals commensurate with the expanding opportunities for entry into the previously male-dominated professions.

Be thorough and exacting in your documentation of the problem. The information in this section will be referred to when you detail the consequences of adopting your proposal.

PROPOSAL

Present the solution to the problem by providing all of the particulars of your proposal. As already suggested, you may wish to consider the following subheadings:

Equipment/Procedure
Capabilities
Costs
Personnel
Timetable

If the proposal involves the purchase of new equipment, describe it accurately and explain its function and capabilities. If the proposal involves a new procedure or policy, explain exactly how it will work. Under costs include initial purchase price, financing, installation, labor, and training costs. If new or transferred personnel are involved, include the qualifications for the position or the qualifications of the employee. Summarize duties and salaries. If you use the traditional format, include a timetable or work plan for making your proposal operable in this section. If you use the streamlined format, place your schedule of tasks to be accomplished in the final action section. An effective proposal not only seeks action but also becomes the blueprint by which the action will be performed. Because the actual proposal section of your report is often lengthy, samples are not given here but may be reviewed in Figure 15.6.

CONSEQUENCES

This section may be divided into *Advantages and Disadvantages* or *Strengths and Limitations.* Here you refer to the details in your problem section and explain how each problem will be solved or alleviated.

If your proposal costs money, show in graphics how implementation will save the company money over projected periods of time. Often a large expenditure will ultimately save money over a few years by increasing efficiency or production. If you propose a new system, number and list all its benefits. Employee morale is an important concern of management; if adoption of your proposal will improve employee morale, detail these benefits.

Do not ignore disadvantages (temporary work stoppage, layoffs of employees, or limited capabilities). Discuss these disadvantages or limitations in a positive manner.

Following are brief sections from proposals outlining consequences. The first example is from a proposal to pave an American Legion Hall parking lot:

Consequences:

By paving the existing parking lot, we will be able to accommodate 30 or more cars in addition to the present 140. By eliminating the problems of dust and erosion, the physical appearance and value of the property will be enhanced. During the two weeks of construction, Sunset City officials have agreed in writing to allow our 57 employees and approximately 100 daytime visitors to park in the Sunset City Hall west parking lot.

The second example is from a proposal to adopt new displacement and furlough rules for an airline pilots' association:

Advantages:

Adoption of the system-wide seniority rules in the pilot displacement and furlough process results in the following advantages:

- Senior pilots are able to displace at any base.
- The total number of displacements arising out of a curtailment situation is small, and the displacements are mostly confined to the base where the curtailment situation occurs. Thereafter, significantly fewer pilots are affected by curtailment.
- A maximum of two base moves exists for every pilot curtailed.
- Protection of captaincy is guaranteed for all but the junior captain at base. The junior captain has base protection.
- The indicated cost saving is 30 percent over the current operating agreement.

Disadvantages:

These benefits are achieved at the cost of limiting pilots' freedom of choice in the following manner:

- A pilot's displacement choices are limited.
- If a captain desires to revert to first officer status to displace in a particular equipment, he or she may not be able to do so.

CONCLUSIONS

The body of your proposal has examined the problem and presented a blueprint for action. You have anticipated the questions and possible objections to your proposal and objectively, but persuasively, responded to them. Your closing section is essentially an urge to action. In the traditional format, summarize your proposal features and reemphasize the advantages in numbered statements. Close with a persuasive statement, such as

> I urge you to give serious consideration to this proposal and am available to discuss its particulars with you.

In the streamlined format, title your section *Action* and present your schedule, work plan, or timetable. Figure 15.6 illustrates the optional closing section for proposals.

Your signature, position, and date are appended to the final page in the right-hand corner of the paper:

Janet Gorky
_____

Janet Gorky, Assembler   11/15/0X

Figure 15.6 illustrates all of the particulars of a thorough proposal.

A PROPOSAL

THE DEVELOPMENT OF A

REIMBURSABLE EXPENSE POLICY

AND

MONITORING AND CONTROL SYSTEM

Prepared for

Mr. Duane P. Morton

President

Ace Electronics Company, Inc.

Fresno, California

By

Charles E. Smith

Senior Partner

Smith Business Consultants

San Francisco, California

24 February 2000

FIGURE 15.6   *Sample traditional proposal with formal report features*   (Courtesy of Charles E. Smith; adaptation)

SMITH BUSINESS CONSULTANTS
1000 Plaza Court
San Francisco, California 94536
213-659-1222

24 February 2000

Mr. Duane F. Morton, President
Ace Electronics Company, Inc.
2120 West Broward Boulevard
Fresno, California 93710

Dear Mr. Morton:

We have completed the study that you authorized in November 199X concerning the control of reimbursable business expenses at Ace Electronics Company and are pleased to submit the enclosed proposal to make expense practices more equitable.

We appreciate the cooperation of your officers and other employees in conducting our study. All transcribed interviews and records of research findings will be kept in confidence for six months should you need to review this material.

We thank you for this opportunity to be of assistance.

Very truly yours,

Charles E. Smith

Charles E. Smith
Senior Partner

DES:jsv

Enclosure

ii

FIGURE 15.6   *continued*

TABLE OF CONTENTS

FIGURE 15.6   *continued*

SUMMARY

This report for Ace Electronics Company, Inc. proposes the development of a reimbursable expense policy along with a monitoring and control system to make expense practices more equitable, to allow for abuse detection, and to reduce costs.

Extensive interviews and expense records research reveal that expenses are excessive, existing policy is inadequate, and methods of approval and control are deficient. By implementing this proposal, the approximately yearly $3.3 million reimbursable expenses may be reduced by as much as $1 million.

We propose the development of a written reimbursable expense policy that incorporates a matrix of all allowable expenses, categorizes employees, sets conditions and restrictions, and clearly states approval individuals. Further, we propose procedural regulations that establish periodic report periods and deadlines, entail new report forms, and institute a definite system of review and analysis. A work plan to carry out these proposals includes the appointment of a director, the establishment of committees, and a listing of chronological duties.

By implementing this proposal reimbursable expenses shall be reduced, and adoption may well lead to cost reduction sensitivity in other expense areas.

FIGURE 15.6   *continued*

PROPOSAL TO DEVELOP A REIMBURSABLE EXPENSE POLICY AND MONITORING AND CONTROL SYSTEM

INTRODUCTION

Purpose

This proposal to develop a reimbursable expense policy and monitoring and control system is designed to make expense practices more equitable, to establish a system which will detect and prevent abuses, and to reduce costs.

Scope

This report presents a description of our study, a review of our findings, a dual proposal, the advantages of implementation, and the conclusions.

BACKGROUND

Procedure

To determine the facts relating to reimbursable expenses, we used a sample basis. With the assistance of four senior executives, a sample of 25 of the approximately 100 executives at Ace was chosen. Personal interviews were held with 20 of them, and extensive records research was conducted for 15 covering the second quarter of 2000, the period chosen for the analysis.

Our first step was to interview these officers, two financial managers, and the personnel manager and to examine a sampling of expense reports. From that work, we prepared two documents: a list of discussion questions for the interviews, and an outline of the records research requirement. Appendixes A and B exhibit these documents and indicate the extensive nature of the fact finding.

The Smith consultants conducted the interviews, and an Ace team directed by the assistant corporate controller conducted the records research.

1

FIGURE 15.6 *continued*

2

Findings

A. Scope of Expenses

Reimbursable business expenses represent a significant controllable cost at Ace. During the first nine months of 2000 total expenses for officers and others was $2.5 million, an annual expense of over $3.3 million a year. For the officer group alone, the figure was $1 million a year.

B. Policies and Practices

Numerous examples of excessive expense practices and failure to control expenses are evident upon examination of the records and by admission of the officers in confidential interviews. We believe that reimbursable expenses at Ace could be reduced by $1 million a year without adversely affecting the business.

As a result of unclear or inadequate policy or poor communication, most interviewees displayed uncertainty about what was expected of them. During the interview officers were asked: "Can you identify the written expense policy/procedure as it relates to you?" Following are representative answers:

"One might exist, but I'm not familiar with it."

"There's only a memo on transportation."

"There's no written policy."

"Policies are not spelled out. I'm not familiar with any written policy. I do what I did for my former employer."

Further, existing written expense policies lack clarity.

- A limited number of items that may be authorized are listed, but the conditions under which they are authorized are not always clear. For example, statements, such as "when time is a factor," "where necessary for company business," and "only to employees who hold the types of jobs that require such meetings," appear in the policies.
- Some expenses require prior approval, but the approving individual is not always identified.
- Some items of allowable expenses, such as home entertainment and telephone answering services, are not treated.

FIGURE 15.6   *continued*

3

- The frequency of reporting is not covered.
- A few items, such as social club memberships and access to the executive dining room, are inadequately treated. In some cases these are considered executive perquisites and in others reimbursable expenses.

C. Approval and Control

Methods of approval and control are inadequate.

- Officers report they are not comfortable when questioning the expense practices of close associates, and rarely do so.
- Expense documentation and explanation are inadequate. Reporting forms fail to determine such things as class of air travel, cost of overnight accommodations per night, purposes of business meetings, the number of people entertained, and so forth.
- Budgetary control is difficult. Monthly reports do not show expenses by individuals, but by divisions.
- Expense reporting is often tardy. For the sample offices, timeliness ran from prompt (2 to 3 days after the reported period) to late (6 months). Many officers reported weekly, others monthly, and some on a less periodic schedule.
- The expense report is not a complete record of expenses. Some officers are reimbursed for expenses directly from petty cash, or the invoice for an expense item is paid directly by Ace and not shown on an expense report.

D. Variations

The interviews and the records research reveal a wide range of variable practices.

- In the sample of the 15 officers whose expense reports were analyzed, only 7 charged telephone expenses, 11 had social club expenses, 10 were reimbursed for gifts, and 5 incurred expenses for personal entertainment.
- Air travel practices further illustrate the variety that exists. During the analyzed quarter, 3 of the officers made at least one flight, but 2 of them did not submit ticket receipts. Of the 11 other employees whose flights could be analyzed, 6 flew first class and only 2 paid the difference between first class and coach fare. Seven wives flew with their husbands at company expense, 2 of them first class.

FIGURE 15.6 *continued*

4

- There is considerable variation in the cost of overnight room accomodations. The range is from $60.00 to $250.00 a night.
- The average per meal cost of business meetings range from $50 to $148. Business meetings were usually conducted at lunch.

The great variation in expense practice is the basis for our conclusion that substantial opportunities for cost reduction of at least $1 million a year exist.

PROPOSAL

To correct these deficiencies in the reimbursable expense system, we propose:

1. The development of a written reimbursable expense policy
2. The establishment of firm procedural regulations

Expense Policy

Ace should prepare a written reimbursable expense policy that is clear with regard to each type of expense, that accomodates differences among employees, and that covers certain procedural requirements. To accomplish this:

1. All personnel should be divided into categories. All employees should not be treated uniformly. Their expense spending requirements and privileges should be recognized in the policy. We recommend that the two categories be designated as follows:

   Members of the Board   All Other
   Vice-presidents   Employees

2. A policy matrix should be devised which lists all reimbursable expenses, employee category differences, restrictions and conditions, and approval authority. Appendix C illustrates a sample policy matrix.

FIGURE 15.6   *continued*

5

3. A clarifying policy should be written for wide distribution and inclusion in the Administrative Policy Manual. The policy should include guidelines and standards, procedures for advances, procedures for preparation, processing, and approval of the expense report, and specific guidelines and standards for each type of reimbursable expense in the matrix. Appendix D shows a sample administrative policy page.

Monitoring and Control

The following procedural regulations should be established:

1. The report period should be monthly; deadline for submission should be one week thereafter.
2. Expense report forms should be revised to require documentation and explanation.
3. The Expense Report must be established as the sole vehicle for recording and reimbursing expenses.
4. Responsibilities of the reporting individual, the controller's department, and the approving individual should be developed into a definite system.
5. The role of the controller's department should be expanded. An editing function there should first evaluate the adequacy of the documentation and explanation, and the Expense Report should be returned to the reporting individuals for correction if there are deficiencies. Next the report should be examined for conformance to policy, and any exception should be noted on a Buck Slip which, along with the Expense Report, should be forwarded to the approving individual.

   The controller's department should prepare a summary analysis of each individual's expenses on a quarterly basis. Quarterly and cumulative expenses should be compared to budget.

FIGURE 15.6   *continued*

6

Work Plan

I.   Staffing
     A. Appoint the Vice-president of Finance as Project Director
         with responsibility for implementing all recommendations.
     B. Appoint a Corporate Expense Policy Committee to work
         with the Project Director.
     C. Assign a small staff to work with the Committee without
         interruption.
II. Expense Policy
     A. Establish the basis for assigning all personnel into the
         recommended two categories.
     B. Prepare an expense matrix and policy for each type of
         reimbursable expense for each category of personnel.
     C. Submit the policy to officers for discussion and
         modification.
     D. Approve the policy.
     E. Disseminate the policy and conduct familiarization training.
III. Monitoring and Control
     A. Revise Expense Report format and create Daily Expense
         Diary, Buck Slip, and Quarterly Analysis.
     B. Establish the edit and analysis functions in each division.
     C. Train expense report editors and analysts.
     D. Test the system and make adjustments.
     E. Commence live operation of the system.

ADVANTAGES

By developing a written reimbursable expense policy and
establishing firm procedural regulations, the following advantages
should accrue:
     1. Reimbursable expense practices should become more
         equitable over the broad organization due to clarification
         and control.
     2. A system for detection and prevention of abuses will be
         established.
     3. Superiors shall exercise more control of expenses.

FIGURE 15.6   *continued*

7

4. Quarterly analysis should encourage discipline and facilitate budget preparation.
5. Implementation should reduce costs.

CONCLUSION

An attractive opportunity for cost reduction exists and provides the basis for achieving economies through innovations in policy and control. In our opinion, the recommendations in this report are well worth implementing and, in addition, will lead to sensitivity towards cost reduction in other areas.

Charles E. Smith   *2/24/00*

Charles E. Smith   2/24/00
Senior Partner

CES:jsv

FIGURE 15.6   *continued*

APPENDIX A

REIMBURSABLE EXPENSE INTERVIEW TOPICS

1. Position held during second quarter, 2000
   a. Title
   b. Superior, subordinates
   c. Nature of position
   d. Approval authority for expense reports
2. Whose expense reports do you approve?
   a. Are summaries prepared?
   b. How do you control subordinates' expenses?
   c. Do you ever question their expenses? Details.
   d. Have you ever had a drive to reduce expenses? Explain.
3. Who approves your expense reports?
   a. Are summaries prepared?
   b. Are your expenses ever questioned? Details.
   c. Do you ever seek prior approval for expenses? Details.
   d. Do you have any prior understanding with regard to your expenses?
4. What is the budgetary control over your expenses?
   a. In what account in what cost center are your expenses accumulated?
   b. How do your expenses compare to budget?
5. Have you ever been told what the expense policy is as it relates to you? Details.
6. Can you identify the written expense policy/procedure as it relates to you? Details. What are the salient points?
7. Identify the forms that are used: expense reports, petty cash disbursements, report of outstanding advances, summaries.
8. What is the frequency of reporting and approval and what is your timeliness? Any difficulties?
9. What are the controlling policies/procedures with regard to advance accounts, and what are your practices?
10. List all of the categories of your reimbursable expenses.

8

FIGURE 15.6   *continued*

9

11. Do you incur reimbursable expenses that are not shown on your expense report? If so, where are they recorded?
12. What are the significant restrictions on your expense spending practices?
13. Do you ever misrepresent expense items, e.g., combining minor items to recover excessive charges?
14. What is the influence of your status on your expense practices? Is there discrimination by rank? Amplify.
15. How do you view the company attitude concerning expenses? Have there been any changes? How does it compare with other companies in which you have worked?

FIGURE 15.6   *continued*

RECORDS RESEARCH OUTLINE

I. Sources of information
    A. Expense report
    B. Summary business expense report
    C. Petty cash disbursement form
    D. Report of outstanding advances
    E. Approval authority
    F. Quarterly Budget Report for appropriate cost center
II. Information required
    A. Controlling policy/procedure
    B. Approval authority
    C. Summary
    D. Type of form
    E. Completeness and deficiencies
    F. Date and total expenses for period
    G. Documentation check
    H. Assumptions
III. Quarterly Summary
    A. Transportation
    B. Hotel rooms
    C. Meals
    D. Business meetings
    E. Other
IV. Analysis
    A. Personnel category differences
    B. Direct payments
    C. Nature and amount of petty cash disbursements
    D. Timeliness
    E. Discrepancies
    F. Significance

10

FIGURE 15.6 *continued*

SAMPLE REIMBURSABLE EXPENSE POLICY MATRIX

Type of Expense	EMPLOYEE CATEGORY President Vice-pres and above	All Other Employees	Restrictions & Conditions	Approval Required
Airline and Railroad	First Class Reimbursable	a) Coach reim- bursable b) First Class Restricted	Restrictions: a) Traveling in company of Pres., V-P, or above b) Duration of trip exceeds 7 hr c) Traveling with cus- tomer, etc.	President Vice-Pres or above
Buses	Reimbursable	Reimbursable		
Rented Cars	Reimbursable	Reimbursable	Condition: a) Other forms of public transport cannot meet business requirements b) Standard size low- priced car	
Personal Cars	Reimbursable	Reimbursable	Condition: Must not be used on long trips when other trans- port is less expensive and obtainable	
Taxi Cabs	Reimbursable	Reimbursable	Condition: Other forms of public transport cannot meet business requirements	
Limousines and Company Cars	Restrictively Reimbursable	Restrictively Reimbursable	Restriction: a) Essential b) Mileage log main- tained	President Vice-Pres or above
Air Charter	Reimbursable	Reimbursable	Restriction: Extreme emergency situation only	President Vice-Pres or above
Lodgings Single Rooms First Class Accommodations	Reimbursable	Reimbursable		
Suites	Restrictively Reimbursable	Restrictively Reimbursable	Restriction: Essential for the conduct of business	President Vice-Pres or above

11

FIGURE 15.6   *continued*

SAMPLE ADMINISTRATIVE POLICY PAGE

ACE Administrative Policy Manual

Section	Dist. List	Date Issued	Policy No.
REIMBURSABLE BUSINESS EXPENSES 3		12/30/0X	RB-1

Subject

Ace Reimbursable Business Expense Policy   Page 3
                                                                 of 39

III. TYPES OF REIMBURSABLE BUSINESS EXPENSES
      For ease of reference, a matrix summarizing reimbursable
      expenses has been included in Section XI of this Policy.

   A. **Transportation Expenses**
       All requests for transportation should be submitted
       through the Transportation Department.

       Where use of the Transportation Department is
       impractical, an employee may arrange his own
       transportation.

       The restrictions and documentation required for
       transportation expenses are as follows:

       1. **Airlines and Railroads**—Coach type accommodations
           are to be used. All expenses in this category must be
           documented by a ticket stub when submitting the Ace
           Expense Report. When a ticket purchased by Ace is
           not used, it should be returned to the Controller's
           Department with the expense report for a refund. The
           amount of the unused ticket should be included on the
           expense report.

           When traveling to and from the airports and Ace,
           public transportation should be used wherever
           practical.

       2. **Buses**—The use of long-distance buses is permitted
           when they can reasonably meet the needs of the
           Company.

                                 12

FIGURE 15.6   *continued*

3. **Rented Cars**—Use of rented cars should be limited to those situations where public transportation is not available or cannot meet business requirements. When the situation requires a rented car, short-term arrangements should be made by the individual. On rentals exceeding one month, arrangements should be made through the Ace Purchasing Department. On all rentals, a standard size, low-priced car should be used. In all cases optional insurance should not be pur-chased and where national car rental services are used, the discount should be obtained. When submit-ting this expense for reimbursement, a copy of the itemized invoice is required.

13

FIGURE 15.6   *continued*

CHECKLIST

Longer Reports and Proposals

ALL LONGER REPORTS

❑ **1.** Have I identified the type of longer report I am writing?

 ❑ Quarterly or annual report?

 ❑ Long-range planning document?

 ❑ Systems evaluation?

 ❑ Proposal?

 ❑ Other?

❑ **2.** Have I designed an appropriate title page? Does it include

 ❑ Name, position, and affiliation of the receiver?

 ❑ My name, position, and affiliation?

 ❑ Date?

❑ **3.** Have I written an appropriate (letter or memorandum) transmittal correspondence?

❑ **4.** Does the transmittal correspondence contain all of its required sections?

 ❑ Title and purpose of the report?

 ❑ Identity of the commissioner of the report and when it was requested, or why it is being submitted?

 ❑ Comments on encountered problems?

 ❑ Acknowledgment of others who assisted in the report preparation?

 ❑ Elicitation of feedback?

❑ **5.** Have I written an appropriate abstract or summary?

❑ **6.** Have I included a table of contents and list of illustrations?

❑ **7.** Have I attached appropriate back matter (glossaries, survey instruments, survey tabulations, financial projections, job descriptions, personnel resumes, equipment brochures, list of references, or other material)?

❑ **8.** Have I designed an attractive document?

PROPOSALS

❑ **1.** Have I developed items 1 to 8 above, which are relevant to all longer reports, in my proposal?

❑ **2.** Have I determined the nature of my proposal?

   ❑ Internal?

   ❑ External?

   ❑ Solicited?

   ❑ Unsolicited?

❑ **3.** Have I included a descriptive or informative abstract for a scientific or technical proposal?

❑ **4.** Have I included an executive summary for management that is persuasive, causally organized, and of appropriate length?

❑ **5.** Have I used a thorough and readable format? Does it include

   ❑ Statement of the objective and purpose?

   ❑ Explanation of the background and problem?

   ❑ Discussion of advantages and disadvantages?

   ❑ Conclusions, recommendations, and/or an action schedule?

❑ **6.** Does the proposal cover all necessary aspects?

   ❑ Equipment/procedure?

❑ Capabilities?

❑ Costs?

❑ Personnel?

❑ Timetable?

❑ Other?

❑ **7.** Have I signed and dated the proposal?

EXERCISES

1. **Descriptive Abstract.**   Locate and photocopy a scientific or technical article from a professional journal. Write a short *descriptive abstract* that presents an overview of the purpose, scope, and methodology of the article. Do not confuse the abstract with an executive summary.

2. **Informative Abstract.**   Using the same article, write a brief *informative abstract* for the same material. Include the major conclusions.

3. **Executive Summary.**   Write an executive summary for the following brief proposal:

    Subject: Modification of the North Campus perimeter road and parking lot access road speed bumps.

    Objective: To alleviate complaints and to prevent further damage to students' and visitors' automobiles.

    Problem: Since the installation of 80 speed bumps on the campus perimeter road and parking lot access roads in March 200X, more than 100 students and other campus visitors have registered written complaints to the Security Department regarding damage to their automobiles. The height of the bumps is 8 in.; compact cars have only an 8-in. clearance. Further, 47 percent of the complaints are by drivers of large and intermediate automobiles. Reported damages include front end misalignments, rear end leakage, and muffler dents.

    Inspection reveals that the bumps are excessively gouged and scraped. Although some automobile damage may be a result of excessive speed, a State Road Department inspector agreed, following his 15 June 200X visit, that the bumps are too high and too abrupt.

    Appendix A exhibits a copy of his letter.

Proposal: Therefore, we propose that the 80 speed bumps be modified by grinding the peaks to a 6-in. height and sloping the sides with asphalt to a 35-degree slant. These modifications will reduce the shock that automobiles are experiencing yet still deter speeding. Figure 1 shows a cross-section of the speed bumps before and after modification:

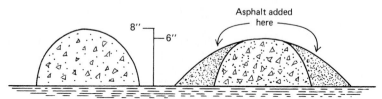

FIGURE 1   *Cross-section of a typical speed bump before and after modification*

Equipment: Because the college does not own the equipment necessary for the modifications, the equipment must be rented. Necessary rental equipment includes

- One hand-held, gas-operated, heavy duty grinder with emery stone disk
- One gas-operated hot asphalt mixer
- One water-filled, hand-operated 500-pound roller

Labor: No special skills are needed to operate the equipment. Three Maintenance Department personnel can complete the modifications in an estimated 120 hours over two weeks.

Procedure: One man can grind the bumps at the rate of 1½ hours per bump. Two men can mix and apply the asphalt in five 6-hour shifts.

Cost: Modifications can be completed at an estimated $380.00 as shown in the following table:

Speed Bump Modification Cost Estimate

Item	Cost ($)	Time	Quantity	Total ($)
Grinder	55.00/wk	2 wk	—	110.00
Mixer	140.00/wk	1 wk	—	140.00
Roller	10.00/wk	1 wk	—	10.00
Asphalt Mix	2.00/55	lb bag	25	50.00
Pebbles	70.00/load	—	1	70.00
			TOTAL	380.00

Advantages: 1. Speed bumps will still deter speeding.

2. Damage to automobiles should decrease.

3. Project can be accomplished by existing staff.

4. Project can be accomplished prior to Term I traffic.

Action: 1. Approve proposal by 15 July.

2. Arrange for rental equipment and supplies by 1 August.

3. Modify speed bumps 1–15 August while the campus is closed to regular classes.

4. **Title Page.** Title the proposal in Exercise 3 and compose a title page. You are the Supervisor of the Maintenance Department of your college, presenting the proposal to the campus provost or dean.

5. **Memo of Transmittal.** Write a memo of transmittal. Include a brief summary of the proposal. Acknowledge an individual or company for assistance in obtaining equipment and supply prices.

WRITING PROJECTS

1. **Individual or Collaborative—Long Proposal.** Write a long proposal to solve a specific problem on your campus or at your workplace. Select a subject of sufficient complexity to warrant a proposal in longer report format, but do not tackle something beyond your scope. Chapter 5 discusses information access, including library research, surveys, the Internet, and subsidiary sources. Your proposal should require some research by survey, interviews with authorities, journal articles, equipment brochure perusal, and related files. To clarify the problem, explain what procedures are being used now and how they are faulty. Document excessive costs, inefficient time use, damages, accidents, thefts, security problems, and so forth. Use survey results, interviews, and other research results to establish the severity of the problem. Present your actual proposal along with capabilities, costs, features, personnel, and a time schedule. Relate the points in your consequence section to the points you made in the background or problem section, explaining how each problem will be alleviated. Be persuasive. After you complete your proposal draft, prepare all of the appropriate longer report features: title page, transmittal correspondence, a table of contents and list of illustrations, abstract and/or summary, and supplements, such as survey instrument, interview questions, pages from equipment brochures, and the like, a glossary, and a bibliography. Submit your proposal in an attractive folder. Suggested subjects are

On the Campus	*At Work*
new equipment	new equipment
new system of registration	new uniforms
improved parking facilities	improved systems of duty roster
picnic or stone tables	improved method of displaying wares
snack bars by classrooms	index for wage increases
additional computer lab time	improved lounge facilities
additional study carels	improved working conditions
new special interest club	new in-service training program
Saturday or 5:00 P.M. classes	sports team program
a women's center	fire evacuation procedures
a campus jogging path	an advertising program
student entertainment project	new staff orientation program
free movie program	grievance procedures
improved bomb scare procedures	dental insurance program
other?	other?

NOTES

Producing Professional Papers

ZITS by Jerry Scott and Jim Borgman

Reprinted with special permission of King Features Syndicate.

S K I L L S

After studying this chapter, you should be able to

1. Understand the nature of professional scientific and technical papers.
2. Understand the difference between scientific writing and science writing.
3. Locate and be familiar with journals in your field.
4. Discover the editorial policy of three or more technical journals in your field.
5. Understand how to submit an article for publication.
6. Critique the overall organization and content of a scientific/technical paper.
7. Select and limit scientific/technical and academic subjects.
8. Formulate a suitable thesis for a professional paper.
9. Prepare a working and polished outline.
10. Brainstorm a report by considering specific expository development methods.
11. Develop a suitable introduction and closing to a professional paper.
12. Write a brief, documented professional paper.
13. Begin to recognize different styles of documentation by examination of the sample papers.

INTRODUCTION

Professional papers include documented scientific and technical articles intended for publication in special interest journals and documented academic papers. The latter include undergraduate term papers, graduate theses, and dissertations, intended to hone the writer's research skills and to present experimental results and original insights into existing knowledge in a given field. Both types, scientific/technical articles and academic papers, are characterized by documentation, accuracy, and clarity.

Science writing and scientific writing are not quite the same thing in that they address different audiences. *Science writing* explains scientific matters, such as new discoveries, theories, and research, to laypeople in magazines and journals aimed at a general audience. Excellent examples of science writing can be found, for example, in the magazine *Discover, The New Yorker* magazine section "Annals of Science," *The New York Times,* and other major newspapers. Courses in science writing are taught in journalism schools. *Scientific writing* is done by scientists writing about science (those same discoveries, theories, and research) for other scientists in specialized journals aimed at specific audiences. Technical writing involves not only all of the strategies covered in previous chapters but also, as in scientific writing, preparing articles for publication. Moreover, these technical articles are designed for peers, rather than laypeople. Generally, science writing is not documented and uses more expressive language than scientific/technical and academic papers. Compare the differences in language in the following two examples of writing about photographs from the Hubble Space Telescope:

Example 1—Scientific Writing

The photograph taken inside the Eagle Nebula depicts three columns of dark dust and gas rising into a blue sky, which reveals new-forming stars. The columns are illuminated from the heat of the new stars; a red turns to gold at the edges. Spirals of interstellar material form into steep peaks and disperse into weblike forms.

Example 2—Science Writing

The photograph [taken inside the Eagle Nebula] must be the Hubble Space Telescope's most emotion-laden image yet—an icon to rival the Apollo photograph of the Earth taken from the surface of the moon. It depicts three eerie pillars of dark dust and gas thrusting up into a blue iridescence sprinkled with newborn stars. The pillars glow from the heat of star birth: a deep red fades to halos of gold at their margins. Whorls of interstellar dust alternately billow into cliff-top peaks and trail off in wispy webs of stellar Spanish moss.[1]

Scientific/Technical Articles

Professional papers written for publication are prepared by researchers to inform other specialists and those in similar areas of current developments in the field. These papers report on new research, discoveries, new methodology, and inventions to keep colleagues abreast of the latest advances. Because the articles have been reviewed by an editorial board of specialists, readers may rely on the authority of the material. Students should seek out the journals that print papers in their field and read at least one or two of them regularly. The most recent findings in any field will be in the journals, not in textbooks.

---

[1] © 1996 *The New Yorker* magazine. Davis Sobel, "A Reporter at Large: Among Planets," December 9, p. 90.

Academic Papers

As a student, you probably think that you will not be required to write scientific and technical papers until much later in your career. However, it is likely (especially if you are majoring in one of the sciences or technologies) that you will be required to write professional-level papers while still in college. The papers you may write as an undergraduate student are most likely to be confined to either class-directed laboratory reports or papers reporting on secondary research you have done for a particular class. On the other hand, papers written at the graduate level are often the result of original, primary research and experimentation. These papers are often submitted for publication by the student, the professor, or both and must be in perfect form, as your reputation (and future) may be greatly hindered by a poorly written paper. Conversely, an excellent paper will enhance your employment and advancement potential.

THE SCIENTIFIC/TECHNICAL PAPER

The professional writing a scientific/technical paper must consider four elements for successful submission and/or publication:

- Designing, researching, and completing the project
- Selecting the intended journal for publication
- Writing the paper
- Submitting the paper for publication

Designing, Researching, and Completing the Project

The scientific/technical paper usually requires primary and secondary research (see Chapter 5). The interpretation or thesis of a scientific/technical paper often comes from reading conflicting technical reports and articles by other authors or from one's own intuition. Unlike secondary research in which the writer proves a premise with facts determined by other scholars' research, primary research involves using original sources, such as experimentation, surveys, or other means, to obtain first-hand data.

Some projects may involve preapproval and funding for primary research. One may seek approval for the project from an institution, a targeted journal, or a variety of funding sources in the field. If the research is primary and is funded by a university or other outside source, accurate records must be maintained along with any combination of laboratory reports, progress reports, and even EPA (Environmental Protection Agency) and OSHA (Occupational Safety and Hazard Act) reports. Once the research has been completed and verified, the writer needs to compose the paper for submission to the university or the appropriate journal.

Selecting the Journal

Review journals in your field in order to select one for article submission. Determine which are the most prestigious, which are most likely to publish your type of material, and which are aimed at your desired audience.

INSTRUCTIONS TO AUTHORS

Most journals publish statements in each issue that define the intended audiences, comment on their editorial policy for selecting unsolicited articles, specify style and documentation guidelines, and include mailing directions. You may review the instructions to authors from more than 3,000 journals in the health and science fields by accessing **<http://www. mc.edu/lib/instr/libinsta.html>**. Even if you are not considering any of these particular journals, look up a few to familiarize yourself with the variety of requirements. If you have some other journal in mind, you can likely locate it on the Internet using its title in a keyword search. Of course, these journals are available at libraries, too. Look for their statements on editorial policy and guides for authors, which detail manuscript formats (margins, pagination, headings, title pages, and abstract requirements), stipulate text requirements (length, number of copies and disks to submit, and so on), specify documentation styles, and suggest methods for handling tables and figures. Some journals require your credentials in a formal or informal resume, your institutional affiliation, and other support materials. Figure 16.1 is an example of author submission guidelines for the *Journal of Dental Research* and the *Technical Communication Quarterly*, respectively.

QUERIES

If you are seeking to publish in a journal, you may wish to approach those journals that seem most promising in advance. In a brief letter, you should ask if the journal is interested in your intended paper by clarifying the specific subject, emphasizing the importance of your article, explaining the approach you have used to develop the paper (for example, testing and research), and summarizing the particulars (length, graphics, and other features). You should also include your credentials, organization affiliation, address, phone number, fax number, and e-mail address.

Association of Teachers of Technical Writing

Home   ATTW Information   Publications   ATTW Conference   News
Jobs   Calls for Papers   Teaching   Academic Programs   Resources   Search

Technical Communication Quarterly
Submission Guidelines

TCQ publishes theoretical articles with industrial and pedagogical applications as well as practical articles with a sound basis in theory. Articles cover a range of professional writing topics, including pedagogy, rhetoric, linguistics, ethics, organizational communication, business/industrial communication, intercultural communication, text design, graphics, audience analysis, electronic communication, and documentation issues as they pertain to technical communication.

Follow these guidelines when preparing manuscripts for submission:

- Send an original and four copies of the manuscript.
- Prepare a floppy disk for submission when the article is accepted. Place the entire manuscript in one document using ASCII, Word, WordPerfect, or Macintosh text files.
- Prepare a cover sheet with the title of the manuscript, author's name, place of affiliation mailing address, phone number, and a 35-word biographical sketch.
- Do not make references to the author in the text or on any page besides the cover sheet.
- Center the title on top of the first page of text.
- Do not exceed 8000 words in the text.
- Provide a 50-75 word informative abstract.
- Tables and figures should appear on separate pages at the end of the text.
- Provide camera-ready quality copies of all figures.
- Include no footnotes.
- Indicate italics by underscoring or italicizing.
- Indicate a first-level, or primary, header by boldface or all capital letters.
- Indicate a second-level, or secondary, header by indenting an boldface or all capital letters.
- Use the parenthetical method for citing a reference in the text, according to The MLA Style Manual (1999). Provide a list of works cited at the end in a section called "Works Cited." Follow The MLA Style Manual in the "Works Cited" section. Include first names of all authors in the "Works Cited" and in the first mention in the text.

FIGURE 16.1   *Another submission guidelines* (TECHNICAL COMMUNICATION QUARTERLY's Author Submission Guidelines found online at www.attw.org are reprinted by permission of Technical Communications Quarterly.)

Writing the Report

The professional scientific/technical paper usually includes the following elements:

- A precise title
- Definition of the problem
- Review of the literature
- Methods and materials
- Results
- Discussion
- Conclusions and implications
- Documentation

TITLE

Often, a journal will require both a short title, such as "Improving Testing Components," and a longer title, such as "Environmental Stress Testing: Improving the Quality and Reliability of Testing Components." Essentially, devise a precise title that clearly indicates the focus and the content to follow.

DEFINITION OF THE PROBLEM

Defining the problem is the major task of the introductory material. What weaknesses or shortcomings in previous techniques and methods have prompted the investigation that has led to the preparation of the scientific/technical paper? State the purpose of the research that focuses the resultant paper. State your thesis clearly.

REVIEW OF THE LITERATURE

Review the results found in other literature on the subject. Summarize the findings to provide a context for your new discussion; the summary lends authority by indicating that you have a command of the subject.

METHODS AND MATERIALS

Next, explain the methods and materials that you have used in your research. Decide whether to discuss your methods in the passive voice ("Next, the leg bone was connected to the thigh bone"), in the active voice ("The leg bone connects to the thigh bone"), or in the instructional mode

("Connect the leg bone to the thigh bone"). Carefully explain the steps taken in the investigation.

RESULTS

Discuss the exact results of your investigation. The methods section and results section are the major components of your article.

DISCUSSION

Explain any steps, results, or information about the product or procedure that have not yet been clarified.

CONCLUSIONS

Finally, draw valid conclusions and suggest implications for future use of the product or procedure. These closing strategies are similar to the conclusions and recommendations of other reports.

DOCUMENTATION

Include the internal references and a final bibliography, which may be titled *Endnotes, Bibliography, References,* or *Works Cited,* depending on the documentation style you are using (see Chapter 6).

Figure 16.2 shows a professional technical article written for the *Journal of Dental Research* and includes the four headings: Introduction (the problem and review of the literature), Materials and methods, Results, and Discussion. It is documented in the CBE System style with variations in the reference listings requested by the publishing journal.

SUBMITTING THE REPORT
FOR PUBLICATION

Once you have written the paper and determined the submission guidelines of the selected journal, you should mail two copies of the article with a cover letter and either a self-addressed, stamped envelope (SASE) or loose postage, to ensure that the editors will acknowledge receipt of your manuscript. Usually, the manuscript is not bound, is in proper manuscript form, and has the author's name at the top, right-hand corner of each page.

Increasingly, journals request a computer disk of the document, either along with the hard copy or in place of it.

MATERIALS SCIENCE

Microscopic Study of Smooth Silver-plated Retention Pins in Amalgam

Y. GALINDO,* K. McLACHLAN,** and Z. KASLOFF***

*Université Laval, Ecole de Médecine Dentaire, Ste-Foy, Québec, Canada G1K 7P4; **University of Manitoba, Faculty of Civil Engineering, Winnipeg, Manitoba, Canada R3T 2N2; and ***University of Colorado, School of Dentistry, Denver, Colorado 80262

A silver-plating technique was developed in an effort to produce good mechanical bonding characteristics between stainless steel pins and amalgam. Metallographic microscope and scanning electron microscope (SEM) studies were made to assess the presence, or otherwise, of such a bond between (a) the silver layer plating and the surface of the stainless steel pins, and (b) and silver plating and the amalgam. Unplated stainless steel and sterling silver pins were used as a control and as a comparison, respectively. A "rubbing" technique of condensation was devised to closely adapt amalgam to the pins. It is concluded that there is strong evidence for the existence of a good bond between the plated pins and amalgam. The mechanical performance of the bond is discussed elsewhere.[1]

J Dent Res 59(2):124-128, February 1980

Introduction.

Metallic pins are used in dentistry to retain large amalgam restorations and to oppose tilting forces when the restoration is subjected to occlusal loads. The inability of these pins to bond with amalgam contributes to the formation of high stress concentrations around the top of the pin.[2,3] Fracture of the amalgam is likely to ensue.

Previous investigations[2] have shown that stress concentrations around smooth pins possessing bonding properties with amalgam are significantly less than is the case with non-bonded pins. Likewise, such pins increase the retention of the restoration[3] and markedly reduce the occurrence of fracture and crack propagation.[4,5] The reduction of strength, due to the inclusion of pins in amalgam specimens, has been shown to be much less when bonded, rather than unbonded pins are used.[2,4,6,7,8]

Received for publication November 29, 1978
Accepted for publication February 29, 1979
This study was supported in part by the Medical Research Council of Canada, Grant MA-4065.
*Present address: Canadore College, Health Sciences, P.O. Box 5001, North Bay, Ontario, Canada P1A3X9

This evidence led to the investigation and development of smooth pins likely to provide a metallurgical bond with amalgam, and be clinically acceptable in all other respects. A microscopic evaluation was selected for establishing the goodness of the pin-amalgam metallurgical adaptation which, in turn, would indicate the likely quality of any bond.

Materials and methods.

Smooth, silver-plated, stainless steel pins were the primary subject of this study. They were compared with sterling silver and unplated stainless steel pins. Sterling silver pins were used only as a mode for comparison, since it has been shown that they adapt well and form a metallurgical bond with amalgam.[1,4,5,7,8] Unplated stainless steel smooth pins were also included as a control, because it is well known that they do not form a bond with amalgam.

Silver-plated stainless steel pins were made with a core of 18-8 stainless steel orthodontic wire. In order to achieve the plating, the passive layer which forms on stainless steel was first eliminated by pickling the previously cleaned wire in an acid chloride solution. To prevent the passive layer from forming again, a very thin, microscopically non-detectable flash of nickel was applied by a plating process. This process consisted of immediately transferring the wire to an acid nickel-chloride solution at room temperature, and using a nickel anode for the electroplating procedure. Once the passive layer had thus been eliminated, and its formation avoided by the nickel flash, successful electroplating of silver onto the surface was readily performed. In this case, an electrolytic, low concentration, silver-cyanide plating solution was used with a stainless steel strip anode. This was followed by the final plating process, involving a highly con-

124

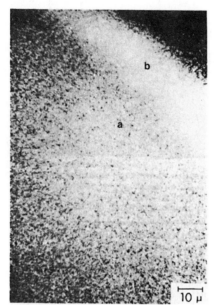

Fig. 1 – Nickel-plated stainless steel wire (X-ray microscope, scanning for nickel, 1250 X). (a) nickel contained in stainless steel wire (white dots), (b) nickel contained in layer of nickel plate.

in sterling silver, using standard methods. Sterling silver pins, as cast, yielded properties which were poor in strength and in modulus of elasticity. These properties were greatly improved by heat-treating the pins at 650°F for one hour.[3,9]

Amalgam specimens containing each type of pin were prepared and all specimens were constructed in the same way. The mold utilized was similar to that used for dental amalgam specimens, described in the American Dental Association Specification No. 1. During the course of the investigation, a technique for achieving good adaptation of the amalgam to the pin was developed. It consisted of thoroughly rubbing the first portion of the freshly mixed amalgam around the pin with a plastic instrument. This procedure assured a reaction between the whole surface of the pin and the amalgam, thus eliminating voids around the pin. All specimens were between 24 hours and five days old when viewed microscopically.

The three special nickel-plated stainless steel pin specimens were mounted to be analyzed under X-ray microprobe. Four specimens of each type of pin used in this research were mounted on bakelite bases for metallographic microscope examination, and similar ones were mounted on stubs with a conductive silver com-

centrated silver plating solution and a pure silver anode. In this manner, the good bonding characteristics expected from pure silver and amalgam could be combined with the desirable physical properties of stainless steel as a pin material.[3] This plating method is the one used for certain airplane sections, and was adapted for pins in dentistry.*

Because the thickness of the nickel flash was too minute to be observed directly, three separate stainless steel specimens were made. Each of these specimens was subjected to the nickel plating procedure for a period ten times as long as that used in preparing pins for the silver plating process. Under these conditions, the thickness of the nickel layer could be measured. Thus, an indirect measure of the thickness of nickel under the silver plating of the experimental pins could be obtained.

The method of making sterling silver pins is now described. Pin patterns were first made from the blue inlay wax and then cast

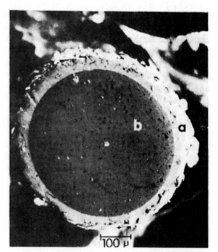

Fig. 2 – Silver-plated stainless steel wire (SEM, 170 X) (a) silver plate, (b) stainless steel wire. Transversal section.

*Bristol Aerospace, Winnipeg, Manitoba, Canada

FIGURE 16.2 *continued*

126   GALINDO ET AL.   J Dent Res February 1980

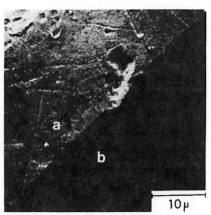

10 μ

Fig. 3 – Silver-plated stainless steel wire (SEM 3000 X), (a) silver-plate layer, (b) stainless steel wire. Transversal section.

Fig. 4 – Silver-plated stainless steel pin in amalgam. Conventional method of amalgam condensation (metallographic microscope, 350 X). (a) stainless steel pin, (b) silver-plate layer, (c) void, (d) amalgam. Transversal section.

pound** for scanning electron microscope (SEM) study. All specimens within each group presented similar, reproducible characteristics. Therefore, only microphotographs of typical specimens of each type were made.

Results.

Results obtained from the pin-amalgam specimens used as controls confirmed two expected and well known facts: a total lack of bond between stainless steel and amalgam, and a very good adaptation of amalgam to sterling silver. The latter strongly implies the presence of a bond, and is consistent with findings of other researchers mentioned earlier. It was, therefore, judged unnecessary to present, in this paper, micrographs of these specimens.

An illustration of the nickel coating on a stainless steel pin is shown in Figure 1. This picture was obtained with an electron probe scan for nickel on one of the stainless steel specimens that was subjected to ten times the normal nickel-plating period. From this it can be seen that complete continuity between the nickel in the pin and that of the plated layer exists.

Silver-plated stainless steel wire specimens as seen under SEM are presented in Figures 2 and 3. The results of this examination were consistent and confirmed that the

**Silver Dag - Acheson Colloids Canada Ltd., Brantford, Ontario, Canada.

plated layer of silver adapts very well to the stainless steel.

The effectiveness of the "rubbing" technique prior to condensation is confirmed in Figures 4 and 5, which are views under the metallographic microscope. A continuous, intimate adaptation of the amalgam to the pin and an absence of voids at the interface is evident. Similar results were obtained with the sterling-silver pin and were consistent with all specimens of both types used in this experiment.

Pin-amalgam specimens were likewise viewed under SEM. Figures 6 and 7 illustrate the close adaptation of the amalgam to the surface of the silver plating, so that a definite boundary between them is not readily discerned. In order to confirm the above results, a longitudinal pin-amalgam section, seen under a metallographic microscope, is shown in Figure 8.

Discussion.

In establishing the quality of any bond for the stated purposes, mechanical tests will be the final arbiter. However, microscopic study of the interface regions was necessary to confirm that metallurgical conditions existed, and that satisfactory adaptation had been achieved.

From the indirect evidence of the X-ray microscope (Fig. 1), the thickness of the nickel flash upon which the silver plating was performed was about one micrometer. The same evidence also shows that the nickel contained in the stainless steel and that from the nickel-plated flash are

FIGURE 16.2   *continued*

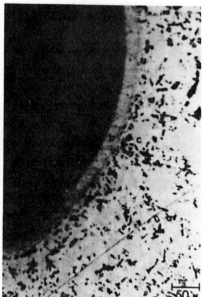

Fig. 5 — Silver-plated stainless steel pin in amalgam. Rubbing method of amalgam condensation (metallographic microscope, 305 X). (a) stainless steel pin, (b) silver-plate layer, (c) amalgam. Transversal section.

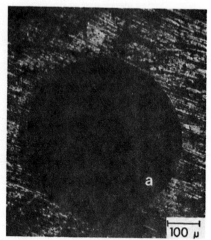

Fig. 6 — Silver-plated stainless steel pin in amalgam (SEM, 175 X). (a) stainless steel pin, (b) silver-plate layer, (c) amalgam. Transversal section.

continuous and, in all probability, chemically bonded.

Microscopic examinations (including SEM) of the layer of silver-plating on stainless steel (Figs. 2 and 3, typical) showed extremely good adaptation with no voids, therefore, strongly suggesting the presence of a good mechanical bond. Because silver is known to bond with amalgam, it was expected that the silver-plating would act as a "soldering" agent between stainless steel and amalgam. Micrographic evidence (Figs. 4 to 8) confirmed this expectation, particularly when compared with that from sterling silver pins in amalgam. It is, then, reasonable to suggest that good mechanical bonding between stainless steel and amalgam can be achieved through the plating method presented here.

When amalgam is condensed without using the "rubbing" technique, large voids are readily seen at both the silver-plated and sterling silver pin and amalgam interface. On the other hand, when rubbing is used, close adaptation is obtained, and only the small and expected voids of the amalgam

mass can be detected at the surface of the pin.

Conclusions.

Specimens utilized in this study confirmed the bonding potential between smooth pins and amalgam under the conditions of this research. The plating technique

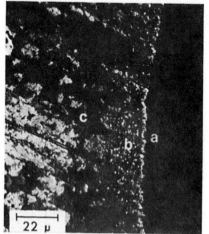

Fig. 7 — Silver-plated stainless steel pin in amalgam (SEM, 1050 X). (a) stainless steel pin, (b) silver-plate layer, (c) amalgam. Transversal section.

FIGURE 16.2   *continued*

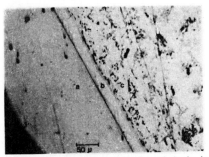

Fig. 8 – Silver-plated stainless steel pin in amalgam (metallographic microscope, 350 X). (a) stainless steel pin, (b) silver-plate layer, (c) amalgam. Longitudinal section.

used produced evidence of the strong possibility of a bond being present between silver and stainless steel. The technique of rubbing amalgam on the pin during condensation proved beneficial in achieving an excellent adaptation between the surface of the pin and amalgam. This technique is also useful in keeping stress concentration influences around the pin to a minimum, when retaining an amalgam restoration. The results of mechanical tests are needed to confirm the clinical usefulness of these conclusions.

Acknowledgments.

The authors are indebted to Mr. G. Freedman and Mr. Gordon Richardson from Bristol Aerospace, Winnipeg, Manitoba, for their guidance regarding the plating technique. Special thanks to Mr. B. Bergman for his help in constructing the special devices used in this research. We wish to acknowledge Dr. Peter Williams from the School of Dentistry and Dr. J. R. Cahoon from the Faculty of Engineering of the University of Manitoba for their advice.

REFERENCES

1. GALINDO, Y.; McLACHLAN, K.: and KASLOFF, Z.: Mechanical Tests of Smooth Silver-plated Retention Pins in Amalgam, *J Dent Res,* (in press).
2. DHURU, V.: A Photoelastic Study of Stress Concentration Produced by Retention Pins in an Amalgam Restoration, MSc Thesis, University of Manitoba, 1972.
3. GALINDO, Y.: The Development and Testing of Retention Pins which Metallurgically Bond with Dental Amalgam, MSc Thesis, University of Manitoba, 1973.
4. MOFFA, J. P.; GOING, R. R.; and GETTLEMAN, L.: Silver Pins: Their Influence on the Strength and Adaptation of Amalgam, *J Prosth Dent* 28:491-499, 1972.
5. LUGASSY, A. A.; LAUTENSCHLAGER, E.P.; and HARCOURT, J. K.: Crack Propagation in Dental Amalgam, *Aust Dent J* 16:302-306, 1971.
6. CECCONI, B. T. and ASGAR, K.: Pins in Amalgam: A Study of Reinforcement, *J Prosth Dent* 26:159-169, 1971.
7. DUPERON, D. F.: The Effect of Selected Pin Retention Materials on Certain Properties of Dental Silver Amalgam, MSc Thesis, University of Manitoba, 1970.
8. PETERSON, E. A. and FREEDMAN, G.: Laminate Reinforced Dental Amalgam, *J Dent Res* 51:70-87, 1972.
9. METALS HANDBOOK COMMITTEE: Metals Handbook, American Society for Metals, 6th ed., 1960.

FIGURE 16.2 *continued*

Pack the cover letter and manuscript (or disk) between cardboard covers and mail it in a padded envelope (jiffy bag) to protect the contents. Do not despair if your article is not accepted; fewer than 5 percent are. Instead, use the suggested revision notes (if included) as free advice and keep trying.

ACADEMIC PAPERS

As an undergraduate student, you are probably not involved in developing new technical products and materials on which you can report. Nevertheless, you can research the latest findings within your field and develop a research paper of publishable value. As a graduate student, you will be required to write a number of documented papers throughout your studies. The academic paper process requires a review of general expository writing skills along with the documentation techniques that will help you to write clearly. The skills include

1. Selecting and limiting the subject
2. Formulating the thesis
3. Preparing the preliminary bibliography
4. Brainstorming the methods of development
5. Consulting with professors
6. Preparing the working outline
7. Conducting the research and taking notes
8. Writing the rough draft
9. Writing suitable introductions and closings
10. Providing the documentation

Selecting and Limiting the Subject

In the professional world, the selection of a subject is usually yours, but as a student, a supervisor or professor may assign you a research report topic, such as "New Findings of the Influence of Temperature on Viruses," "The Practicality of Laser Surgery on Heart Patients," or "Secure Financial Electronic Transactions on the Internet." If the subject selection is entirely yours, consider the following three principles:

1. Select a subject that can be thoroughly investigated within the confines of your length limitations. "Computers" is too vast a subject and fails to suggest a particular focus. But subjects such as "Methods to Combat Computer Viruses" or "The Advantages of Vocal Software" are both more narrow and focused.

2. Select a subject on which you either are an authority or are able to access a wide variety of published research material, interviews, or survey responses. "Cures for the Ebola Virus" will not be productive because research is inconclusive. "Procedure for a Magnetic Circuit Test on Transformer Performance" may be too specialized for you and your audience. "Methods to Reduce Pollution in the Aral Sea (Uzbekistan/Kazakhstan)" is too far removed for you to obtain published, up-to-date material.

3. Select a subject that will allow you to formulate a judgment. A study on "Genetically Engineered Crops: A Technological Leap Forward or Destruction of the Ecosystem?" will yield information on which you can formulate an informed judgment.

Formulating the Thesis

The next step is to develop a preliminary thesis, a statement that focuses the purpose of your paper. This statement of the central idea to be developed in your paper will help you to organize facts, limit your note taking, and eliminate needless research.

Consider what you know about your limited subject. You probably already have some opinions or new information, or you would not have selected the subject in the first place. State in one sentence (or two if necessary) an opinion, conclusion, generalization, or prediction about your subject. For instance, if you have narrowed your subject to "The Effectiveness of Computer Access Control Systems," you may tentatively state, "Computer access control systems are improving." If you are an engineer who has been working on an improved system, your thesis may state, "A fail-safe method is now available to eliminate computer viruses."

Often, the thesis of a technical paper refutes the benefits of a present system and points out new findings. Such theses frequently contain the words *however, instead, nevertheless,* or *consequently.*

Examples

For decades Medic Alert neck tags or bracelets have provided instant medical history; however, laser optic technology can now produce revolutionary medical data-memory cards containing up to 800 pages of information.

The consumer no longer has to withdraw money from a wallet or suffer through a time-consuming check approval process; instead, she can pay a retailer directly out of her checking account via an electronic fund transfer with a debit card.

Genetically engineered crops can marshal a plant's natural defenses against weeds and viruses and will flourish with only a minimal application of chemical fertilizer; however, such biotechnology could threaten the biosphere.

While you are taking notes and extending your research, you should revise or refine your thesis. After you have assembled all of your data, reduce the thesis to one sentence to unite your findings. You may wish to add an organizational clue to the structure of your paper. For instance, your final thesis may read, "Despite advances, computer access control systems demand constant upgrading due to the increasing amounts of sensitive computerized data, the expansion of networking systems, and the growing number of sophisticated, computer-competent criminals." This thesis not only contains an overall inference on the need for constant improvement, but also provides a clue to the sections of the report: the problems of data proliferation, networking, and unethical programmers.

Preparing the Preliminary Bibliography

Once you have formulated a thesis for your report, the next logical step is to locate possible sources of information. Determine if you can locate an adequate number of timely resources and provide a working list of sources from which you can choose. Chapter 5 discusses in detail how to access your sources both in the library and through a personal computer, and indicates methods of primary research.

Brainstorming the Methods of Development

There are several major strategies for approaching the organization of material. These include *narration* (telling of events), *description* (recounting in precise detail), and *exposition* (explaining certain points). The professional paper usually requires expository strategies, although it may contain elements of narration and description.

There are eight methods of developing expository material, which may be used singularly but are likely to be used in combination:

- Definition
- Description
- Classification and division
- Exemplification
- Comparison and/or Contrast
- Cause and/or Effect
- Instructions
- Process analysis

Chapters 8 through 11 examine in detail strategies for definition, description, instructions, and process analysis. The student writer should employ a number of these expository strategies in producing effective academic papers. Ask yourself these questions:

1. Which terms will I need to define?
2. What descriptions are necessary?
3. How will I classify (place into major sections) and divide (separate into subsections) my material?
4. What specific examples will support my thesis?
5. What comparisons (likenesses) and/or contrasts (differences) will help to explain my material?
6. Are cause and/or effect factors an essential part of my evidence?
7. Are instructions for operation of any procedure necessary?
8. Will I need to analyze a process in my paper?

Search for material (see Chapter 5) that will support the pertinent strategies in the preceding lists. Then organize the strategies into a logical presentation.

Consulting with Professors

Once a preliminary bibliography of sources has been developed and the methods of development decided upon, it is wise to consult with one or more professors prior to continuing with the research. Professors are experts on the subjects they teach and can advise you on whether you are on the right track and offer help with sources and development. At the graduate level, a written proposal and preliminary bibliography should be submitted for the professor's approval.

Preparing the Working Outline

Develop a preliminary or working outline of your materials. Consider the writing strategies that you have brainstormed. Your first draft may be sketchy, but it will serve as a guideline to your note taking. Without an outline, you are likely to take notes on materials irrelevant to your purpose or overlook an area that should be explored.

You should not be totally bound by your working outline. As you think over your subject and take notes, you will want to expand, rearrange, discard, and subordinate your outline topics. Your final research report should include a final outline.

FORMAT

Select a traditional or decimal outline format:

I. A.
        1.   1.0
        2.   1.1
    B.   1.1.1.
        1.   1.1.2.
        2.
                                            1.2
            a.   1.2.1.
            b.   1.2.2.
                (1)   2.0
                (2)   2.1
                    (a)   2.1.1.
                    (b)   2.1.2.

Although the Roman-numeraled, traditional outline allows for five levels of subordination, you will seldom find it necessary to use subheads beyond two subordinates. The decimal outline format is popular for technical presentations.

TOPIC OUTLINE

Although each entry in an outline may be a sentence, a topic outline is not only easier to develop in the preliminary stage but is also more common in the final draft. Major topics name the divisions of your paper. Next, subtopics are subordinated and indented under each major topic, and subsequent developmental topics are subordinated beneath each subtopic.

Partial Preliminary Outline

    1.0 Introduction including thesis
            1.1 Genetic crop engineering—definition
            1.2 Genetic crop engineering—process
                    1.2.1 Golden rice
                    1.2.2 Corn
                    1.2.3 Additional subtopics, as necessary
    2.0 Causes of engineering
            2.1 Worldwide hunger
            2.2 Improvement of natural qualities
            2.3 Additional subtopics, as necessary
    3.0 Effects of engineering
            3.1 Resistance to insect pests

    3.2 Fewer polluting chemicals
    3.3 Additional subtopics, as necessary
   4.0 Ecologists' concerns
    4.1 Sierra Club
    4.2 Friends of the Earth
      4.2.1 Allergens
      4.2.2 Genetic pollution
      4.2.3 Additional subtopics, as necessary
   5.0 International opposition
    5.1 Britain
      5.1.1 Immune system damages
      5.1.2 Additional subtopics, as necessary
    5.2 France
      5.2.1 Tainted foods
      5.2.2 Additional subtopics, as necessary

Continue working on your outline as you take notes. Your professor may require you to include your polished outline with the final paper. Figure 16.3 shows a complete outline for a student academic paper.

Conducting Research and Taking Notes

With your thesis in mind, your outline to guide you, and your preliminary bibliography, you now proceed with the research and note-taking tasks. Chapter 5 covers this material thoroughly.

Writing the Rough Draft

Using your outline as a structural guide and your organized note cards (Chapter 5) as the content guide, write a rough draft of your research paper. Do not forget to use graphics and other visuals when they clarify your data. Follow these guidelines for your rough draft:

1. **Double space your draft** in order to permit expansion or revision of your draft. Determine if your professor wants you to single or double space your final draft.

2. **Write your introduction.** Be sure your limited subject and polished thesis are clearly stated. A clue to the overall organization of the paper is needed if it is not indicated in your thesis. Clearly define your key terms before beginning to present the evidence that supports your thesis. (The next section of this chapter contains additional information on writing introductions.)

OUTLINE

Thesis: Tomography, the innovative creation of a combination of scientific minds of our generation, is used to help cardiologists analyze the heart, to aid brain surgeons in the detection of cerebral disorders, to assist geologists in the study of earth's composition, and to permit aerospace technicians to inspect the MX missiles for malfunctions.

  I. Introduction
 II. Background
     A. Definition
     B. Origin
III. Cardiological applications
     A. Heart
        1. Disorders
        2. Normalities
     B. Cardiovascular system
        1. Arteries
        2. Veins
IV. Cerebral applications
     A. Brain scans
        1. Normal children
        2. Impaired children
     B. Oxygen/glucose loss
        1. Causes
        2. Effects
  V. Geological applications
     A. Earth's core
     B. Earth's surface
VI. Aerospace applications
     A. Study of MX missiles
        1. Disorders
        2. Remedies
     B. Approval of MX missiles

FIGURE 16.3   *Sample student review article outline*

3. **Write the body paragraphs.**   Each body paragraph or grouping of paragraphs should begin with a topic sentence (a sentence that clarifies the topic of the paragraph and states your opinion or generalization about it) based on the organizational parts as included in your thesis or revealed in your introduction. Each paragraph must have unity (address itself to the same subtopics of the overall subject), coherence (logical progression of evidence), and transitions (words and phrases that connect the evidence, such as *for instance, in the first place, furthermore, finally*). Review Chapter 2. Include your graphics and visuals, but do not use them just to "decorate" your paper; use them only where appropriate.

4. **Include the author's name within the text** when paraphrasing opinions and conclusions or directly quoting any source material. All of your researched data will require internal and end documentation (Chapter 6). Check the five safeguards in Chapter 5 in order to avoid a plagiarism charge and to ensure that all of your evidence is documented.

5. **Insert your internal reference notes in parentheses** as you write, and circle each in order not to overlook one when you prepare your final draft and bibliography listing (see Chapter 6).

6. **Write your conclusion.**   This may be one or several paragraphs that summarize your findings, present solutions, or recommend a particular action. See the next section in this chapter for some other tips on closings.

7. **Edit all mechanics** (spelling and punctuation) and style considerations (sentence construction and variation, grammar, usage, and other language concerns).

Writing Introductions and Closings

Consider the introduction and closing of your paper. The introduction of the truly professional paper is serious and forthright. If you have a particular broad-interest journal written with laypeople in mind, you may want to write a catchier introduction. The major purpose of the introduction is to state your thesis, but you may consider including

- A reader-attention device
- A context in which to consider your ideas
- Definition of key terms
- The thesis itself
- Organizational clues

Reader-attention devices include

- A direct quotation of a well-expressed statement
- A startling fact or statement
- A rhetorical question
- A related anecdote

Here are some sample reader-attention devices

1. One of life's great excuses—"The check is in the mail"—may be heading for extinction. Financial institutions and retailers want to stop handling cash and checks, so they are switching to cheaper, faster electronic point-of-sale (POS) transactions. Point-of-sale is the electronic transfer of money from a consumer's bank account to a merchant's bank account with the use of a plastic card and merchant computer terminals. The new system is creating a gradual revolution in the way people buy items and deal with their banks.

2. "In our opinion it is impossible to exist in the record industry without some sort of computerized point-of-sale system," says Irving Heisler, president of Montreal-based Discus Music World. Discus operates more than 100 units across Canada. The major advantages of the point-of-sale system lie in being able to serve customers better while at the same time lowering operating costs by making both inventory management and labor scheduling more efficient.

3. A shopper in a Dublin, California-based Lucky Store wants to pay cash for her purchases. She no longer has to suffer a time-consuming check approval process. Instead, by using a debit card, in just seven seconds she can pay the retailer directly out of her checking account with an electronic funds transaction. This point-of-sale system effectively speeds the sale, reduces paperwork, and costs less than traditional credit card transactions.

Here is an introduction fulfilling the five expository requirements:

Will genetically engineered crops revolutionize farming and better feed the world's poverty-level inhabitants? (*rhetorical question*). Transgenic gardening can fight off worms that infect cotton crops, add enhancing beta-carotene to rice, and reduce the need for pollutive chemical sprays; however, ecologists fear that some valuable insects will be wiped out, certain killer weeds will be made hardier, and other crops will be inadvertently contaminated (*context for consideration of ideas*). Genetic engineering is the science of experimental techniques of manipulating an organism's DNA endowment by introducing or eliminating specific genes through molecular biological procedures (*definition*). The potential benefits of such biotechnology far outweigh the ecological backlash that can be overcome by more research and experimentation (*thesis and organization clue combined*).

Your closing will usually restate your thesis in its original or in paraphrased terms and reach beyond it to conclusions and suggested further logical implications of your findings. Your final few sentences may

- Draw a stronger conclusion than already expressed
- Artfully summarize your material or emphasize just one strong aspect of it
- State a climactic fact
- Urge your reader to action

A few brief examples follow:

1. **Emphasis of one aspect.** Beta-carotene enhanced golden rice will feed at least a million children in Asia who would otherwise die every year because they are weakened by vitamin-A deficiency.

2. **Stronger conclusion.** There is no question that agricultural biotechnology can be harnessed to the good of humankind. It can solve the hollow-bellied hunger of world citizens whose plight is worsening with the international population explosion.

3. **Climactic fact.** The point-of-sale system enables management to reduce losses by thousands of dollar. Discus Music World increased its sales by 20 percent in the first year of use. You can't ask for more than that.

4. **Urge to action.** Quicker consumer services and lower operating costs have persuaded merchants to use the POS system. Shouldn't you install the system at your place of business?

Providing Documentation

Determine in advance the required documentation style in order to label your note cards, write your notes, write your draft, and polish the paper. Review the internal and end documentation particulars in Chapter 6 for a variety of styles for appropriate subjects.

Figure 16.4 shows a student academic paper, a study of hospice care versus hospital care. It is documented in the CBE (Council of Biology Editors) numbering style.

Venci 1

Cathy Venci

Professor M. Minnassian

English Comp. 2210-09

December 12, 1989

Hospice Versus Hospital: An Examination of Contrasting
Approaches to Patient Care

Medicine and health care have made enormous life-saving advancements in this age of technology. Such procedures as organ
transplants and laser surgery are now used routinely. Due to scientific research, many diseases that once were life threatening no
longer pose a serious threat to our health. However, despite scientific advancements, there comes a point when technology reaches its
limits. This is when the health-care professional must face the reality of the terminally ill patient. This patient has very special needs
that go beyond medical treatment. The hospice movement is meeting
the needs of these dying patients. *Mosely's Medical Dictionary* defines *hospice* as "a program offering continuous supportive care to
dying people enabling them to live out the final days of their lives
comfortably and as fully as possible" (1). The hospice concept of
care differs from the traditional hospital in the approach to the
treatment of the patient and his/her family, the role of the health-
care professional, and the development for care in the future.

FIGURE 16.4 *Sample paper documented in the number system style (CBE)*   (Courtesy of
student Cathy Venci)

Venci 2

A brief history of hospice care will further define the philosophy behind the modern hospice care movement. In the Middle Ages, the hospice was a place of refuge for the traveler. The religious sects that operated the hospices offered food and shelter to the poor. In the middle 1800s these shelters developed into retreats for people who were dying from incurable diseases such as tuberculosis (2). According to Hamilton and Reid, the advancement of hospice care came in 1967 when Cicely Saunders opened St. Christopher's Hospice in London (3). They state that St. Christopher's Hospice was formed with the purpose of helping the terminally ill patient to remain comfortable and relatively free from pain without any artificial means to prolong dying (3). These same principles are the foundation of the hospice concept in America today.

The hospice program has taken a different approach to the treatment of the patient. Although the hospital and hospice are both committed to medical needs of the patient, the hospital's emphasis is on the care of victims of acute illness. The goals of the hospital, according to Michael Hamilton and Helen Reid, authors of *A Hospice Handbook,* are to diagnose and cure disease through modern technology (3). Another hospice expert, Kenneth P. Cohen, states that since technology is the main source of treatment, it is used to the extent

FIGURE 16.4   *continued*

Venci 3

of prolonging life, even when there is no known cure (4). He adds that life support systems and aggressive treatment are normal procedures in the hospital setting (4). Health reporter Jane Toot notes that the treatment of the terminal patient includes

- Intravenous fluids
- Forced feedings
- Nasogastric tubes
- Laboratory examination (5)

Cohen says that the most important treatment the dying patient needs is pain control therapy (4). According to Cohen, hospital procedures have a fixed routine of medication every four hours or so. The patient may be experiencing serious pain before relief is given (4).

There are differences between hospital and hospice attention to the dying patient's family. According to Toot, the hospital is not equipped to deal with the family (5). The patient is often in an acute-care ward that has very strict visiting regulations, which tend to create a sense of isolation for both the patient and the family (5). Toot notes that the family is often confused over the treatment of the patient. She states that this confusion is due to the hospital excluding the family in making decisions concerning the treatment of the patient (5).

FIGURE 16.4   *continued*

Venci 4

On the other hand, the hospice is developed to accommodate the terminally ill patient. The emphasis is on caring for the patient's symptoms and not curing his/her disease (5). Hamilton and Reid refer to the hospice as "a place for dying" (3, p. 48). Accordingly, there are no life support systems in the hospice program, there are no last-minute life saving attempts (3). Toot notes the major treatment of the patient is pain control; pain-killing drugs, such as morphine, are given to keep the patient free from pain (5). Other forms of treatment, according to Toot, consist of

- Massage for relaxation
- Heat therapy for circulation
- Exercise for flexibility (5)

Family participation is encouraged by hospice staff. Patients go home frequently to spend time with the family (5). Further, the hospice gives the patient access to

- Unlimited visiting hours for family
- Personal belongings and pets
- Around-the-clock medical care and pain relief
- Counseling and spiritual guidance for both the patient and the family (5)

FIGURE 16.4 *continued*

Venci 5

This daily care emphasizes the quality of life at all times (5). The whole hospice program can be summarized in one statement: "Dying patients are human beings. These people have real needs and cares that need to be attended to" (6). A survey (see attached instrument with numbered tabulations) of 100 hospice patients in Moore County reveals the value patients place on hospice services (7). Figure 1 shows the number one ratings of four services by percentages:

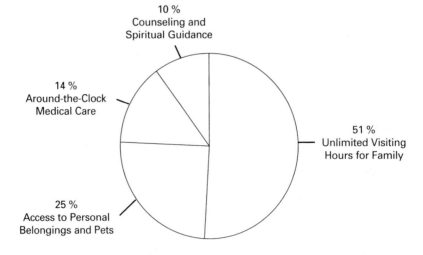

Figure 1   Number one ratings for four hospice services (7)

In addition, 41 percent selected access to belongings and pets as the second most highly valued. Thirty percent rated around-the-clock medical care as third highest, and 40 percent rated counseling and

FIGURE 16.4   *continued*

Venci 6

spiritual guidance least highly. Clearly being in their own homes
with family, belongings, and pets nearby is the premier desire of terminally ill patients (7).

    Another area of contrast between hospice and hospital care con-
cerns the role of the health-care professional. Cohen notes that the
hospital staff is increasingly involved in the challenge of technical
care and diagnosis; he states

> Hospital staff do not necessarily ignore the dying patient, but
> they give priority to patients for whom they can provide life-
> saving measures. This is not unreasonable, considering their
> training and emphasis on curative functions rather than the
> caring functions needed by the dying (4).

Table 1 shows the contrast of involvement in hospital and hospice
staff members:

Table 1   Contrast between staff members in hospice and hospital

Hospital staff	Hospice staff
Hierarchy system of care	Team approach to sharing
physician	responsibility
nurses	
aides	

FIGURE 16.4   *continued*

Table 1 Continued

Hospital staff	Hospice staff
Rotation of wards and shifts by staff members	Daily care of same patient
Little communication concerning patient cases	Daily discussions and meetings concerning patient cases
Few volunteers on staff	Many volunteers on staff

Based on Toot (5)

The cost of hospital and hospice care differs. The hospital trend will continue to focus on technology and research. Due to the high cost of these components, the cost of hospital care will continue to increase. Another study (8) states that the cost of a bed space in a hospital is now $240 and upward a day (8). The study also notes that this same bed space in a hospice costs as little as $34 per day (8). This difference in medical expenses is one of the major factors in hospice growth (8). A *New York Times* reporter (9) points out that another factor in hospice care growth is the age of the population. With an aging population, there is a greater number of people who need the kind of care hospice programs offer (10). Two-thirds of hospice patients are over sixty-five years of age (10).

In the past the majority of hospice funding came from donations by individuals, civic groups, and church organizations (9). Medicare

FIGURE 16.4   *continued*

is now paying some of the expenses (10). Fifteen years ago there were only 200 hospices in the United States (10). There are now thousands of groups, and with the help of Medicare funding there will be even greater numbers (10).

In conclusion, hospitals are doing an excellent job in providing a cure for the victim of acute illness and accident. The hospital should continue to do everything possible to advance research and technology to meet the needs of the patient who has a chance to return to society. On the other hand, with the increase of the age of our population, it is apparent there will be a greater need for the care and counseling that hospice provides to those in their final days of living.

FIGURE 16.4   *continued*

Venci 9

Hospice Services Survey

The purpose of this survey is to determine the value you place on benefits received through hospice care. Results will be included in a brief paper examining hospital versus hospice care.

Please rate the following benefits of hospice care in your opinion. (Place a 1 before the service you deem most valuable, a 2 before the second most valuable service, a 3 before the third most valuable service, and a 4 before the least valuable).

_____ a.   Unlimited visiting hours for family
_____ b.   Access to belongings and pets
_____ c.   Around-the-clock medical care
_____ d.   Counseling and spiritual guidance

Tabulated Survey Results

Most valuable selections:	Second most valuable:
51 selected a.	22 selected a.
14 selected b.	41 selected b.
25 selected c.	10 selected c.
10 selected d.	27 selected d.

Third most valuable:	Fourth most valuable:
17 selected a.	10 selected a.
30 selected b.	15 selected b.
30 selected c.	35 selected c.
23 selected d.	40 selected d.

FIGURE 16.4   *continued*

Venci 10

References

1. *Mosley's Medical Dictionary,* 1993 ed.

2. Hospice care. *Encyclopedia Americana,* 1993 ed.: 436–438.

3. Hamilton, M., Reid, H. *A Hospice Handbook.* Grand Rapids, MI William H. Eerdmans Publishing, 1980.

4. Cohen, K. P. *Hospice: Prescription for Terminal Care.* Germantown, MD, Aspen Publications, 1996.

5. Toot, J. Physical therapy and hospice. *Physical Therapy Journal* 64 (1994): 665–670.

6. Coor, C., Coor, D. *Hospice Care Principles and Practices.* New York: Springer Publishing Company, 1983.

7. Venci, C. Survey on value of hospice services. Moore County, FL, 1997.

8. Mudd, P. High ideals and hard cases. *Hastings Center Report* (April 1982): 11–14.

9. Boundy, D. Growth of hospice programs is cited. *New York Times* 20 May 1994, sec. 22:6.

10. Friedland, S. Hospice benefit off to slow start. *New York Times* 22 Nov. 1994, sec. 11: 1, 7–8.

FIGURE 16.4 *continued*

C H E C K L I S T

Writing Scientific/Technical Articles and Academic Papers

SCIENTIFIC/TECHNICAL ARTICLES

❏ **1.** Have I designed the research article as a study of a product or a procedure?

❏ **2.** Have I received approval from superiors (or funding sources)?

❏ **3.** Have I completed the research?

    ❏ Primary sources?

    ❏ Secondary sources?

❏ **4.** Have I used multiple sources and checked the research for accuracy?

❏ **5.** Have I selected a journal in the field most likely to publish the paper?

❏ **6.** Have I obtained the instructions to authors for the selected journal?

❏ **7.** Have I provided a precise title?

❏ **8.** Does my introduction clarify the purpose of the paper?

    ❏ Does it define a problem?

    ❏ Does it review the existing literature?

❏ **9.** Have I thoroughly discussed the methods/materials used in the investigation of the problem?

❏ **10.** Have I included a thorough discussion of other aspects of the investigation?

❏ **11.** Have I discussed the results of the investigation?

❏ **12.** Does my conclusion make valid judgments and suggest implications for the future use of the product or problem solution?

❏ **13.** Have I provided the appropriate internal and end documentation (see Chapter 6)?

❏ **14.** Have I submitted the article for publication appropriately?

 ❏ Have I provided a cover letter?

 ❏ Have I included two copies and/or a disk of the paper?

 ❏ Have I included a self-addressed, stamped envelope for acknowledgment of receipt of the manuscript?

 ❏ Have I packaged the manuscripts and/or disk carefully?

ACADEMIC PAPERS

❏ **1.** Have I selected a scientific or technical subject and limited it sufficiently for length considerations?

❏ **2.** Have I formulated a preliminary thesis to focus the purpose of the paper?

❏ **3.** Have I prepared a preliminary bibliography and cards or computer file (see Chapter 5)?

❏ **4.** Have I decided on the appropriate method(s) of development?

 ❏ Definition?

 ❏ Description?

 ❏ Classification and division?

 ❏ Exemplification?

 ❏ Comparison and/or contrast?

 ❏ Cause and/or effect?

 ❏ Instructions?

 ❏ Process analysis?

❏ **5.**   Have I consulted with my professors on my progress?

❏ **6.**   Have I developed a working bibliography that I will polish for inclusion?

❏ **7.**   Have I conducted the research and taken notes (see Chapter 5)?

❏ **8.**   Have I written the rough draft?

    ❏ Have I included an introduction?

    ❏ Have I written the expository body paragraphs?

    ❏ Have I incorporated partial and sentence direct quotations appropriately?

    ❏ Have I written a closing?

❏   **9.** Have I provided all of the appropriate documentation (see Chapter 6)?

❏ **10.** Have I edited all of the mechanics of the paper?

❏ **11.** Have I assembled the paper correctly (title page, outline, paper, bibliography)?

EXERCISES

1. **Professional Journals**   Compile a list of at least five journals in your field.

2. **Journal Guidelines**   Obtain the instructions to authors from three of the journals in your field.

3. **Thesis Statements**   Rewrite the following broad thesis statements to indicate a more limited focus and purpose for each:

   a. The Internet has revolutionized academic research.

   b. Many diet prescriptions now available have harmful side effects.

   c. Macroengineering will improve continental travel in the future.

   d. The twenty-first century is ushering in changes in the job market.

   e. New medicines to lower cholesterol are emerging.

   f. Desalinization systems will provide additional usable water.

   g. Mars exploration is providing benefits here on Earth.

    h. Computers are changing the nature of international telephone calls.

    i. Our knowledge is increasing in the study of humanity's origins.

    j. Global warming is threatening our coastal cities.

4. **Outlining.** Revise the following outline from a traditional sentence format to a decimal, topic outline.

> Thesis: Changes in automobile, rapid transit, and air travel are revolutionizing the twenty-first century.

I. Automobiles are undergoing dramatic changes.
  A. Cars are being totally computerized.
    1. Voice commands can lock and unlock the car, operate the radio, and operate the climate control system.
    2. Vehicles will all be equipped with telephones.
    3. All systems will be electronically monitored.
  B. Designs are increasingly aerodynamic.
    1. Future cars will have sharply raked windshields and aerodynamic bodies.
    2. Undercarriages and wheels will be enclosed.
    3. Windows will be flush with the body.
    4. Grills and outside ornaments will become obsolete.
  C. Alternative fuels are replacing gasoline.
    1. Hydrogen fuel is being developed.
    2. Electric cars are now on the market.
II. Rapid transit is easing the burden of traffic in cities.
  A. Urban rail systems between cities are expanding.
  B. Bus systems and people movers are expanding within cities.
  C. Bullet trains will curb air travel.
    1. Fast trains will replace airplanes for moderate trips.
    2. The high cost of air travel will become prohibitive.
III. High-speed jetliners will make long-distance travel easier.
  A. Supersonic jets will continue to replace slower jumbo-jets.
  B. The number of flights to various points will become greater.
  C. New airlines will offer long-distance and intrastate flights.

WRITING PROJECTS

1. **Scientific/Technical Article** Using the information in the Scientific/Technical Articles section of this chapter, write a one-page evaluation of the style, clarity, and organization of a published article of your choice. Include a copy of the article with your evaluation.

2. **Collaborative—Scientific/Technical Paper**   Select two class-mates in your field to collaborate on an article theoretically for publication. After selecting a broad subject from the following list and narrowing your subject to a focused topic, select a journal most likely to publish your paper and obtain its author guidelines for inclusion with your paper. Indicate which journal you have selected in prefatory material to your paper:

> This paper is intended for publication in *Name of Journal* with the attached author guidelines.

Use the organization of a scientific/technical article; research and write the paper on some problem of the twenty-first century. In your paper, delineate the problem by reviewing the existing literature. Discuss the problem, the methods being explored to alleviate the problem, and the results obtained by those methods. Draw conclusions on the findings and point out future implications. Following is a list of broad subjects that can be narrowed to focused topics.

education	pollution
transportation	fuels
merchandising	the shape of cities
endangered species	new computer applications
the economy	travel modes
satellites	electronic marvels
genetic engineering	weather modification
architecture	home appliances
population	space exploration
new occupations	health care
e-commerce	medicine
the American family	crime
life expectancy	food production
minerals and materials	other?

3. **Collaborative—Academic Research Paper.**   Select two class-mates in your academic field of study to collaborate on an academic research paper concerning a breakthrough discovery in your field. Develop a bibliography that includes a variety of sources (encyclopedias, magazines and journals, newspapers, brochures and pamphlets, and Internet sources). Consider primary sources, too, such as interviews, surveys, letters and their responses, public forums, and direct observation. Take notes. Develop a thesis, a preliminary outline, and methods of development. Investigate who made the discovery, how it was made, how it was tested, what benefits it offers, what problems it poses, and what future applications are possible.

Include graphics and other visuals as appropriate and helpful. Use the documentation style for internal referencing and final bibliography appropriate to your subject or as assigned by your professor. If your style requires a title page, abstract, headings, and so on, adhere to that style.

NOTES

PART five

The Presentation Strategies

CHAPTER 17

Performing Verbal Communications

CATHY   by Cathy Guisewite

S K I L L S

After studying this chapter, you should be able to

1. Identify potential professional situations in which oral communication skills are necessary.

2. Name three considerations in preparing for a successful information-gathering interview.

3. Prepare for and role play an information-gathering interview with appropriate strategies.

4. Name and practice eight strategies for productive telephone communications.

5. Name and practice three strategies for reaching group decisions and understand the drawbacks of each.

6. Name and practice six strategies for worthwhile appraisals.

7. Name and practice four strategies for productive reprimands.

8. Name five strategies for a convincing persuasive oral report.

9. Present both persuasive and informative oral reports.

10. Discuss the significance of at least five nonverbal messages that are unconsciously sent and received.

INTRODUCTION

Effective verbal communications are important in every work setting. Every organization requires its employees to develop competent and productive speaking skills. Initially, to gain employment, you must be able to speak persuasively about your own potential and to reveal yourself to be knowledgeable about the organization that has granted you an interview (see Chapter 13).

Once you are employed, you will be expected to demonstrate careful and useful interpersonal communication skills in a variety of tasks including information-gathering interviews, telephone transactions, group decision-making discussions, and informal and formal speeches. You should even be aware of the nonverbal communications you purposefully and inadvertently send and receive.

INTERVIEWS

Employment Interviews

In addition to covering the writing strategies of the resume and cover letter necessary to obtain an employment interview, Chapter 13 covered employment interview strategies. Review those strategies now because they offer valuable information about all of the verbal communications that this chapter discusses.

Information-Gathering Interviews

You may often find yourself acting as an interviewer in order to obtain information from your coworkers, superiors, and people in other organizations. You need to determine in advance the exact purpose of the interview. In addition, you need to preplan by jotting down the topics that need to be covered and organizing them logically. Be prepared to take notes or to tape the interview with the permission of the interviewee.

PURPOSE

Write out your purpose statement so that you identify the overall purpose and the subcategories of information that you are seeking. You are not going to read this purpose, but you are going to state it clearly both in seeking the interview appointment and in beginning the actual interview.

Sample Purpose Statement 1

The purpose of my interview with you, the twelve members of my department, is to obtain from each of you your opinions and answers to twenty questions regarding our needs for updating computers, software, and printers in order to draft a proposal for necessary improvements.

Sample Purpose Statement 2

The purpose of my interview with you, John Avery, Manager of the Media and Technology Communications Center of Motorola Corporation, is both to discuss the changing role of the industrial technical writer and to obtain sample user guides and other tiers of materials for publication in the fourth edition of my textbook.

The next step is to establish the appointment by phone call, memo, letter, or personal contact. At this time, state the purpose of the interview and set a specific date giving the interviewee(s) plenty of time to prepare for the actual interview.

CONTENT PREPARATION

Decide on the specific questions you are going to ask and organize them logically. You may want to prepare a very brief topic outline to serve as your visual guide during the interview, but your most effective interview will be realized without much reference to personal notes (unless you are actually using a survey or questionnaire). Your questions should be specific, open-ended, and unbiased. Give the person(s) plenty of time to respond. Some sample questions for the interview on the changing roles of industrial technical writers may include the following:

1. In the past, the technical writer was given only a month or so after the development of a new product to write a user manual. I understand that technical writers now operate within a different time frame and that their tasks involve much more than just the written document. Would you please tell me at what point the writer is now involved and describe for me these new tasks?
2. Why has the time schedule changed?
3. Does your staff do all of the document design and desktop publishing or do you use professional printers? If so, please differentiate those tasks you undertake and those you job out.
4. You mention preparation of online materials. Would you please elaborate on the nature and purpose of these materials?

Continue along this line, taking your cues for further questions from the responses given to you. If you do not fully understand a response, rephrase your question or ask for amplification. Avoid questions that can be answered with a simple *yes* or *no* or are too vague, such as, "What do you do?" Do not ask biased questions, such as, "Would you agree that technical writers are underpaid?" Ask, "In view of the increased number of tasks, are modern technical writers being adequately compensated?" Be prepared to ask follow-up questions, such as, "Why? Why not? Do you foresee even more tasks being required? What recommendations would you make to cope with the increasing tasks of the writer?"

Your conduct should be similar to that you exhibit when you are the interviewee. Review the tips in Chapter 13 about being on time, dressing professionally, shaking hands and making eye contact, showing your personality, and being courteous. Remember that the actions that will nullify the value of your interview include "winging it," over-familiarity, nervousness, offensive language, failure to respond to questions, and abrupt endings. In addition, consider these suggestions for a successful information-gathering interview:

1. Thank your respondent at the beginning of the interview for granting you the time.

2. Restate your purpose and elaborate on how you intend to use the interview responses.

3. Let your interviewee do most of the talking. Don't "lead" the respondent or state your own opinions.

4. Do not tape an interview without the explicit consent of the interviewee. If you take notes, keep them to a minimum, jotting down words and phrases, statistics, and other specific information to jog your memory when you write up the proceedings.

5. If you interview more than one person on the same subject, standardize your questions and ask them in the same order to each respondent.

6. Offer to provide a copy of any reports, documents, or other materials that will incorporate the interviewee's answers and comments when your study is complete.

7. Arrange to write out your interview results very soon after the interview in order to remember clearly what transpired.

When interviewing a number of people about the same subject, you may want to devise a survey or questionnaire. Review the material on surveys in Chapter 5.

TELEPHONE CONVERSATIONS

At all times, you are a representative of your organization. Many transactions are conducted by telephone both within and outside your company. Your telephone etiquette reflects not only your own verbal skills but also the company's image. Certain basic strategies will help you to receive messages in a professional manner. Talking on the phone involves both talking and listening skills. Consider these tips:

- Pay attention to the volume, tone, and clarity of your voice
- Use a warm and pleasant tone
- Do not shout or whisper
- Be sure that your mouth is about an inch or two from the mouthpiece—no closer and no farther
- Avoid using a speaker phone because it degrades sound quality
- Enunciate clearly; avoid slurring of "Hullo," "Whajasay?," "Yeah," and the like
- Avoid distracting background noises, such as another conversation, computer printers, radios, tapping, blowing, or chewing
- Be prepared to take notes; have a notepad, pens, and pencils near your telephone

- Have on hand, too, any reports, letters, or other printed matter that may need to be referenced during your conversations; do not make your listener wait while you locate such materials

PLACING A CALL

When placing a call, identify yourself by a courteous "Hello" and state your name and other identifying comments, such as, "Hello, this is Judy Withrow, the training consultant with Writing Skills Management. I spoke to you last Friday about the possibility of conducting a seminar for your midlevel management people." After an acknowledgment, proceed to state the purpose of your call, such as, "I'm calling to determine if you received the brochures I mailed to you and to set a date for an interview." After transacting your business, specify what it is you want your listener to do: call you back in an hour or a week, mail you material, or confirm details in writing. If it is appropriate, thank the listener for the information, interest, or time.

RECEIVING A CALL

When receiving a call, say "Hello" courteously and identify yourself, your department, or your organization: "Hello, Tom Brown speaking," "Hello. Personnel Department, Miss Shipley speaking," or "Ace Company. This is Bob Martin, Sales Manager. May I help you?" End the conversation courteously, too. It may be appropriate to say, "Thank you for calling," "I'll send the materials to you in today's mail," or "I hope I've assisted you." Say "Goodbye" in response to the caller's closing and hang up carefully. A receiver that is banged down or dropped on its cradle may unduly irritate the caller who has not yet hung up.

If you answer the telephone for someone else, be doubly courteous. Let the caller know that the right number has been reached: "Hello, Ms Steven's office; this is Jerry Thomas, her assistant, speaking. May I help you?" Do not make lame excuses for another's absence, such as, "She's busy right now" or "I don't know where she is." State that the individual is engaged, in conference, or out to lunch and offer to take a message. Do whatever you can to have the call returned.

GROUP DISCUSSIONS

From time to time, you will be part of a decision-making group—a committee, a department, or an even larger group. Large groups tend to operate through parliamentary procedure to reach decisions, whereas small groups operate less formally. We are concerned here only with small task groups.

A number of factors affect group processes:

- Purpose for which the group exists
- Personal goals of its members (which may be in conflict with the overall purpose)
- Permanency of the group
- Power and relationships of the members of the group
- Methods (majority rule, compromise, or consensus) used for decision making

The most effective group is one with a clear purpose that is understood and supported by all of its members. Further, the longer a group can work together, the less it tends to be dominated by one or two people. Finally, a group that knows its decisions will be accepted by others will be more successful than a group that perceives an outside threat or an overriding decision maker. There are three major methods for reaching a group decision. Each has its drawbacks.

MAJORITY RULE

Arriving at a decision by majority rule involving a vote is a common method, but not always the best. A vote tends to polarize the camps, making the losers less committed or even antagonistic to the decision.

COMPROMISE

In a compromise decision, both sides give up a little to gain a little. Collective bargaining typically uses compromise decisionmaking methods. Compromise is effective if members of the group have to answer to larger constituencies. The constituents can believe they won something and were not total losers, yet since nobody is 100 percent pleased with the decision, those involved may not work very hard to implement the decision.

CONSENSUS

This method requires that all members agree on the decision. It is the most difficult way to decide but the most effective because all members are satisfied with the outcome. Consensus should lead to better productivity and commitment to implementation of the decision. For a group to reach consensus decisions, follow these steps:

1. All members should participate in clarifying a goal and establishing a procedure.
2. Draw up an agenda of group procedure and appoint an initial leader.
3. Let each member talk freely about the goal and suggest procedures for the group to follow.

4. Hold a group discussion to determine the status and causes of the problem or situation.

5. Establish criteria for evaluating each possible solution or decision.

6. Brainstorm solutions in a creative way to consider all possible solutions:

   a. Have all members offer as many different solutions as possible with no interrupting criticism or evaluation.

   b. List all of the suggestions on a chalkboard or flip chart.

   c. Weigh each potential solution against the criteria to determine which is best in terms of effectiveness, cost, simplicity, possible implementation, and acceptability to the group.

   d. Finally, make a group decision (a consensus) on how to implement the solution with give and take in delegating and accepting responsibilities.

APPRAISALS AND REPRIMANDS

You will probably meet with your supervisor from time to time to assess your job performance. Alternatively, as you move up the career ladder, it may become your responsibility to appraise the performance of your subordinates. Giving and receiving reprimands are also inevitable transactions in any organization. Both appraisals and reprimands tend toward emotional communication, so strategies should be aimed toward objectivity.

APPRAISALS

Certain strategies will lead to an objective and effective interchange, including the following:

1. Schedule an appraisal interview in advance so that both sides may prepare.

2. Limit the interview to a specific time period—15 to 30 minutes—to avoid repetitious harangues.

3. Have both parties prepare separate written assessments of the performance and exchange them in advance of the interview.

4. During the appraisal, concentrate on task-related factors rather than personalities.

5. Offer (or seek) explanations for poor performance.

6. Overall, center the discussion on specific goals and methods for improvement.

These strategies should lead to goodwill, improved performance, and group cohesiveness.

REPRIMANDS

Objectivity is also the key to the effective reprimand situation. A clear company procedure should be in place to handle these touchy interviews. The following steps should help to turn a reprimand into a maturation process:

1. The company should offer written policies and/or procedural manuals to spell out the expected activities of its employees. These written materials will also substantiate that a violation has occurred.

2. The supervisor should attempt to determine if the violation was due to lack of information or was willful. Clear communication is essential.

3. The supervisor must ask questions, use feedback, and be sensitive to the feelings of, and possible penalty to, the violator.

4. The violator should be prepared to ask questions about appropriate recourse, appeals, and corrective actions.

Both the supervisor and the violator must be honest and concerned and work toward the goal of amelioration.

ORAL REPORTS

You will be called upon in your career to make any number of informal or formal oral reports. These may be reports at meetings of your peers and supervisors, presentations at training seminars, speeches at conventions or before civic groups, or presentations to explain proposals and other projects. Oral reports are classified as

- Impromptu speeches
- Memorized speeches
- Manuscript speeches
- Extemporaneous speeches

The impromptu speech is an off-the-cuff presentation likely to occur at a meeting where you are either asked about a project or you decide to present your views on an agenda item. A memorized speech may be appropriate for material which must be communicated many times, but it is difficult to avoid sounding wooden and mechanical in memorized speeches. A manuscript speech, one which is read, may be appropriate to convey very technical or detailed information but calls for exacting practice to avoid a monotone delivery and lack of eye contact. The extemporaneous report is the most widely used and effective oral presentation.

Extemporaneous Reports

The extemporaneous report requires careful audience analysis, clear purpose, logical organization, supportive visual materials, sufficient rehearsal, and skillful delivery.

Audience Analysis

Previous chapters have stressed that an analysis of your audience is essential for effective written reports. The public speaker must also consider the audience who will listen to the oral report. Besides thinking about how much the audience already knows about your subject, what level of technical language is appropriate, and your relationship to the group, you should seek to discover in advance the average age of the group, political persuasions, religious or ethnic affiliations, rural or urban interest, and sex. Knowledge of these factors will help you to infer how the listeners will receive your information. These factors should indicate the degree of formality appropriate to your talk.

It is equally important to continue analyzing your audience during your speech through feedback. You can gauge your audience's reactions to your speech by noting facial expressions, postures, applause, and the like. The alert speaker will make adjustments in delivery based on this feedback.

Purpose

It is important to have a clear purpose in mind: to entertain, to persuade, or to inform. We are not concerned here with the entertaining speech, but as a professional, many of your reports will persuade or inform your audiences.

PERSUASIVE PURPOSE

Actually, you call upon the strategies of persuasion not only in most of your verbal communications but in written communications, as well. Every time you ask for action in a memo, letter, brief report, proposal, interview, group discussion, phone call, or verbal presentation, you consciously or unconsciously employ persuasive strategies. These have been discussed in almost every chapter of the book. Even graphics and visuals help to persuade an individual or a group to your way of thinking by presenting in condensed visual form irrefutable facts, relationships, and abstract concepts.

You need to employ persuasive strategies to bring about overt action or to change beliefs and attitudes. Examples of overt actions are the purchase of products, the election of officers, the adoption of policies, or the alteration of procedures. Persuasion may also change workers' attitudes toward minority and women workers, alleviate difficult working situations, or promote pride in organizational membership. Five strategies will help you to deliver a compelling persuasive speech:

1. **Audience comfort.** Arrange to give your speech in a setting that is comfortable for the audience. Consider whether you should take an authoritative stance up front or arrange the group into a round-table mode.

2. **Identification with the group.** Arrange to be introduced by a person who is well-liked by the group and who will stress your qualifications to speak on the subject. Establish that you are part of the group by the way you dress and act or by actually expressing similarities in ideas, beliefs, or experiences.

3. **Goals and rewards.** State your goals in an honest, friendly, yet assertive manner, emphasizing the rewards (anything that meets the needs and desires of the group) that persuaded listeners will receive.

4. **Credibility.** Use accurate statistics and other evidence along with the *reasons* for believing that the evidence is relevant to your conclusions. Cite the experts whose claims can be readily accepted by the group. If you cite authorities whose expertise is not fully accepted by the group, you will lose credibility.

Above all, remember that change is very difficult for some people to accept. Some members of the group may believe their hierarchal position or very job is threatened; others may believe they are going to have to work harder or fall behind. If you are the spokesperson for the company, you must let everyone know how each will benefit from the change.

INFORMATIONAL PURPOSE

The informative report, or in this consideration, the informative speech, is the most common among professionals. Such speeches may be patterned along the same lines as written reports, for example, analytical and evaluation reports, instructions, analysis of processes, descriptions of mechanisms, and so forth. Do not lose sight of your primary goal, to impart information.

Organization of Speeches

Your evidence or data must be logically organized. A listener is not a reader. A listener cannot back up to review information or skip ahead to the conclusion. Organize your report with the listener in mind.

If your purpose is to persuade, you may consider the problem-solution organizational approach or the advantages-disadvantages approach. Figure 17.1 illustrates the organization of these approaches.

If your purpose is to inform, you will organize along the lines of the type of report which you are presenting orally. Figure 17.2 illustrates an organizational approach to your speech.

Next, prepare an outline of your actual speech and gather your information. Other chapters in this book review outlines and organization for specific types of reports.

Finally, using your outline, prepare $3'' \times 5''$ or larger note cards on the main topics of your speech. Print large enough so that you can read the material easily. Underline key points in red, and indicate by asterisks where you plan to use your visual materials. Use only one side of the note cards, and do not overload a card.

Content Tips

Consider your introduction, transitions, and closings carefully.

INTRODUCTIONS

Create a positive impression by politely addressing the audience with a "Good Morning/Afternoon/Evening." Welcome your audience to the event (conference, seminar, or meeting). Thank them for inviting you to speak. If necessary, quickly establish your expertise, credibility, and professionalism with a short description about yourself and/or your company. Involve your audience by relating the topic to their lives, work, or concerns. Let them know that you will welcome questions at the end of your presentation.

Announce your topic and capture your audience's interest in your subject by beginning with an attention-getting device:

- An anecdote (short story related to your subject)
- Startling facts, figures, or comments
- A suitable joke
- A question
- A compelling direct quotation

State your thesis about your subject.

PROBLEM-SOLUTION APPROACH

I. Approach
   A. Gain attention and goodwill.
   B. Develop credibility, if necessary.
   C. Orient receiver to subject and purpose.

II. Body
   A. Develop problem.
      1. Explain symptoms or results (problem description).
      2. Explain size and/or significance.
      3. Explain cause.
   B. Develop solution.
      1. Explain solution.
      2. Explain how solution eliminates problem.

III. Conclusion
   A. Appeal for action or desired belief.
   B. Allow for discussion.

ADVANTAGES-DISADVANTAGES APPROACH

I. Introduction
   A. Gain attention and goodwill.
   B. Develop credibility, if necessary.
   C. Orient receiver to subject and purpose.

II. Body
   A. Explain disadvantages of present situation.
   B. Explain advantages of new idea, proposal, policy, or situation.
   C. Explain that advantages cannot be obtained without proposed changes.

III. Conclusion
   A. Appeal for action.
   B. Allow for discussion.

FIGURE 17.1 *Persuasive oral report approaches*

INFORMATIVE SPEECH APPROACH

   I. Introduction
      A. Gain attention and goodwill.
      B. Orient receiver to subject and purpose.

  II. Body
      A. Present data in logical organization.
      B. Repeat key terms and provide verbal transitions between parts.
      C. Employ visual aids.

 III. Closing
      A. Summarize.
      B. Emphasize.
      C. Allow for questions and/or discussion.

FIGURE 17.2   *Informative oral report approach*

BODY

Throughout your presentation, consider restating your topic or thesis in the same or slightly different words to help your audience retain your ideas. Use topic sentences to introduce each new point. Use transitional words and phrases *(first, second, finally, furthermore, another reason for,* and so on) to help your audience follow your organization and ideas.

CONCLUSION

End your presentation by reiterating your thesis, main points, and final recommendations or courses of action. Use an expository device as follows to clinch your ending:

- Draw a stronger conclusion than already expressed
- Artfully summarize your material or emphasize just one strong aspect of it
- State a climactic and convincing fact
- Urge your audience to action

Invite your audience to ask questions and to state their opinions and ideas on your presentation, particularly if you are calling for action.

Visuals and Graphics

Visuals and graphics will help you add impact to both persuasive and informational speeches by clarifying and emphasizing your information. The size of the room or auditorium, the kind of people in your audience, the available monies, your particular talents, and the available equipment (chalkboards, projectors, flip chart stands, televisions, computers, etc.), and the nature of your speech are all factors for consideration in planning visual material.

Chapter 3 discussed the value and conventions of graphics in written reports. These same types of graphics can be presented in oral reports. Chapter 18 discusses at length a wide variety of oral report visual props from simple flip charts to video and computer presentations.

Rehearsal

Rehearse your speech a number of times before you actually give it. Practice before a mirror, and use a tape recorder. If possible, give your speech to a small group of friends. Ask them to assess your poise, eye contact, voice, gestures, and rate.

Your voice should be conversational, confident, and enthusiastic. Avoid a monotonous sound by varying the pitch, intensity, volume, rate, and quality of your voice.

Delivery

All of the preceding steps should prepare you for an effective delivery. How your audience perceives you and your information depends on skillful delivery.

General Appearance.   Dress appropriately. Approach the podium confidently and maintain good posture. Gesture naturally. Gestures may be larger in a large room than in a small room. Be animated and maintain eye contact. Do not be ramrod stiff or clutch the podium. Do not jingle keys or make distracting adjustments to your hair, glasses, or clothes.

Audience Interaction.   Pause before you begin your speech to gain the listener's attention. Begin forcefully and engagingly. Keep tabs on your audience. Your listeners will be confirming or contradicting what you say through nonverbal messages. Allow for questions and discussion at the end of your report.

Notes and Visual Usage.   Use your notes and visuals with ease. Place these items in comfortable positions. Do not fidget with your cards or pointers. Stand aside when referring to visual materials and maintain eye contact as you make your points about the materials.

Voice.   Try for vocal variation. Let your voice exude warmth and sincerity. Pronounce your words clearly and distinctly. Do not vocalize "uhs," "ums," and other nervous sounds. Pause when appropriate.

ORAL SPEECH RATING SHEETS

At the end of the chapter are speech rating blanks for your classmates to use to evaluate speeches. Figure 17.3 is a general speech rating sheet useful for any extemporaneous persuasive or informational speech. Figures 17.4 through 17.6 may be used to assess oral descriptions of a mechanism, instructions, and analyses of a process, all of which are good subjects for informational speeches. Use these rating sheets in addition to the chapter checklist on preparing verbal communications.

NONVERBAL COMMUNICATIONS

Verbal communications involve more than the words being spoken. Our senses of sight, hearing, touch, and smell also receive messages that can color our own and others' perceptions of the spoken word. These perceptions are called *nonverbal messages*. A number of message carriers have already been mentioned—posture and position, voice modulation, eye contact, handshaking, background noise, and the like. For effective communications, you should be alert both to the nonverbal messages you send and those you receive. Ideally, nonverbal messages should be consistent with verbal messages, but such consistency is often violated.

Overt Actions.   Actions often do speak louder than words. Obvious actions, such as jabbing with a finger, pacing back and forth, waving papers around, and pounding on desks quite obviously signal aggression and anger. But subtle body actions, such as slouching, staring at the ceiling, standing with crossed arms, and the like, convey messages, too. Slouching conveys boredom; staring at a ceiling conveys disinterest; and crossed arms convey coldness or aggression. Open arm and body positions convey trust, warmth, and sincerity.

Covert Actions.   Some actions, such as blushes, tight jaws, smiles, and the like, reveal how a person is feeling. Who has not been confused by the person with furrowed brow and a tight-mouthed smile asserting, "No, I'm not mad at you"? Be alert to these signals in interpersonal communications, and learn to control your own covert actions to convey consistent verbal and nonverbal messages.

Eye Contact.   Eyes are the primary conveyors of nonverbal messages. Use your eyes to your advantage. An eyebrow flash by both parties often occurs within seconds of making eye contact with another person. Both you and the other person's eyebrows will lift briefly in a visual handshake. If you do not receive the warm eyebrow flash, consider whether the person is signaling hostility or is just shy, anxious, or even nearsighted. Intermittent three-second, direct eye contact conveys interest and receptivity, whereas reduced eye contact may convey concentration, disinterest, or deceit. Between communicators, eye contact regulates interactions and monitors the verbal messages. Breaking eye contact sideways conveys distraction or lack of interest and generates discomfort in the other person. Looking up will also throw the other person off balance.

Appearance.   Physical appearances also convey messages, although often inaccurately. We tend to stereotype persons according to three basic body types: athletic, frail, or obese. We expect the athletic person to be strong, mature, or self-reliant, whereas we expect the frail person to be indecisive, bookish, or pessimistic. Examine your personal tendency to stereotype people by their body type and actively rethink the validity of these prejudices. Be aware of how your own body type influences the perceptions of others. Changeable factors in appearance—length of hair, mustaches, eyeglasses, uniforms, jewelry, cosmetics, hem lengths, and so on—convey impressions, too. We attribute skills, personality traits, and abilities based on how people look.

Space.   Space is also an important nonverbal message carrier. The arrangement of office furniture or conference facilities can enhance or detract from open communication. Also, individuals have a sense of territory; to encroach on another's space is perceived as a threat or an aggression. Women are generally more comfortable conversing face-to-face, while men tend to prefer a side-on position that moves into a more frontal one. If you remain standing while your conversation partner is seated, you signal dominance, which may make the seated party uncomfortable. Even the distance that we set between ourselves and those with whom we are communicating conveys messages. Arabs and Latins tend to communicate at much closer distances (18 to 36 inches) than do North Americans or other Westerners who unconsciously prefer 30 to 48 inches. Arabs may perceive Western colleagues as remote and unfriendly, whereas the Westerners feel intimidated as their hosts keep moving in closer.

Extemporaneous Speech
Rating Sheet

Speaker _____ Subject _____ Evaluator _____

ITEMS	COMMENTS	SCORE
ORGANIZATION: Clear arrangement of ideas? Introduction, body, conclusion? Pattern of development adapted to ideas and audience?		
LANGUAGE: Clear, accurate, varied, vivid? Appropriate standard of usage? In conversational mode?		
MATERIAL: Specific, valid, relevant, sufficient, interesting? Properly distributed? Adaptive to audience? Personal credibility? Use of evidence?		
DELIVERY: Poised, at ease, communicative, direct? Eye contact? Aware of audience reaction to speech? Do gestures match voice and language?		
ANALYSIS: Approach to subject original, interesting? Central idea, purpose clear, divided into significant, interesting, subordinate ideas?		
VOICE: Pleasing, adequate, distracting? Varied or monotonous in pitch, intensity, volume, rate, quality? Expressive of logical emotional meanings?		

TOTAL _____

SCALE:

10	7	4	1
Superior	Average	Inadequate	Poor

FIGURE 17.3   *Extemporaneous speech rating sheet*

Mechanism Description Rating Sheet

Speaker _____ Mechanism _____ Evaluator _____

Did the speaker	Yes	Somewhat	No	Comment
1. Make necessary preparations before starting?	—	—	—	_____
2. Define intended audience?	—	—	—	_____
3. Name mechanism precisely?	—	—	—	_____
4. Define and/or state purpose of mechanism?	—	—	—	_____
5. Provide an overall description?	—	—	—	_____
6. Discuss the operational theory?	—	—	—	_____
7. State by whom, when, and where the mechanism is operated?	—	—	—	_____
8. Provide a list of the main parts?	—	—	—	_____
9. Describe the parts in the order listed?	—	—	—	_____
10. Define and/or state purpose of each part?	—	—	—	_____
11. List subparts of assemblies?	—	—	—	_____
12. Describe each part adequately?	—	—	—	_____
13. Avoid wordiness?	—	—	—	_____
14. Assess the advantages and disadvantages?	—	—	—	_____
15. Describe optional uses?	—	—	—	_____

FIGURE 17.4   *Rating sheet for oral description of a mechanism*

Did the speaker	Yes	Somewhat	No	Comment
16. Compare the mechanism to other models?	—	—	—	————
17. Discuss the cost and availability of mechanism?	—	—	—	————
18. Avoid reading the presentation?	—	—	—	————
19. Use adequate, well-organized notecards?	—	—	—	————
20. Show evidence of rehearsal?	—	—	—	————
21. Maintain effective eye contact?	—	—	—	————
22. Speak clearly so all could hear?	—	—	—	————
23. Use appropriate graphics?	—	—	—	
a. Number and title clearly?	—	—	—	————
b. Print legibly?	—	—	—	————
c. Keep graphics simple and uncluttered?	—	—	—	————
d. Lay out logically?	—	—	—	————
e. Credit sources of graphics?	—	—	—	————
24. Handle graphics with ease?	—	—	—	————
25. Do you feel you can judge the reliability, practicality, and efficiency of the mechanism?	—	—	—	————

SUGGESTED GRADE ——————————

FIGURE 17.4 *continued*

Instructions Rating Sheet

Speaker ——————— Instructions for ————— Evaluator —————

Did the speaker	Yes	Somewhat	No	Comment
1. Make necessary preparations before starting?	—	—	—	————
2. Define intended audience?	—	—	—	————
3. Provide a specific, limiting title?	—	—	—	————
4. State the instructional or behavioral objective?	—	—	—	————
5. Stress the importance of the instructions?	—	—	—	————
6. Define key terms?	—	—	—	————
7. State preliminary warnings or cautions?	—	—	—	————
8. Divide the process into main steps?	—	—	—	————
9. Provide a precise list of tools, materials, etc.?	—	—	—	————
10. Divide the main steps into substeps?	—	—	—	————
11. Provide warnings and notes for individual steps?	—	—	—	————
12. State each step as a command?	—	—	—	————
13. Avoid combining steps and substeps?	—	—	—	————
14. Express all steps in parallel, grammatical terms?	—	—	—	————
15. Avoid wordiness?	—	—	—	————

FIGURE 17.5   *Rating sheet for oral instructions*

Did the speaker	Yes	Somewhat	No	Comment
16. Avoid reading presentation?	——	——	——	————
17. Use adequate, well-organized notecards?	——	——	——	————
18. Show evidence of having rehearsed?	——	——	——	————
19. Maintain effective eye contact?	——	——	——	————
20. Use appropriate graphics?	——	——	——	————
a. Number and title clearly?	——	——	——	————
b. Print legibly?	——	——	——	————
c. Keep graphics simple and uncluttered?	——	——	——	————
d. Lay out logically?	——	——	——	————
e. Credit sources of graphics?	——	——	——	————
21. Handle graphics with ease?	——	——	——	————
22. Invite questions from the audience?	——	——	——	————
23. Do you feel you could perform the set of instructions accurately and efficiently?	——	——	——	————

SUGGESTED GRADE ————————————

FIGURE 17.5   *continued*

Process Analysis Rating Sheet

Speaker _____ Process _____ Evaluator _____

Did the speaker	Yes	Somewhat	No	Comment
1. Make necessary preparations before starting?	—	—	—	_____
2. Define intended audience?	—	—	—	_____
3. Name process precisely?	—	—	—	_____
4. Select a process primarily involving human action?	—	—	—	_____
5. Define or state purpose of process?	—	—	—	_____
6. Define or explain terms?	—	—	—	_____
7. Explain theory on which process is based?	—	—	—	_____
8. Indicate by whom, when, and where the process is performed?	—	—	—	_____
9. Indicate special conditions, requirements, preparations, precautions which apply to the entire process?	—	—	—	_____
10. Precisely list materials, tools, and apparatus?	—	—	—	_____
11. Divide process into five or six main stages?	—	—	—	_____
12. Arrange steps in numbered, chronological order?	—	—	—	_____
13. Provide a flow chart of main steps?	—	—	—	_____
14. Discuss steps in order listed?	—	—	—	_____
15. Define or state purpose of each main step?	—	—	—	_____

FIGURE 17.6   *Rating sheet for oral analysis of a process*

Did the speaker	Yes	Somewhat	No	Comment
16. Divide main steps into sub-steps?	——	——	——	—————
17. Describe special conditions for each step?	——	——	——	—————
18. Explain theory which applies only to one step?	——	——	——	—————
19. Adequately analyze each main step with emphasis on why, to what degree, to what extent, etc.?	——	——	——	—————
20. Avoid excessive detail?	——	——	——	—————
21. Evaluate effectiveness of process?	——	——	——	—————
22. Discuss advantages/ disadvantages?	——	——	——	—————
23. Discuss importance of process?	——	——	——	—————
24. Evaluate results of process?	——	——	——	—————
25. Explain how process is part of larger process?	——	——	——	—————
26. Compare process to similar processes?	——	——	——	—————
27. Assess cost and time factors?	——	——	——	—————
28. Avoid reading presentation?	——	——	——	—————
29. Use adequate, well-organized note cards?	——	——	——	—————
30. Show evidence of rehearsal?	——	——	——	—————
31. Speak clearly so all could hear?	——	——	——	—————
32. Avoid shifting to instructions?	——	——	——	—————

FIGURE 17.6   *continued*

Did the speaker	Yes	Somewhat	No	Comment
33. Use appropriate graphics?	—	—	—	————
a. Number and title clearly?	—	—	—	————
b. Print legibly?	—	—	—	————
c. Use color effectively?	—	—	—	————
d. Keep graphics simple and uncluttered?	—	—	—	————
e. Lay out logically?	—	—	—	————
34. Handle graphics with ease?	—	—	—	————
35. Invite questions from the audience?	—	—	—	————
36. Do you feel you can judge the reliability, practicality, and efficiency of the process?	—	—	—	————

SUGGESTED GRADE ————————————

FIGURE 17.6   *continued*

CHECKLIST

Verbal Communications

INFORMATION-GATHERING INTERVIEW

☐ **1.** Have I reviewed the **do's** and **do not's** of all interviews (see Chapter 13)?

☐ **2.** Have I determined a clear-cut purpose that I can present to the interviewee(s)?

☐ **3.** Have I decided on the questions I am going to ask?

　　☐ Are these questions organized logically?

　　☐ Specific?

　　☐ Open-ended?

　　☐ Unbiased?

☐ **4.** Have I reviewed, so that I can practice the courtesies of interviews?

　　☐ Will I remember to thank the respondent in advance for granting me the time?

　　☐ Will I inform the respondent how I intend to use the information?

　　☐ Am I prepared to listen rather than to "lead" the respondent?

　　☐ Have I received the respondent's permission to tape the interview?

　　☐ Am I prepared to take notes?

　　☐ Have I prepared to offer a copy of my report, document, or other material that will incorporate the interviewee(s) responses?

　　☐ Have I prepared a survey or questionnaire for multiple respondents?

TELEPHONE CONVERSATIONS

❑   **1.** Have I practiced modulating the volume, tone, and clarity of my voice?

❑   **2.** Am I prepared to enunciate all words correctly?

❑   **3.** Am I aware of the need to avoid distracting background noises?

❑   **4.** Have I provided a note pad, pencils, and pens by my telephone?

❑   **5.** Will I need any reference materials for a particular telephone conversation?

GROUP DISCUSSIONS

❑   **1.** Have I assessed the group that will participate in the discussion?

❑ Have I made clear the purpose of the discussion?

❑ Have I evaluated the members' personal goals that might be in conflict with the overall goal?

❑ Have I considered the permanency or impermanency of the group?

❑ Have I considered the power and relationships of the group members?

❑   **2.** Have I decided on the method (majority rule, compromise, or consensus) that will guide the group's decisions?

❑ Will a vote of the majority suffice?

❑ Will the members be able to compromise to reach decisions?

❑ Have we the time for a consensus decision?

❑   **3.** If I am looking for a consensus, am I prepared to give the group the power to establish a consensus?

❑ Has the group collectively established a goal and a procedure?

❏ Has the group devised an agenda?

❏ Have I allowed time for each member to contribute freely?

❏ Have I planned for a group discussion to determine the status and causes of the problem or situation?

❏ Has the group established criteria for evaluating each possible solution or decision?

❏ Will the group employ the four strategies of brainstorming?

   ❏ Suggestions without criticism or interruption?

   ❏ Listing of all the suggestions so that everyone can see them?

   ❏ Weighing the potential solutions against the criteria?

   ❏ Making a group decision and delegating and accepting responsibilities?

APPRAISALS

❏ 1. Am I prepared to participate in a performance appraisal?

❏ 2. Is the appraisal interview scheduled far enough in advance for both parties to prepare?

❏ 3. Has a specific time limit been established?

❏ 4. Am I ready to concentrate on task-related factors rather than personalities?

❏ 5. Am I ready to seek or offer explanations for poor performance?

❏ 6. Will I be able to concentrate on steps for improvement?

REPRIMANDS

❏ 1. Am I prepared to participate in an objective reprimand situation?

❏   **2.** Are there written policies and/or procedural manuals for the type of behavior to be discussed?

❏   **3.** Am I prepared to evaluate if violations are due to lack of information or willfulness?

❏   **4.** Am I prepared to discuss the violations with questions, feedback, and sensitivity?

❏   **5.** Am I prepared to ask or to answer questions about recourse, appeals, and corrective actions?

ORAL REPORTS

❏   **1.** Am I prepared to give an impromptu, memorized, manuscript, and extemporaneous report?

❏   **2.** Have I assessed my audience?

    ❏   Have I considered their prior knowledge of the subject?

    ❏   Have I considered their average age, sex, and rural or urban interests?

    ❏   Have I assessed their political and religious persuasions as well as their ethnic affiliations?

❏   **3.** Have I organized my report logically?

❏   **4.** Have I outlined my speech?

❏   **5.** Have I crafted my introduction, body, and closing to be convincing?

❏   **6.** Have I prepared note cards that are brief, readable, and highlighted for key points and visuals?

❏   **7.** Have I prepared supporting visuals and made the necessary preparations for handling them?

❏   **8.** Have I rehearsed my speech?

❏   **9.** Have I considered my appearance, audience interaction, and voice?

❑ **10.** Have I considered strategies for a **persuasive** speech?

  ❑ Have I arranged for the audience's comfort?

  ❑ Have I arranged for an introduction to stress my qualifications?

  ❑ Have I considered my dress, actions, and other possible expressions of similarities with the group?

  ❑ Have I stated my goal emphasizing the groups' rewards should such emerge?

  ❑ Have I included convincing statistics and the statements of acceptable authorities?

  ❑ Have I considered my nonverbal communications?

❑ **11.** Have I considered strategies for an **informative** speech?

  ❑ Besides employing the above strategies, have I ensured that my audience will be able to judge the reliability, practicality, and efficiency of the subject of my speech?

EXERCISES

1. **Information-Gathering Interview.**   Working in groups of two or three, devise interview questions about a specific subject that will affect your classmates or the entire student body. Possible subjects are the need for computer lab changes, the establishment of snack booths on campus, class registration procedures, campus bookstore policies, cafeteria procedures, and bike racks. Be sure to have a clear purpose, logical organization, and specific, open-ended, unbiased questions. Review the relevant interview strategies discussed in Chapter 13 and the seven suggestions in this chapter. Have each member of the group participate either as the interviewer, the respondent, or the critic who will assess the verbal and nonverbal communication skills of the other two. Encourage the rest of the class to participate in a critique.

2. **Telephone Communications.**   Working in pairs, role play a business telephone transaction. Devise your own transaction or try the following situation: The caller has mailed a cover letter and resume to the director of personnel. The caller is trying to find out if the letter arrived and seeks to arrange an interview for the next

week. The callee, the director of personnel, is eager to arrange an interview but will ask a few questions. Both should be prepared to take notes. The class will assess the skills of both the caller and the callee.

3. **Appraisals.**   Working in pairs, devise a performance appraisal situation between a supervisor and a subordinate. Each will provide a written assessment to be exchanged before the role playing in class. The supervisor will discuss five elements of the subordinate's work: timeliness, productivity, attention to detail, enthusiasm, and peer relationships. The supervisor will discuss two areas needing improvement. The class will assess the skills of both the supervisor and the subordinate in terms of objectivity, concentration on task-related factors, explanations for negative appraisal, and discussion on goals and improvement.

4. **Reprimands.**   Working in pairs, devise a reprimand situation between a supervisor and a subordinate. The pair should devise a brief, written policy or procedural statement that, presumably, the company already had in writing. The reprimand may be over repeated tardiness, low productivity, lack of attention to detail (as in a written report), general demeanor, or peer relationships. The class will assess the skills of both the supervisor and subordinate in terms of determination of cause, questions, feedback, sensitivity, and agreement on appropriate recourse, appeal, or corrective action.

5. **Group Discussion for a Consensus Decision.**   Role play a group discussion among five or six classmates. The group is to reach a consensus on one of the following situations:

    a.   A donor has offered $10,000 to your school for a scholarship in a professional field. The group consists of one member each from the computer department, the nursing department, the architectural department, and the dentistry department (or if these are separate schools in your university, the members are from each of the schools). The group is to decide how to implement the scholarship. Following the discussion, the class will assess the verbal and nonverbal skills of the participants.

    b.   A department in a business has $5,000 budgeted for employee travel to professional conferences and seminars around the country for the fiscal year. There are 10 members in the department; each member has two or three conferences he or she would like to attend. Expenses would range from $200 for short, in-county seminars to $2,000 for week-long, out-of-state conferences. The group must decide on the criteria for spending the money. Following the discussion, the class will assess the verbal and nonverbal skills of the participants.

6. **Extemporaneous Oral Reports.**   Prepare a 6- to 10-minute extemporaneous oral report with visual materials for one of the following situations:

a. *A persuasive speech.* Choose a controversial subject (four-day work week, class registration procedures, bookstore policies, degree requirements in your field, etc.). Prior to your speech, pass out index cards to the class and ask each member to write opinions on this subject. Deliver your speech. Each classmate will evaluate your presentation by rating you according to the rating sheet in Figure 17.3. Following the speech, ask your classmates to read their opinion cards and discuss how you have or have not changed their opinions.

b. *An informative speech.* Refer to Chapters 9, 10, and 11 on instructions, descriptions of mechanisms, and analysis of a process. Select one of the report strategies; prepare and deliver an informative oral report. Each classmate will assess your presentation by filling out the appropriate rating sheet in Figures 17.4, 17.5, or 17.6.

Notice: An assignment for another extemporaneous oral report follows Chapter 18 after coverage of accompanying visual communication alternatives.

7. **Nonverbal Communications.**   Select an employer, a coworker, a professor, or a friend. Tell that person you desire just a five-minute interview about a problem you are having. Devise a situation. Hold the informal interview, paying close attention to the nonverbal messages the interviewee sends, such as overt actions, covert actions, eye contact, appearance, and space. Be mindful of your own nonverbal messages during this interchange. Following the interview, quickly write down your observations. Write a two- or three-paragraph paper about your observations. Do not use the actual name of the interviewee.

NOTES

CHAPTER 18

Managing Visual Communications

MR. FUMBLE   by PeJay Ryan

By permission of PeJay Ryan.

S K I L L S

After studying this chapter, you should be able to

1. Identify potential professional situations in which visual aids are necessary.
2. Name twelve visual communication aids along with their individual strengths and weaknesses.
3. Practice the attribution of sources for your visuals to avoid plagiarism.
4. Understand and practice effective design, with attention to simplicity, unity, emphasis, balance, and legibility.
5. Construct at least six visual aids from among the following: handouts, demonstrations or exhibits, chalkboard displays, flip chart displays, posters, transparencies, slides, filmstrips, videotapes, and computer presentations.
6. Handle visuals properly and with ease.
7. Understand the value and production steps of videotaped employment interviews and job performances.
8. Participate in a videoconference meeting or employment interview.

INTRODUCTION

Throughout this textbook, the power of visual communications has been stressed. As you have designed documents, incorporated emphatic features (boldface, italic type, and so on), explored various fonts and type sizes, and devised graphics and other visuals for your documents, you have experienced first hand the value of visuals. Visuals can be further put to use by incorporating them into your oral presentations by the following methods:

- Printed or photocopied handouts
- Demonstrations and exhibits
- Chalkboards
- Flip charts
- Posters
- Transparencies
- Opaque projections
- Slides
- Filmstrips

- Videotapes
- Computer presentations
- Videoconferencing

Each of these tools, used alone or in combination, may enhance your formal presentations. The key to success is to select and use them carefully and intentionally where they best fit the situation in terms of your purpose (to instruct, motivate, sell, or entertain), the audience's prior knowledge about your subject, the size of the room, the size of the audience, and the equipment readily available to you. Consider your options as you organize your oral presentation and practice in front of friends in the facility you will be using, if possible.

TYPES OF VISUALS

Print or Photocopy Handouts

Computer-printed or photocopied handouts are very easy to use, especially with larger audiences that may have trouble seeing the physically distant chalkboard, exhibit, flip chart, or poster. If you prepare too much material on the page, your audience will tend to read it as you speak, thereby missing part of what you say.

Keep the art and words to a minimum. Make your handouts clear and concise as well as relevant to the material you are discussing in your presentation. Handouts offer the advantages of being prepared in advance, being simple to transport, and giving your audience something to take home for review.

Demonstrations/Exhibits

Occasionally your purpose will be to demonstrate a technique, a piece of equipment, or a process, such as coronary pulmonary resuscitation, a blood pressure cuff, or a laboratory experiment. You may consider using exhibits such as architectural models, products, and equipment. These visuals must be easy to use to avoid long and complicated pauses during your presentation and large enough for your audience to see clearly.

Limit the use of demonstrations and exhibits to small audiences and facilities, such as a conference room or small classroom. Do not use demonstrations and exhibits unless everyone in the room can clearly see the exhibit plus your use or manipulation of it. You may require your audience to cluster around you or the exhibit.

Chalkboards

Most classrooms and many conference rooms contain chalkboards. You may wish to present an outline, simple text, graphics, and the like on the chalkboard; however, if you do plan to use a chalkboard, prepare the material ahead of time and use multicolored chalk. Your audience will be greatly distracted if it has to wait while you write and draw. Consider whether everyone in the facility will be able to see and to read all of your material.

Flip Charts

Use flip charts (large, ring-bound pads) on an easel for informal presentations. You can prepare some pages of graphics or other visuals in advance. Write in any color felt-tip marker and flip the paper over the top to move on to the next sheet. Flip charts are very handy if you are going to be listing topics in a situation such as a brainstorming session. New electronic flip charts are now on the market that will allow you to write on their surface, print out the messages, or download information to a personal computer.

Flip charts have advantages over chalkboards. You may easily transport them and place them closer to your audience; a chalkboard is usually fixed and generally behind you. Flip charts, like exhibits, should be used in small rooms and with small audiences of fewer than 25 people.

Posters

Posters are very useful but are limited in scope. Complicated posters with typewritten pages and small diagrams, often used by high school students for science fair exhibits are functionally useless for an oral presentation. They cannot be seen clearly from more than a few feet away. The rule is simple: your poster must be clearly visible and easy to read from the farthest seat in the room and from all angles. This requirement limits the amount of information you can include on the poster.

Masking tape will hold posters firmly on to a blackboard without marking the slate. Posters should not be propped up against the podium or on chairs; they are sure to fall. Use easels if at all possible to hold them. Both posters and easels are highly transportable.

Greatly enlarged photographs, usually mounted on foam core board along with brief text, are widely used for seminar and sales table displays. Use photographs of text, persons, situations, or graphics. Two drawbacks to the technique of visual presentation can be high cost and loss of picture quality.

For regular posters, use sturdy white or pale-colored poster material. If you use more than one poster, use the same color for each; multiple color poster displays are distracting. Sketch artwork and lettering directly onto the poster. Letter stencils or paste-on letters are useful. You may, of course, also use felt-tip pens, crayons, grease pencils, translucent color tapes and color sheets, and many other art supplies.

Transparencies

Overhead transparencies are simple sheets of clear plastic that can be written on by hand, prepared by photocopying printed or art material, or printed directly on your computer printer. Transparencies are cheap and allow you to apply artwork and lettering directly onto the plastic for quick, temporary use. You may also prepare your material on plain white paper and photocopy it onto a transparency. The most popular method of preparation, however, is to prepare your material on a computer, using at minimum 35-point type size, a variety of fonts, art, and color. If you have an inkjet or laser printer, you may print directly onto transparencies. Copies of your transparencies can be made to use as handouts in conjunction with your presentation.

Ensure ahead of time that your facility has an overhead projector and a white wall or screen of the appropriate size for your group. Ensure, also, that you have extra bulbs available in case one burns out during your presentation.

Opaque Projections

An opaque projector is used to project images from paper onto a screen. This works well in a small classroom or boardroom situation to project pages from a bound volume or any printed document for discussion and analysis. The advantages are that no art preparation is required ahead of time; therefore, there is no expense. The disadvantages are that the room must be dark, magnification may be limited, and the machines tend to be a bit noisy. Nevertheless, for impromptu or brief visual enhancement, opaque projectors work well.

Slides

Anyone with a knowledge of photography may produce slides. Effective presentations can be created using only a 35-mm camera and a small amount of ingenuity. You may photograph people, scenery, and artwork; journal, magazine, and book illustrations; graphics; and computer-printed clip art, icons, and lettering.

Once your slides have been developed and returned, you need only to select those you intend to use in your presentation, place them in the order you intend to use them, and arrange them in the tray of a projector.

Two projectors are often used simultaneously to make comparisons and contrasts. Determine ahead of time, of course, if a slide projector(s) and screen(s) are available in the facility you intend to use. A light-projection pointer will help you to direct the audience's attention to specific detail within the slide. Many facilities provide pointers, but you may decide to buy your own reasonably priced, pen-sized pointer to carry with you.

Slides require a dark room, which denies you eye contact with your audience and may limit audience note-taking. Computer software will allow you to create small paper copies of your slides (six to nine per page) to distribute to your audience.

Filmstrips

Filmstrips are extremely useful because of their ease of handling; however, they are difficult to make, expensive to purchase, and are becoming difficult to find. More versatile slide presentations and videotapes are replacing filmstrips; however, a check at your library might reveal a selection on file which may be suitable for your presentation. They, too, require a dark room, limiting eye contact and note-taking.

Videotapes

A videotape is without a doubt the most popular currently used visual aid. It is used by executives, teachers, lawyers, students, and others. There are countless videos available for purchase, or you may prepare your own using a video camera, or even copy material from a television program. Some understanding of photography, cinematography, and videotape editing will enable you to produce professional videos with voice and other sound effects. Your camera user manual will include helpful information.

You will require a VCR for your presentation facility. Remember that the size of the screen will dictate the acceptable size of your audience. Videos offer the advantages of instant replay and freeze-framing for emphasis. Because they may require a reasonably dark room, they may deny you eye contact with your audience and limit note-taking.

VIDEOTAPED EMPLOYMENT
INTERVIEWS AND JOB PERFORMANCES

In addition to videotapes that you prepare for an oral presentation, some companies may request a videotape of you expressing your qualifications for employment or performing the duties of your job. You may prepare a video that shows you either talking about yourself and your goals or actually working at your job. This allows potential employers to get a much clearer image of what you are like as a person and how you present yourself to the public.

The following elements need to be addressed if you decide to produce your own video:

- Designing the presentation
- Preparing the script
- Designing and preparing the set
- Rehearsing the presentation
- Recording and editing the interview
- Copying the presentation

In designing the presentation, consider your strengths and weaknesses and anticipate any questions that an interviewer may ask about the information you have offered. For the job performance video, tape yourself actually doing your work and interacting with others. Prepare a script not to exceed 20 minutes. Design a set that reflects your field of work. If the position involves desk work, use a desk with books, possibly a computer, and other appropriate paraphernalia arranged on it. If you are looking for an architectural or engineering position, stand or sit next to a drafting table and carry out some actual work. Let the setting demonstrate that you are a part of the profession and comfortable within its surroundings.

If you design the set, consider color, lighting, and camera placement. Avoid clashing colors or strong patterns that may detract from you, the main focus. Use soft light coming directly toward your face to eliminate the blemishes, wrinkles, and shadows that overhead lights may underscore. Determine the best side of your face and place the camera about 30 degrees to the opposite side, which will allow you to turn your head slightly, giving a more relaxed look and allowing the front light to highlight the best side of your face. The incidental shadow falling on the opposite side of your face provides modeling and dimension.

Rehearse your presentation on tape so that you can retain what looks good and eliminate what looks bad. Check the sound and lighting. Tape your final presentation and make two or three initial copies from the master, saving the master for further use. Copies from copies tend to lose their contrast.

Mailing boxes specifically designed for videocassettes are available from postal supply stores. Be sure to label both the tape and the box appropriately, include the correct postage, and include your cover letter and resume in an envelope attached to the package.

Computer Presentations

With the right software, you can create drawings, graphics, and text on slides for a computer presentation. You may also create your own materials, incorporate clip art and online downloads, or use a scanner to import materials from other sources. Templates and tools allow you to select background and layout layers. You may add titles and text in any size type or font within the presentation and add arrows, banners, and textures, and even pictures within pictures. With a few clicks, you can alter the spacing around objects. You may easily change the order of slides, add more slides, or import existing ones from another presentation. Further, you can turn any drawing or text into a 3-D object or reshape text or objects to special images such as waves, bow ties, and pennants. You can even rotate (or flip) a selected object to change its orientation. In addition, animation, movie files (digitally encoded live action), and unique transition effects can be added, making the slide seem to roll to the side or to open out of the center of the screen. You can open a page on the Internet, play a sound file, or skip to another part of the slide show. The possibilities are almost endless.

Furthermore, you may project your computer slide presentation onto a larger screen for a larger audience and print out copies of your slide show (six to nine per page) to allow your audience to take notes on them and to take away as handouts.

Videoconferencing

Videoconferencing is becoming increasingly popular because it saves thousands of hours lost in travel time. In the videoconference, you share live video, audio, and data over wired or wireless telephone lines—meeting face-to-face on television or computer monitors and by video telephones. You can even transmit graphics, slides, or videotape during the meeting. If you do not have the capability for a personal computer videoconference or an on-site videoconference facility, you may use the videoconferencing rooms and state-of-the-art equipment in hundreds of Kinko's and other multiservice office centers around the world.

In the videoconference, you may share any number of other visual aids already discussed, such as demonstrations, exhibits, flip charts, posters, slides, or videotapes. You may also distribute premeeting materials, such as handouts, agendas, and the like. You can even hold multipoint conferences (more than two sites).

An increasing number of potential employers may request an interactive videoconference employment interview, saving the company and you travel expenses. To execute an interactive videoconference interview, both you and the interviewer will require special software and hardware (cameras mounted on top of your computer monitors, microphones, and so on). Kinko's and your college learning resource center will be able to help you undertake such interviews. You may also store the interview on your hard drive and send it as an e-mail attachment to a prospective employer.

Set the conference or interview time and date well in advance so that all participants can plan to be on hand. Dress conservatively and avoid white, plaids, prints, and stripes, which tend to vibrate or glare. If you are using props, set them up where they are accessible and can be easily seen on camera.

If the videoconference is of a meeting, have each location chair introduce all participants, even if they are off-camera at the time. Review your meeting agenda and goals. Be yourself and speak naturally and clearly. In a group, identify yourself as you speak from time to time to help your audience. Pause slightly after finishing a comment to allow for a brief lag in the video image transfer. Avoid coughing into the microphone, drumming your fingers, shuffling papers, or holding side conversations. Allow time to review questions and decisions, and end on time.

If the videoconference is an employment interview, consider the above and review interview materials in Chapter 13. Remember that you can "stage" such an interview, store it on your hard drive, and e-mail it as an attachment to any number of prospective employers.

SOURCE ATTRIBUTIONS

Plagiarism is as much a concern with visual materials used in oral presentations as it is with written documents. Thus, your visual presentations need to be credited to their sources. These credits may be printed directly on the visual aid, but this is not always possible. Therefore, you should *always* verbally acknowledge the source of a graphic when using it in a presentation.

HANDLING VISUALS

Visuals will lose all of their inherent power if you fail to use them carefully. Follow these tips:

1. Ensure ahead of time that your facility has the required equipment (easels, videocassette recorders, slide projectors, overhead projectors, screens, microphones, and pointers), electrical outlets, and accessible light switches. Check the furniture (desks, podiums, tables) and seating arrangement, and rearrange if necessary.

2. As a rule, limit your exhibits, flip chart pages, posters, and transparencies to no more than four or five in a 20-minute presentation to avoid a circus atmosphere.

3. Try not to display your visuals until you refer to them in your presentation in order not to distract your audience.

4. Do not get in the way of your visuals. Stand aside and avoid turning your back to your audience.

5. Interpret or discuss each visual. Do not just let it "speak" for itself.

6. After discussing a visual, remove it or step away from it and refocus attention on yourself.

7. Distribute handouts *after* a presentation in order to focus your audience's attention on you. Sometimes you will want to distribute handouts ahead of time to help the audience follow your agenda.

MATERIALS

Other than computer-generated art and lettering, basic supplies for visual aid construction are available in college bookstores and art and paper supply shops, and include

- Felt-tip pens in various widths and colors
- Grease pencils in a variety of colors
- Color projection pens and pencils for transparencies
- Prepared cut-a-color and shading sheets, both translucent and transparent
- Chart pax tapes, both translucent and transparent
- Lettering templates
- Letter, symbol, and shape stencils
- Dry transfer letters
- Adhesive letters in black and colors

Poster boards of various qualities, transparencies (acetate and thermals), flip charts, photocopiers, blank videotapes, and all of the other possible materials you may consider are also readily available.

EFFECTIVE DESIGN

Review the design information in Chapters 3 and 4, which include document page design, graphics and visual constructions, drawings, illustrations, maps, photographs, text art, clip art, and icons. Those chapters include tips on the amount of detail, clarity, texture, color, size, and orientation on a page, which also apply to oral presentation visuals. If you are preparing handouts, chalkboard art, flip chart visuals, posters, transparencies, or computer slides, consider the following key design elements:

- Simplicity
- Unity
- Emphasis
- Balance
- Legibility

1. **Simplicity.**   Fewer elements are more pleasing to the eye than a hodgepodge of detail. Bold key detail has more impact than does complex art. If the material contains words, limit them to 15 or 20. Figure 18.1 illustrates both a simple design and a complex design:

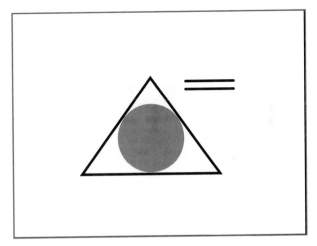

Simple, pleasing design

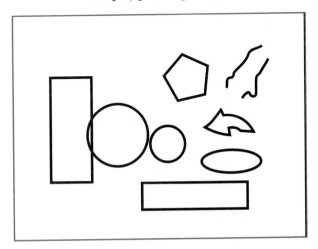

Too complex design

FIGURE 18.1   *Simplicity of design versus complexity of design*

The circle within the triangle with two parallel lines indicating text exemplifies simplicity of design, whereas the multiple figures with squiggly text lines illustrate a complexity of design that is difficult on the eye (and the nervous system).

2. **Unity.** Unity may be achieved by overlapping elements and arrows, and by a conformity of shapes and sizes on any one visual. Random layout reduces unity. Figure 18.2 illustrates unity:

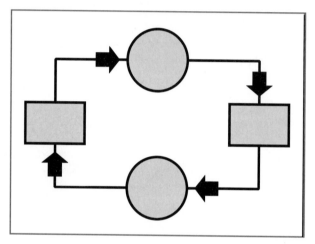

FIGURE 18.2   *Unity of design*

The double circles, double rectangles, and connecting arrows produce the unifying result.

3. **Emphasis.** Use color, bold sizes, and white space to achieve emphasis.

4. **Balance.** Balance may be formal or informal. Formal balance is achieved when an imaginary axis divides the design into two mirror halves horizontally or vertically. Informal balance is asymmetrical but logical. It is more dynamic and more attention-getting. Figure 18.3 illustrates some principles of balance:

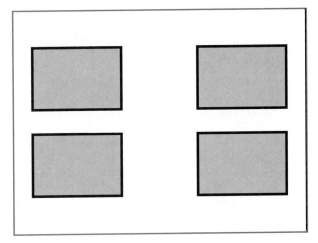

Symmetrical design

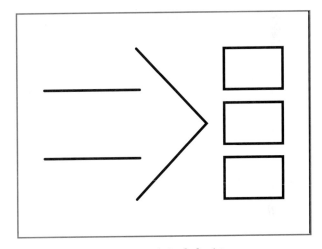

Assymmetrical design

FIGURE 18.3   *Formal balance and informal
(asymmetrical) designs*

The first design is formal, with the halves mirrored both horizontally and vertically, while the second design is informal, having asymmetrical elements. Both are effective, but perhaps the informal design is more sophisticated.

5. **Legibility.**   Wording on visual materials should be minimal. Sanserif or gothic letters are easier to read than script and other fancy fonts. Figure 18.4 illustrates a variety of fonts.

TIMES NEW ROMAN	**FUTURA**
CLAIRVAUX	GOUDY

FIGURE 18.4   *Illustration of plain and fancy fonts*

Use capital letters for short titles and labels, but use a combination of upper- and lowercase letters for verbal content of six or more words. Figures 18.5 and 18.6 show examples of all capital titles and upper- and lowercase combination titles.

SHORT

TITLES

FIGURE 18.5   *All capital letters used for short titles*

Combinations of Uppercase and Lowercase Titles Are Easier to Read

FIGURE 18.6   *Combination letters for titles of six words or more*

The maximum viewing distance is generally accepted as being eight times the horizontal dimension of the graphic. Thus, a four-foot-wide poster has a maximum viewing distance of 32 feet. Minimum lowercase letter sizes are 1-in. high on standard posterboard and ¼-in. high (35-point font) for transparencies. Capital letters are correspondingly larger. Thick letters are more legible than thin letters.

Application of these design principles will turn your oral assignments into dynamic presentations.

CHECKLIST

Visual Communications

GENERAL

❏ **1.** Have I carefully planned the purpose, objectives, and organization of my presentation?

❏ **2.** Have I selected appropriate visuals for the content of my presentation and the facilities in which I will present them?

❏ **3.** Have I considered handouts, chalkboard material, flip charts, posters, transparencies, opaque projections, slides, filmstrips, videotapes, and computer presentations?

❏ **4.** Will I be using these visuals in a face-to-face situation or in a videoconference?

❏ **5.** Have I collected the appropriate art materials and/or computer software for visuals execution?

❏ **6.** Have I considered for each aid that I am constructing the principles of effective design?

    ❏ Simplicity?

    ❏ Unity?

    ❏ Emphasis?

    ❏ Balance?

    ❏ Legibility?

❏ **7.** Do I need to attribute verbally the sources of my visuals?

❏ **8.** Have I practiced handling my visuals effectively?

HANDOUTS

❏ **1.** Have I limited my art work and words to a minimum?

❏ **2.** Have I used thick enough lettering and appropriate fonts and art work to distinguish this from a normal page of print?

❏ **3.** Will I distribute my handouts before or after my presentation?

DEMONSTRATIONS AND EXHIBITS

❏ **1.** Should I consider giving a demonstration?

❏ **2.** Should I use an item as an exhibit?

❏ **3.** Can I handle a demonstration and exhibit with ease and without long pauses?

❏ **4.** Will everyone in the room be able to see my demonstration or exhibit?

CHALKBOARDS

❏ **1.** Will I be able to prepare my chalkboard material in advance?

❏ **2.** Do I have multicolored chalk for emphasis?

❏ **3.** Will everyone in the room be able to see and to read my material?

FLIP CHARTS

❏ **1.** Would a portable flip chart be more appropriate than a chalkboard?

❏ **2.** Should I prepare some of the graphics or print ahead of time?

❑ **3.** Do I have color felt-tip markers for emphasis?

❑ **4.** Will everyone in the room be able to see and to read my material?

POSTERS

❑ **1.** Would posters be appropriate for my visual aids?

❑ **2.** How will I display my posters—by taping to a wall or chalkboard, or using an easel?

❑ **3.** Would enlargements of photographs be appropriate for my posters?

❑ **4.** Will all of my posters be uniform in color and size?

❑ **5.** Will everyone in the room be able to see and to read my posters?

TRANSPARENCIES

❑ **1.** Should I create transparencies?

    ❑ By hand?

    ❑ On plain white paper for photocopying?

    ❑ By computer printout?

❑ **2.** Shall I make copies to distribute as handouts?

❑ **3.** Do I have the necessary acetates, reproduction equipment, and projection equipment?

OPAQUE PROJECTIONS

❑ **1.** Have I organized the book pages or other print materials to be projected?

❑ **2.** Do I have access to an opaque projector, a blank wall or screen large enough for my projections, and a light-projection pointer?

SLIDES

❏ **1.** Do I have the equipment and photographic ability to produce slides?

❏ **2.** Have I organized my slides into a logical order?

❏ **3.** Do I have the necessary equipment (slide projector and tray, light-projection pointer, and wall or screen large enough for my projections)?

❏ **4.** Have I printed out copies of the slides (six to nine per page) for the audience to retain?

FILMSTRIPS

❏ **1.** Is it possible to purchase a filmstrip that will serve as an appropriate visual aid?

❏ **2.** Do I have access to the machinery to produce a filmstrip?

❏ **3.** Do I have the necessary projection equipment and screen?

VIDEOTAPES

❏ **1.** Do I have the necessary equipment to use a videotape (video camera, VCR, television screen, and blank tapes)?

❏ **2.** Should I add voice and other sound effects?

❏ **3.** Will everyone in the room be able to see the screen?

❏ **4.** Should I consider making a video of myself for an employment interview or job performance review?

  ❏ Have I carefully designed the presentation?

  ❏ Have I prepared a script?

  ❏ Have I arranged an appropriate set?

  ❏ Have I rehearsed my presentation?

❏ After recording, have I carefully edited the tape?

❏ Have I made copies and saved the master tape?

❏ Have I obtained the appropriate mailing box?

❏ Should I include my resume, cover letter, or other written material?

COMPUTER PRESENTATIONS

❏ **1.** Does my facility have a computer and screen?

❏ **2.** Do I have the software to develop a computer presentation?

❏ **3.** Do I have the know-how to use the software?

    ❏ Have I practiced incorporating clip art, online downloads, and scanned material into the presentation?

    ❏ Have I considered backgrounds and layout layers?

    ❏ Have I used appropriate fonts, type sizes, arrows, banners, and textures?

    ❏ Have I spaced around the material artistically?

    ❏ Should I turn any object into a 3-D object or reshape text into special images?

    ❏ Should I change the orientation of any of my slides?

    ❏ Would animation, movie files, rolled slides, or center-opening slides add to the presentation?

    ❏ Should I incorporate words and music into my presentation?

VIDEOCONFERENCING

❏ **1.** Would a videoconference presentation be appropriate?

❏ **2.** Do I have access to videoconferencing facilities?

❏ **3.** Should I develop visual aids for the presentation?

❏   **4.** Have I set the date and time well in advance with my groups?

❏   **5.** Have I arranged for the appropriate introductions at each site?

❏   **6.** Does everyone understand the agenda and goals?

❏   **7.** Have I practiced speaking naturally and clearly and avoiding distractions?

❏   **8.** Have I dressed conservatively and avoided white, plaids, prints, and stripes?

❏   **9.** Have I paused after my comments to allow for the brief lag in image transfer?

❏   **10.** Have I encouraged the participants to ask questions and add comments?

❏   **11.** If I have taped a videoconference on-camera interview, have I stored it on my hard drive in preparation for e-mail attachment to other prospective employers?

EXERCISES

1. Prepare at least three visuals for a subject of your choice. Your visuals may include any combination of a handout, a chalkboard visual, a flipchart visual, a poster, or a transparency.

2. In class, present your visuals and orally evaluate and critique each other's visuals prepared for Exercise 1.

3. For each of the following, determine the best type of visual aid for the conditions presented:

   a. A stockholder's meeting with an audience of 500

   b. A board meeting with 10 participants

   c. A classroom oral report on describing a mechanism

   d. A demonstration on emergency care to a group of 25

   e. A sales presentation in a person's home with 6 attendees

ORAL/VISUAL PROJECTS

1. **Oral Extemporaneous Speech with Visuals.**   Prepare an oral extemporaneous speech for classroom presentation (review Chapter 17) and include at least three visual aids (handouts, demonstration/exhibit, chalkboard material, flip charts, poster, transparencies, or opaque projections in your presentation.

2. **Collaborative Presentation.**   With a group of three, prepare an oral extemporaneous speech for a classroom presentation with visual aid selection of slides, videotape, or computer presentation on a subject of your choice. Execution of these options will earn you extra credit because they are more difficult than the aids assigned in Project 1.

NOTES

CHAPTER *19*

Designing and Managing Web Sites

PAGAN THOUGHTS   by Pejay Ryan

By permission of Pejay Ryan.

S K I L L S

After studying this chapter, you should be able to

1.   Define Web sites and explain how they are accessed on the Internet.
2.   Name and execute the nine main steps for creating and maintaining a Web site.
3.   Analyze the anticipated audience and the purpose of your prospective Web site.
4.   Define a home page and linked pages.
5.   Use an inviting, concise, and informative style throughout your site.
6.   Decide on a site design, page layouts, graphics, and link designs and placement.
7.   Convert the text to HyperText Markup Language by using the tags or conversion software.
8.   Add preselected graphics, color, dividers, sound, animation, and other multimedia options.
9.   Test and revise the site for effectiveness.
10.   Place the site on an Internet host server.
11.   Advertise the existence of your site.
12.   Maintain and update the site.

INTRODUCTION

Designing and managing Web sites is increasingly the job of the technical writer in business and industry. Businesses, industries, manufacturers, organizations, schools, government agencies, and individuals are hosting sites on the World Wide Web to promote their products and services or to disperse information. The Web, as you know, is an Internet service that displays and links together hypertext documents via a special protocol called Hypertext Transfer Protocol, or HTTP. A Web site consists of the home page and a set of linked pages. As a student, you may want to create your own Web site to provide information about yourself, to share pictures, to post resumes, and to advertise your skills, activities, and views of interest to your friends, prospective employers, and the general public.

Web pages can be built quickly and easily. A host of books are available on site design. The journals *Web Week* and *InfoWorld* are excellent sources for Web site developers. Moderately priced software packages are also widely available to teach you how to create a Web site, step by step. Such software includes America Online's Web Pages Made Easy ™, Microsoft's FrontPage, Softquad's HoTMetal, Spanner's Web Site Design, and more. Even Microsoft Word and Corel WordPerfect will instruct you on how to create impressive Web sites. All of the software programs and manuals are great tools that allow you to make even the most daunting of designs simple, with drag-and-drop editing, formatted templates, and graphics.

Further, there are hundreds of Web sites available that contain design tutorials, graphics, and various other tools to make your Web design and creation experience quick and painless. The World Wide Web Developer's Virtual Library, **<http://www.stars.com>**, will link you to over 1,000 sites containing design aids and free downloads. There are currently over 45,000 sites on the Internet to help you create Web sites.

WEB SITE STEPS

There are nine basic steps to creating and maintaining a Web site. They are

1. Analyzing your purpose and audience.
2. Creating the text, graphics, and links.
3. Designing the site.
4. Designing the pages.
5. Converting content to HTML.
6. Testing and revising the site.
7. Placing the files on an Internet server.
8. Spreading the word.
9. Maintaining the site.

Figure 19–1 (page 632) shows a flow chart of these steps.

Analyzing Purpose and Audience

Step one is to define your Web site purpose. Is it to inform? to entertain? to promote products, services, and activities, such as conferences, tours, cultural events? to post a resume? Who will visit your web site? What will be their ages, interests, perceptions, cultural biases, and technical knowledge? Answers to these questions will guide you in design and content decisions. Chapter 2 discusses audience analysis in depth and is worth reviewing here.

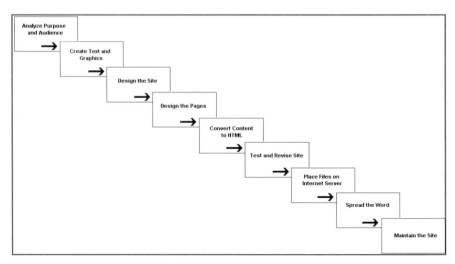

FIGURE 19.1   *Nine steps for designing and maintaining a Web site.*

Creating the Text, Graphics, and Links

Step two is to write the text (title and body paragraphs), select the graphics (photographs, drawings, clip art, diagrams, and the like), and decide on the links you want to include (to your personal e-mail, to separate pages within your site, or to other Web sites).

HOME PAGE

Your site will consist of a *home page* and *linked pages.* The home page should provide identification information (the name of the product or service, company or organization name, your name if the site is a personal Web site, e-mail addresses, etc.) The home page should also provide informative and appealing graphics. Your home page graphics should be simple so that they download quickly; otherwise, viewers may become impatient and leave your site before it is finished loading. The object of the home page is to entice the viewer to the linked pages where larger graphics can be stored.

A welcoming, informative introduction should be included on the home page; your tone should be personal and positive. If it is a personal site, it may well begin, "Welcome to the Home Page of (Your Name)." Hypertext links or buttons, or both, should be added to connect the reader to subsequent screens or back to the home page as well as to your e-mail address and other relevant Web sites. You may wish to place these across the top, frame them down the right-hand side, or locate them in other key spots with the appropriate underlining, icons, or pictures for clicking to other pages.

Your tutorial will instruct you on how to create the links, but the copy must be typed out in advance, using brief noun phrases, e-mail addresses, or other Web addresses with the complete URL (Uniform Resource Locator), including the transfer protocol (http://), the domain (i.e., www.aol.com/), and file names with proper punctuation (i.e., yourname/tour.html/).

LINKED PAGES

The *linked pages* require headings or titles to indicate to the reader which screen is being viewed. The linked pages must include text that thoroughly develops ideas with specific detail. Graphics, sound, animation, and more links may enhance the information.

STYLE

Keep your text concise by using short words (one to two syllables), short sentences (10 to 12 words per sentence), and short paragraphs (about four typed lines). Line length should not exceed two-thirds of the screen because it is difficult for the screen reader to track longer lines. Try to limit each Web page to one or two viewable screens to eliminate excessive scrolling. It is far more difficult to read extensive text on a screen than on a hard copy page due to glare and distractions of sound and animation.

Designing the Site

As already explained, your site will consist of a home page and linked pages. As a rule, you will not want more than three levels of links. Figure 19.2 shows a three-tiered site design. Large, complex sites, such as government and scientific research sites, may contain dozens of pages. Each page will contain links to the other pages as well as back to the home page. Keep your site as easy as possible to navigate, so that your viewer has a number of options at the end of each page, including a link to your e-mail address. Review the information on headers and footers in Chapter 4.

Another way to design the site is to use frames, a division of each page into sections or windows. One frame will display the master list of all linkable pages, and the other frame will display the page with text and visuals that the visitor is viewing. The site map of linkable pages will always be visible on any linked page. Using frames, however, will reduce the size of the screen displaying the information on each page. Because of the need for conciseness, this is not necessarily a drawback. Figure 19.3 shows a Web page with frames.

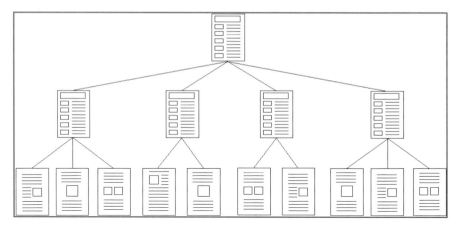

FIGURE 19.2   *Three-tiered site design depicting first-tier home page, second tier with 4 linked pages, and third tier with 10 linked pages.*

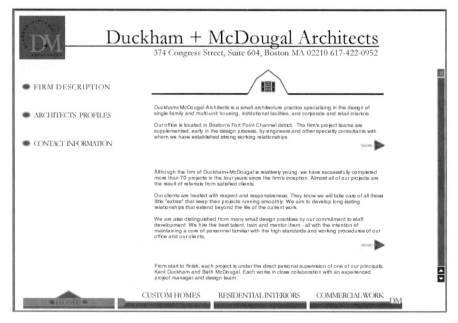

FIGURE 19–3   *Sample Web site home page with frames for linkable pages* (Courtesy of Duckham & McDougal Architects. By permission.)

Designing the Pages

The fourth step is to create the actual page designs. Carefully review all of the document design elements discussed in Chapters 4 and 18. You will want to consider space, columns, titles and headings, line length, leading, kerning, fonts and type sizes, margins, indentations, justifications, lists, numbers, bullets, emphatic features (boldface, all capitals, small capitals, drop capitals, windows, symbols, icons, inverse text, and color). Consider also borders, fills, and watermarks, plus headers and footers. Review the material in Chapter 18 on simplicity, unity, emphasis, balance, and legibility.

After word-processing your text, decide on the placement of your graphics. Your graphics must supplement the delivery of your message, not just decorate it. Too many graphics will slow the downloading of the site. You may crop your original graphics, increase contrast and brightness, or present them in color or grayscale (black and white). Consider adding sound, animation, and other multimedia.

The key is to keep each page simple and readable. Do not overdo any page because, remember, added graphics and possibly sound and animation may distract from the text. Ideally, such additions will be added only to enhance the text, not to confuse the viewer. Try to contain each linked page set to no longer than two or three pages to avoid the demands of endless scrolling.

JUMP LINKS

For long text, use separate files, or *jump links,* a series of heading links at the top of the material to allow viewers to jump from the top of the page directly to the screen they wish to read. Include footers at the bottom of each page to allow the viewer to jump back up to the top of the page, to another page set, or to the home page. Easy navigation within your site is essential to pleasing the viewer.

Converting Text, Graphics, Links, Sound, and Animation

The language used to create Web pages is called **HyperText Markup Language (HTML).** This language uses a series of tags to create formatting within the Web browser. While some tags can be difficult to understand at first, many of them are quite simple. For instance, to boldface a series of words, you will enclose them within <b> and </b>; the first tag turns on the browser's boldface function, and the second tag turns it off.

This will not appear as boldface text.
<b>This will appear as boldface text.<b/>

New software programs contain conversion programs to automatically convert your word-processed text to HTML, but knowing how to code HTML yourself gives you strong control over your pages, allowing you more options and the ability to make changes easily. Books and tutorials are widely available on HTML coding. Table 19.1 lists some basic tags.

TABLE 19.1   *HTML Codes*

Purpose	*On*	*Off*
Document setup		
	\<html\>	\</html\>
	Creates a Web page.	The forward slash mark (/) turns all the codes off.
Title	\<title\>	\</title\>
	Title does not appear on Web page.	
Body	\<body\>	\</body\>
	Begins the text.	
Emphasis techniques		
Boldface	\<b\>	\</b\>
Italics	\<i\>	\</i\>
Blinking	\<blink\>	\</blink\>
	Text blinks on and off.	
Font sizes		
Largest	\<H1\>	\</H1\>
	\<H2\>	\</H2\>
	\<H3\>	\</H3\>
	\<H4\>	\</H4\>
	\<H5\>	\</H5\>
Smallest	\<H6\>	\</H6\>
Dividers		
Double return	\<P\>	\</P\>
	Creates a double space.	
Line break	\<BR\>	\</BR\>
	Acts like a hard return.	
Horizontal rule	\<HR\>	\</HR\>
	Places a thin line across the page.	
Center	\<center\>	\</center\>
	Centers text or graphics.	
Itemized lists		
Numbered list	\<OL\>	Looks like this:
	\<L1\>text	1. text
	\<L2\>graphics	2. graphics
	\<L3\>links	3. links
	\</OL\>	
Unnumbered list	\<UL\>	Looks like this:
	\<L1\>text	• text
	\<L2\>graphics	• graphics
	\<L3\>links	• links
	\</UL\>	

TABLE 19.1 *HTML Codes (continued)*

Purpose	On	Off
Backgrounds		
Colors	<body bgcolor="gold"> or any color you desire. The "bg" stands for *background*.	
	<font color="navyblue"> or any other color you choose. There are also thousands of numerical designations for different colors. For instance, black is #000000, pink is #BC8F8F, and red is #FF0000.	</fontcolor>
Patterns	<body backgrounds= "waves.gif"> There are thousands of patterns on the Internet.	
Images/Graphics	You can also find thousands of free pictures, icons, lines, bullets, animations, backgrounds, buttons, clip art, and more on the Internet. You may have art in your personal files for Web site inclusion. Code the graphics or images into your HTML text as follows: <img src="cartoon.gif"> or <img src="cartoon. jpg"> *Cartoon* is just an arbitrary example. Use the file name.	
Hypertext links	Links allow the reader to jump from one linked page to another. The code is <a href= "filename.html"> Use the name of the file you plan to link to.	</a>

Color can be added as page background, boxed backgrounds, for fonts, or as section dividers. Templates for page layout, dividers, and graphic or page borders are available on a number of software programs. Limit your color selections to just two or three colors, or your site will be overwhelmed. Devise a color scheme; for instance, select one color for background, another for your title box (if you use one), and another color, complementary to the background color, for text fonts. Avoid the pastel colors; golds, shades of blues, black, even reds and greens are effective.

You should have by now decided on photographs, drawings, charts, clip art, and other visuals. Graphics will need to be scanned or downloaded and put into a digital format in your graphics files as either a .GIF or .JPG file. The .GIF (Graphics Interchange Format) files will be compressed, can use only 256 colors, allow for only limited file size, and are not always of the highest quality. The .JPG (Joint Photographic Experts Group) files are more popular in that they can contain millions of colors, allow for smaller file sizes, and print with excellent quality.

You can use your own photographs by using a digital camera and the appropriate software or by asking participating photo developers to provide you an online file. With a few clicks, you may format the graphics by size, color, borders, and so forth, and then click and drag these graphics onto the page as you originally designed your model. Too large or too many graphics will slow down the loading of your site onto the viewers' screens, so use them judiciously.

Sound can be added in a number of ways. You can program your site to begin playing music or voice immediately upon accessing *(streaming audio)*, you can require that the sound track be downloaded completely before hearing it, or you can even embed the sound to the page with an interface allowing the viewer to control the sound by clicking buttons to play, pause, stop, or repeat the sound track.

Other multimedia options are movie clips and animations, which can be downloaded from the Internet and then uploaded onto your page.

Testing and Revising the Site

Before placing your site on an Internet server, test and revise it thoroughly. Have a professional (Webmaster, computer consultant, teacher, editor) look at your text and layout, and critique your content, layout, color usage, graphics selection, and all other aspects. Determine if the site loads correctly when you enter the URL on your browser. Check out all of the links to ensure that they work accurately. Test your own e-mail link. If possible, test the site on different kinds of computers, monitors, and browsers. Some browsers may not display your graphics correctly, and you may then wish to add textual explanations.

Placing the File on an Internet Server

There are now hundreds of ways to place your file on an Internet server. If you used a software package, such as America Online's Web Pages Made Easy™, you may automatically register in AOL Hometown, a hosting server. There are hundreds of hosting servers, both local and international. Some major companies with Web sites will host your page for free or sell you hosting privileges on their own sites. For instance, if your site concerns scientific research, you may wish to be included on one of the National Science Foundation sites, or if you have authored a published book, the publisher may link your site on its own Web site. Check out **<http://www.web-site-hosting-allstars.com>** and **<http:geocities. yahoo.com/home/pagebuilder>.** Both offer free hosting plans as well as design plans and tutorials. If you pay for a hosting server, prices will range from perhaps $25 a year to over $100 a year. You can check out other sites by searching "web site hosting servers" on the Internet.

You may either provide a disk of your site for uploading or, if you have the application for FTP (File Transfer Protocol), you may upload your site file by computer codes. Tutorials, books, and Internet sites will guide you in this registration process.

Spreading the Word

Next, you need to publicize the site's existence. If you are involved in a company, add the URL to all products, brochures, letterheads, and advertisements in magazines or on television. Some search engines roam the Web for domain names, names in text, and content keywords, and automatically record the addresses of new sites, but you may notify them directly by looking for links to a registration page on each. Register your site with search engines such as AOL Search, Yahoo, Google, AltaVista, and Infoseek. E-mail your entire address book and key organizations with which you are affiliated to announce your new Web site.

Maintaining the Site

In order to have viewers return to your page, keep your material current and update the material on a regular basis. Actually set calendar dates to maintain the site. If possible, deliver a new twist to the design or the content from time to time. Randomly change graphics or add pictures. Consider changing background, font, and other colors. Alter the layout a bit from time to time. Ensure that your links are current. If you haven't already done so, add a counter to keep track of the number of contacts made to your page.

Finally, check your own site and e-mail frequently to determine its popularity and to answer any inquiries which may arise. If your site is a business Web page, you will have included the steps to place orders, to request additional information, or to make appointments.

Following the steps in this chapter will allow you to devise effective personal or professional Web sites.

CHECKLIST

Designing and Managing Web Sites

AUDIENCE/PURPOSE

❏ **1.** Have I clearly defined the purpose of the site?

    ❏ To inform?

    ❏ To entertain?

    ❏ To promote products, services, and/or activities?

❏ **2.** Have I determined the potential audience?

    ❏ Ages?

    ❏ Interests?

    ❏ Personal/cultural biases?

    ❏ Technical knowledge?

TEXT, GRAPHICS, AND LINKS CREATION

❏ **1.** Have I designed an inviting and informative home page?

    ❏ Title or boxed title?

    ❏ Concise and informative text?

    ❏ Appropriate graphics?

    ❏ Appropriate links?

❏ **2.** Are the linked pages logical and developmental with the above considerations?

SITE AND PAGE DESIGNS

❏ **1.** Have I decided on the numbers of tiers of linked pages?

❏ **2.** Will I use frames, a site map, or other methods of navigation?

❏ **3.** Have I designed my site to enhance visual meaning by considering simplicity, unity, emphasis, balance, and legibility?

❏ **4.** Does the use of space (columns, horizontal lines, title and other boxes, borders, graphics placement) enhance the site?

❏ **5.** Do line lengths, leading and kernings, font selections and type sizes, margins, indentations, and justifications enhance the site?

❏ **6.** Do lists, numbers, bullets, and emphatic features enhance the site?

❏ **7.** Have I decided on appropriate color for title boxes, backgrounds, fonts, graphics, borders, and dividers?

❏ **8.** Have I placed my links appropriately for easy viewer navigation?

    ❏ Frames?

    ❏ Jump links?

    ❏ Footer links?

    ❏ E-mail links?

    ❏ Other Web site links?

COPY CONVERSION TO HTML

❏ **1.** Have I a complete guide to tag the text to HTML?

❏ **2.** Have I HTML conversion software?

❏ **3.** Have I converted my graphics to digital format files?

❏ **4.** Have I executed sound?

    ❏ Streaming audio?

❏ Download-all-first audio?

❏ Embedded interface audio?

❏ **5.** Am I downloading movie clips, animations, or other multimedia?

TESTING AND REVISION

❏ **1.** Have I tested and appropriately revised the site?

❏ **2.** Have I had a professional evaluate the site?

❏ **3.** Does the site load correctly?

❏ **4.** Do the links work accurately?

❏ **5.** Will the site work effectively on different computers, monitors, and browsers?

SERVER SELECTION, ADVERTISING, AND MAINTENANCE

❏ **1.** Have I considered free and automatic, local, international, and other company Web site host servers?

❏ **2.** Have I registered the site on a host server by disk or FTP application?

❏ **3.** Have I publicized the site's existence?

❏ Printed URL on products, brochures, letterhead, advertisements, etc.?

❏ Registered on Internet search engines?

❏ **4.** Have I established a calendar to update and maintain the site?

❏ **5.** Am I considering new copy twists, new graphics, revised layout, new links, and other additions to continue to hook my viewers?

❏ **6.** Have I installed a counter?

EXERCISES

1. **Collaborative—Evaluate Your Own School's Web Site.** What does the home page contain? What are the subjects of the linked pages? Does the site work best for prospective students, enrolled students, faculty, or alumni? Why? Determine if the information is the same as on printed material about the school or if it is designed to take advantage of the unique capabilities of Web design. Is the text inviting, concise, and informative? How could it be improved? Is the layout logical? What graphics are used? Are they appropriate? Logically chosen? Well-placed? Do the pages use frames for the links? If not, how are they included? What devices are used for the links (underlining, icons, pictures, buttons, other)? Do the linked pages allow you to link back to the top of each section and to the home page? How many levels of links does the site contain? Too many? Too few? Just about right? Are they easy to navigate? Is there evidence that the site is updated frequently? Can you e-mail the Webmaster? Do so, and inquire about his or her role in the design of the site and the maintenance of the site. Write an analytical report about the site with recommendations for improvement. Consider carefully your report's document design.

2. **Collaborative—Evaluate a Local Professional or Corporate Organization's Web Site.** Select one that is not too elaborate or involved. What does the home page contain? What are the subjects of the linked pages? Does the site work best for prospective members of the organization or for users of the organization's services, products, or sponsored activities? Explain your conclusions. Determine if the information is the same as on printed material about the organization or business or if it is designed to take advantage of the unique capabilities of Web design. Is the text inviting, concise, and informative? How could it be improved? Is the layout logical? What graphics are used? Are they appropriate? Well-placed? Do the pages use frames for the links? If not, how are they included? What devices are used for the links (underlining, icons, pictures, buttons, other)? Is the site easy to navigate? Do the linked pages allow you to link back to the top of each section and to the home page? How many levels of links does the site contain? Too many? Too few? Just about right? Are there links to other related Web sites? Are the other sites helpful? Is there evidence that the site is updated frequently? Can you e-mail the Webmaster? Do so, and inquire about his or her role in the design of the site and the maintenance of the site. Write an analytical report about the site with recommendations for improvement. Consider carefully your report's document design.

PROJECTS

1. **Individual—Personal Web Site for Yourself or Your Family.**
   Follow the first six steps for designing a Web site, and enter the HTML
   codes by yourself or use a software conversion program that will do the
   formatting for you. Make sure the layout on both the home page and
   the linked pages is attractive; the text inviting, concise, and informative; the graphics sharp; the colors complementary; and the sounds,
   animations, and links workable. Present the site to the class for small
   group or entire class critique. Then consider the next three steps of actually selecting a host server, advertising the site, and maintaining it.

2. **Collaborative Project—Organization Web Site.** Team with at
   least three members of your class to create a Web site for your technical writing class, school, fraternity or sorority, organization (club,
   sport team, music group, theater class, art group, newspaper, tutoring, etc.), church or synagogue, or other organization in which all of
   you are involved. Follow the first six steps of designing a Web site.
   Decide whether to use a software program with automatic HTML
   conversion or to enter the HTML codes yourselves. Make sure the
   layout on both the home page and the linked pages is attractive; the
   text inviting, concise, and informative; the graphics appropriate and
   sharp; the colors complementary; and the sounds, animations, and
   links workable. Present the site on disk to the class for small group or
   entire class critique. Consider the next three steps of actually placing
   the site on a server, advertising, and maintaining the site.

3. **Collaborative Project—Business Web Site.** Team with at least
   three members of your class to create a Web site for a fictitious business. Your site can focus on any service that your group could provide,
   such as house painting, auto detailing, child care, tree trimming, gardening, aerobic training, swimming or tennis instruction, and so on.
   Name the company, invent an address, phone number, company officers, and other pertinent information. Devise a company logo. Follow
   the first six steps to design and test the site. Decide whether to use a
   software program with automatic HTML conversion or to enter the
   HTML codes yourselves. Make sure the layout on both the home page
   and the linked pages is attractive; the text inviting, concise, and informative; the graphics appropriate and sharp; the colors complementary; and the sounds, animations, and links workable. Present
   the site to the class for small group or entire class critique.

NOTES

APPENDIX A

Conventions Of Construction, Grammar, and Usage

INTRODUCTION

Effective communication depends not only on content and format but also on precise adherence to the conventions of sentence construction, grammar, and usage. This appendix will provide you with a brief handbook, exercises, and reference tables of professional and technical writing conventions.

The conventions that govern written English are not prescriptive rules. Rather, they are patterns developed by careful writers over decades and accepted by publishers, professional and technical writers, educators, and the public. Because English is a living language, these conventions alter over periods of time. For instance, it has long been conventional to use a comma before *and* in a written series of three or more items (i.e., *He assembled the nuts, bolts, and rivets*). In recent years some publishers, businesses, and industries have agreed to omit the comma (i.e., *He assembled the nuts, bolts and rivets*). Those who are concerned with language classify the conventions as *Standard English* and *General English*. The majority of professional and technical writers adopt Standard English conventions; therefore, this appendix covers the more formal conventions. Use these conventions to edit your texts and to troubleshoot your writing problems.

THE SENTENCE

Main Sentence Elements

A sentence is a group of words containing a subject and a verb and expressing a complete thought. English sentences are arranged in many word-order patterns that convey meaning. Lewis Carroll's nonsense sentence

> T'was brillig and the slithy toves did gyre and gimble in the wabe.

reveals patterns that make sense to English-speaking people. We recognize a subject and verb *(T'was)*, the conjunctions *(and)*, a helping verb *(did)*, a preposition *(in)*. Using the pattern, we can develop any number of sentences:

> It was Monday, and the electronic technicians did assemble and test in the plant.
>
> It was noon, and the construction workers did toil and sweat in the sun.

There are five basic sentence patterns in English. Subjects, objects, and complements are nouns or pronouns. Verbs are words that express action *(jump)* or state of being *(is, seem)*.

Pattern 1:   Subject + Verb
 S + V
 Water boils.

Pattern 2:   Subject + Verb + Direct Object
 S + V + DO
 Technicians repair computers.

Pattern 3:   Subject + Linking Verb + Subjective Complement
 S + V + SC
 Helium is a gas.
 (A subjective complement renames the subject.)

Pattern 4:   Subject + Verb + Indirect Object + Direct Object
 S + V + IO + DO
 The company gave Ms. Barnes a plaque.
 (An indirect object is the receiver of the direct object.)

Pattern 5:   Subject + Verb + Direct Object + Objective Complement
 S + V + DO + OC
 The union elected him president.
 (An objective complement renames the direct object.)

There are actually many patterns beyond these basic five and many ways to invert or expand a sentence with single, one-word adjectives and adverbs or multiword phrases and clauses. Consider that a Pattern 2 sentence

S + V + DO
The man has a compass.

may be inverted to ask a question

Has the man a compass?

or expanded to

adjective clause
The man who shares my drafting cubicle has
adjective   **prepositional phrase**
a metric compass in his hand.

Recognition of the basic elements allows you to construct sentences that make sense and to punctuate the elements according to the conventions.

Secondary Sentence Elements

Secondary sentence elements are typically used as modifiers; that is, they describe, limit, or make more exact the meaning of the main elements.

ADJECTIVES AND ADVERBS

Single words used as modifiers are related to the words they modify by word order. Adjectives modify nouns or pronouns and usually stand before the word modified.

ADJ   ADJ
He is a precise, concise writer.   (modifies writer)

Adjectives may stand after the modified noun or pronoun or follow a linking verb:

N   ADJ
The temperature made the metal brittle.   (modifies metal)
N   ADJ
The metal was brittle.   (modifies metal)

Adverbs, which modify verbs, adjectives, or other adverbs, are more varied in position. They usually stand close to the particular word or element modified.

> ADV
> He worked late.   (modifies the verb worked)
>
> ADV
> He worked quite late.   (modifies the adverb late)
>
> ADV
> He was rather late.   (modifies the adjective late)
>
> ADV
> He had never been late.   (modifies verb phrase had been)
>
> ADV
> Unfortunately, he was always late.   (modifies whole sentence)

Edit your writing so that the modifiers are clearly related to the words or statements they modify.

Ambiguous: The plasterers have finished the wall almost. (Almost seems to modify wall.)

Clear: The plasterers have almost finished the wall.

Ambiguous: I only need a few seconds.

Clear: I need only a few seconds.

Ambiguous: The red brick doctor's office is closed.

Clear: The doctor's red, brick office is closed.

Usually do not place an adverb within an infinitive phrase (to read, to plan, to construct).

Split infinitive: We have to further plan the assembly.

Clear: We have to plan further the assembly.

EXERCISE

Add adjectives or adverbs within each sentence. If more than one position is possible, explain what change of emphasis would result from shifting the modifier.

1. **Add only.** Last week I ordered a computer package with a spelling checker.

2. **Add definitely.** I am convinced that you are the one to complete the audit.

3. **Add hardly.** Although I adjusted the lever, I could hear the bass resonate.

4. **Add engineer's.**   The well-worn report lay on the shelf.

5. **Add carefully.**   The committee has to read all of the reports.

PHRASES AS MODIFIERS

A phrase is a group of related words without a subject or verb. It cannot stand alone. A phrase is connected to a sentence or to one of its elements by a preposition or a verbal.

PrepositionalPhrases. A prepositional phrase consists of a preposition (*in, at, by, from, under,* etc.) followed by a noun or pronoun plus, possibly, modifiers. It functions as an adjective or adverb, depending on what element it modifies.

> He entered from the door (modifies the verb entered) of my office (modifies the noun door).

Verbal Phrases.   A verbal phrase consists of a participle, gerund, or infinitive (verb forms without full verb function) plus its object or complement and modifiers. A participial phrase functions as an adjective; a gerund phrase as a noun; and an infinitive phrase as either a noun, adjective, or adverb.

Participial phrase:	Circuit boards containing any defects should be scrapped. (modifies boards)
Gerund phrase:	Drafting the blueprint was the next task. (functions as noun subject)
Infinitive phrase:	To conduct a market survey (functions as noun subject) is the easiest way to determine the cost. (functions as adjective modifying way)

Place modifying phrases next to the word modified. Participial and infinitive phrases will give you the most trouble.

Misrelated:	He distributed notebooks to the trainees bound in plastic. (The participial phrase seems to modify trainees.)
Revised:	He distributed notebooks bound in plastic to the trainees.
Misrelated:	The man who was lecturing to emphasize a point pounded the podium. (The infinitive phrase seems to modify lecturing.)
Revised:	The man who was lecturing pounded the podium to emphasize a point.
Dangling participial phrase:	Looking up from my desk, Jane gave me the report. (The phrase seems to modify Jane.)
Revised:	Looking up from my desk, I accepted the report from Jane.

CLAUSES

A clause is a group of words that contains a subject and verb plus modifiers. An **independent** (main) **clause** is a complete expression which could stand alone as a sentence. A **dependent** (subordinate) **clause** also has a subject and verb but functions as part of a sentence. It is related to the independent clause by a connecting word that shows its subordinate relationship either by a relative pronoun (*who, which, that,* etc.) or a subordinate conjunction (*because, although, since, if,* etc.).

Independent clauses: The water tastes brackish because it is contaminated.

The laser beam penetrated the metal plate, and the plate glowed red.

Dependent clauses: If your engine is hot, add antifreeze.

The drive belt, which slipped, shredded.

After I attached the heat sink, the rectifier cooled.

We will make a final test because a dry run was never completed.

Sentence Classification

Sentences may be classified according to the kind and number of clauses they contain as simple, compound, complex, or compound-complex.

SIMPLE SENTENCES

A simple sentence contains an independent clause and no dependent clauses. It may contain any number of modifiers or compound elements.

The capsule exploded.

The tiny, white, plastic capsule expanded and exploded due to the high temperature in the storage bin.

COMPOUND SENTENCES

Compound sentences contain two or more independent clauses and no subordinate clauses. They may be joined by coordinating conjunctions (*and, or, but,* etc.), semicolons, or conjunctive adverbs (*nevertheless, therefore, however,* etc.).

A dot matrix printer is acceptable, but a daisy wheel printer produces easier-to-read copy.

A Radio Shack computer is flexible; it allows you to print hard copy of graphics.

A personal computer is expensive; nevertheless, it is a practical tool for the professional writer.

COMPLEX SENTENCES

A complex sentence contains one independent clause and one or more dependent clauses.

Because it is raining, the slump test will be postponed.

The engineer who originally specified seven pilings changed her mind when she considered the sand content of the soil.

COMPOUND-COMPLEX SENTENCES

A compound-complex sentence contains two or more independent clauses and one or more dependent clauses.

Because the text is illustrated with tables and sample materials, it is an indispensable guide for technical writers, and it may be used by students and professional writers in business and industry.

Complex sentences are used more frequently than simple, compound, or compound-complex sentences in published writing today. Complex sentences allow for more variety than simple sentences and allow the writer to manipulate emphasis. Recognition of independent and dependent clauses is necessary also for adding conventional punctuation.

EXERCISE

1. Combine these simple sentences as directed.
    a. Combine into a compound sentence. A v-t voltmeter measures sine waves. An oscilloscope measures nonsinusoidal voltage.
    b. Combine into a complex sentence. Pressure-sensitive tapes serve the electrical industry well. They insulate all manner of equipment and last longer than other tapes.
    c. Combine into a compound-complex sentence. Magnetic memories can store more digital information on a par with optical disks. Optical disks can recreate visual images at a lower cost than magnetic memories. Optical disks can recreate visual images at a higher speed.
2. Rewrite these complex sentences to shift the emphasis as directed.

    a.   Emphasize the idea that he disliked his car rather than the idea that it uses too much gas. He disliked the car because, as far as I could determine, it used too much gas.

    b.   Emphasize the idea that soaking will separate the components rather than the idea that you can separate them in acids. If you like, you can separate the components by soaking them in acids.

BASIC SENTENCE ERRORS

Sentence Fragments

If a group of words is written as a sentence, but the group lacks a subject or a verb, or cannot stand alone independently, it is called a sentence fragment. A fragment may be corrected by adding a subject or verb, joining it to another sentence, or rewriting the passage in which it occurs.

Fragment:	The siphons, which were described earlier.
Revision:	The siphons, which were described earlier, <u>must be ordered</u>. (verb added)
Fragment:	The company continues to lose money. Although production has increased.
Revision:	The company continues to lose money although production has increased. (joined to another sentence)
Fragment:	Effective writing requires many skills. For example, a command of conventional grammar and the application of correct mechanics.
Revision:	Effective writing requires many skills, such as a command of conventional grammar and the application of correct mechanics. (rewritten and combined into a sentence)

EXERCISE

Rewrite or combine these fragments into complete sentences.

1. Although we are trying to please the personnel of each department by presenting a variety of training programs.

2. The late arrivals having been named in this report along with the reasons for their tardiness.

3. The arrangements should be checked immediately. Registrants to be confirmed immediately.

4. Each month more than $1,500 can be saved if the department buys a copy machine. No increase in quality if the machine is top quality.

5. I recommend we use copper. Not zinc.

Sentence Parallelism

Parallel structure involves writing related ideas in the same grammatical constructions. Adjectives should be parallel with other adjectives, verbs with verbs, phrases with phrases, and clauses with clauses.

Nonparallel: Tungsten steel alloys are tough, ductile, and have strength. (adjective, adjective, verb plus noun)

Parallel: Tungsten steel alloys are tough, ductile, and strong. (all adjectives)

Nonparallel: He must learn the language and to be knowledgeable about his computer. (verb and infinitive)

Parallel: He must learn the language and become knowledgeable about his computer. (both verbs)

Nonparallel: He will be hired if he has the required training, if he has three years experience, and by being recommended by his former employer. (dependent clause, dependent clause, phrase)

Parallel: He will be hired if he has the required training, if he has three years of experience, and if he is recommended by his former employer. (all dependent clauses)

Edit your writing to ensure that parallel ideas are expressed in parallel structures.

EXERCISE

Rewrite the nonparallel structures.

1. The computer is inexpensive, compact, and it is easy to use.
2. Before studying architecture, you should assess whether you have design ability and if you are exacting with numbers.
3. To write well, one must be able to organize materials, have a flexible vocabulary, and one should know grammatical and mechanical conventions.
4. He was well-liked and had training in management skills.
5. She attached the sphygmomanometer by positioning the patient, placing the cup above the elbow, and the clasp was secured.

Run-On and Comma-Spliced Sentences

A run-on (also called a fused) sentence occurs when two or more independent clauses are written as one sentence without appropriate punctuation. A run-on sentence may be corrected by separating the fused clauses with a period or semicolon or by rewriting the sentence.

Run-on:	Ace Company revised its maternity policy men are now eligible for child-rearing leave.
Correction:	Ace Company revised its maternity policy; men are now eligible for child-rearing leave.
	Ace Company revised its maternity policy by allowing men to be eligible for child-rearing leave.

A comma-spliced sentence occurs when two or more independent clauses not joined by a coordinating conjunction or conjunctive adverb are written with only a comma between them.

Splice:	Employees are entitled to eight sick days per year, they may be concurrent.
Correction:	Employees are entitled to eight sick days per year; they may be concurrent.
	or
	Employees are entitled to eight sick days per year. They may be concurrent.
Splice:	The restaurant requires a deposit for our annual dinner engagement, therefore, we must send a $50.00 check.
Correction:	The restaurant requires a deposit for our annual dinner engagement; therefore, we must send a $50.00 check.

Using your understanding of clause structure, edit your writing to avoid run-on and comma-spliced sentences.

EXERCISE

Correct these run-on and comma-spliced sentences.

1. This guide is a product of months of research, compilation was done by specialist Jim Smith.
2. Adult education opportunities are plentiful, moreover, all classes may be offered at our training facility.
3. After three straight days of bargaining, the talks broke down they will resume on Monday.
4. The union refused to consider benefit reductions but it did express willingness to negotiate increased work hours.
5. Mr. Larsen has participated in other civic activities in addition to his involvement in public schools he is a member and past officer of the Chamber of Commerce.

AGREEMENT

The most common grammar errors are subject-verb disagreements and pronoun-antecedent disagreements. For instance, if a subject is plural, then its verb must be plural as well, and if a pronoun antecedent is singular, then its pronoun must be singular as well.

Subject and Verb Agreement

Verbs change form from the singular to the plural. Consider:

Singular	Plural
I am	We are
He is	They are
She was	They were
He edits	They edit

A verb must agree with its subject in number. Generally, English-speaking people make these alterations automatically, but problems arise in a variety of structures.

In a sentence with compound subjects joined by *and,* the verb is plural unless the subjects are considered a unit.

The technician and engineer **are** consulting.   (two persons)

The accountant and auditor **reviews** my books monthly.   (one person)

In a sentence with compound subjects joined by *or, nor, either . . . or, neither . . . nor,* the verb usually agrees with the closest subject.

The diodes or the transistor **is** faulty.

Neither the transistor nor the thermistors **are** operating.

A singular subject followed by a phrase introduced by *as well as, together with, along with, in addition to* ordinarily takes a singular verb.

The president as well as the vice-president **was held** responsible for the mismanagement.

Collective nouns (*committee, jury, crowd, team, herd,* etc.) usually take a singular verb.

The committee **is meeting** Tuesday.

The jury **is arguing** with the judge.

When a collective noun refers to members of the group individually, a plural verb is used.

The <u>jury</u> **are arguing** among themselves.

Expressions signifying quantity or extent (*miles, years, quarts,* etc.) take singular verbs when the amount is considered as a unit.

<u>Ten dollars</u> **is** too much to pay for a tablet.

<u>Six hours</u> **is** too long to work without lunch.

A singular subject followed by a phrase or clause containing plural nouns is still singular.

The highest <u>number</u> of diesel trucks **is** produced in Europe.

The <u>nurse</u> who tends the heart patients **finds** them to be grateful.

When a sentence begins with *there is* or *there are,* the verb is determined by the subject that follows.

There **are** an estimated 100 <u>employees</u> in this building.

There **is** a conflicting <u>opinion</u> over capital punishment.

A verb agrees with its subject and not with its complement.

Our chief <u>trouble</u> **was** (not <u>were</u>) malfunctions in the testing equipment.

EXERCISE

Select the appropriate verb to agree with its subject.

1. Three semesters *(is, are)* not enough to master French.
2. The bulk of our tax dollars *(go, goes)* to defense spending.
3. He is one of those people who *(is, are)* always willing to help.
4. The best benefit *(is, are)* the vacations.
5. Either the man or the woman *(assist, assists)* me with the payroll.

Pronoun Agreement

A pronoun is a word that takes the place of a noun, such as *he, who, itself, their, ourselves,* etc. (A complete list of pronouns is reviewed in Table A.1.)

The <u>technician</u> checked **his** circuit boards.

The <u>woman</u> **who** trained me could assemble the parts **herself.**

TABLE A.1 *Pronouns*

	Subject	*Object*	*Possessive*
PERSONAL PRONOUNS:			
First person			
Singular	I	me	my, mine
Plural	we	us	our, ours
Second person			
Singular & plural	you	you	your, yours
Third person			
Singular			
masculine	he	him	his
feminine	she	her	her, hers
neuter	it	it	its
Plural	they	them	their, theirs
RELATIVE PRONOUNS:	who	whom	whose
	that	that	
	which	which	whose, of which
INTERROGATIVE PRONOUNS:	who	whom	whose
	which	which	whose, of which
	what	what	

REFLECTIVE AND INTENSIVE PRONOUNS:

myself	yourself	himself	herself	itself
oneself	yourselves	ourselves	themselves	

DEMONSTRATIVE PRONOUNS:

this	these	that	those

INDEFINITE PRONOUNS:

all	both	everything	nobody	several
another	each	few	none	some
any	each one	many	no one	somebody
anybody	either	most	nothing	someone
anyone	everybody	much	one	something
anything	everyone	neither	other	such

RECIPROCAL PRONOUNS:

each other	one another

NUMERAL PRONOUNS:

one, two, three . . .	first, second, third . . .

A pronoun must agree in number with the word for which it stands, its antecedent.

Faulty: Ace Company is furloughing 20 of **their** employees.

Correct: Ace Company is furloughing 20 of **its** employees.

Faulty: Send the receipts to the bookkeeping department. **They** will issue the refunds.

Correct: Send the receipts to the bookkeeping department. **It** will issue the refunds.

Faulty: Anyone can take **their** accrued sick leave when necessary.

Correct: Anyone can take **his** (or **his or her**) accrued sick leave when necessary.

When a pronoun's antecedent is a collective noun, the pronoun may be either singular or plural, depending on the meaning of the noun.

The committee planned **its** next meeting. (the unit)

The committee gave **their** reports. (the individual members)

Usually a singular pronoun is used to refer to nouns joined by *or* or *nor*.

Neither Jane nor Judy did **her** share.

When the antecedent is a common-gender noun (*customer, manager, instructor, student, supervisor, employee,* etc.), the traditional practice has been to use *he* and *his* as in

A manager routinely evaluates **his** employees.

However, writers who are sensitive to sexist elements of our language are more prone to use both *his* **or** *her* if the gender of the antecedent is not known.

If *his* **or** *her* must be repeated frequently, the cumbersome usage may be avoided by changing the singular antecedent to a plural construction.

Managers routinely evaluate **their** employees.

Some indefinite pronouns (*some, all, none, any,* etc.) used as antecedents require singular or plural pronouns, depending on the meaning of the statement.

Everyone, everybody, anyone, anybody, someone, no one, and *nobody* are always singular.

Everyone must turn in **his or her** timesheet.

Somebody erased **his or her** floppy disk.

All, any, some, or *most* are either singular or plural, depending on the meaning of the statement.

All of the employees received **their** payroll deduction forms. (All refers to employees and is plural; all is the antecedent of their.)

All of the manuscript has been typed, but **it** has not been proofread. (All refers to manuscript and is singular; all is the antecedent of it.)

In Standard English usage, *none* is usually singular unless the meaning is clearly plural.

Standard: None of the men finished **his** work.

General: None of the men finished **their** work.

Clearly plural: None of the new computers **are** as large as their predecessors. (The sentence clearly refers to all new computers.)

When a pronoun is used, it must have a clearly identified antecedent.

Ambiguous: The CRT fell on the keyboard and broke **it.** (It could refer to CRT or keyboard.)

Clear: The CRT broke when **it** fell on the keyboard.

Ambiguous: The consultants recommended a new method of shipping parts. This is the company's best alternative for the future. (It is not clear if this refers to method or to the implied word recommendation.)

Clear: The consultants recommended a new method of shipping parts. This recommendation is the best alternative for the future.

EXERCISE

Correct these misused pronouns.

1. The company had high hopes for the new research program, but they encountered financial problems.
2. Neither the supervisor nor the laborers want his pay reduced.
3. Everybody supported their union.
4. An instructor should encourage his students to ask questions.
5. Electrical engineering is an interesting field, and that is what I want to be.

Pronoun Case

Review Table A.1. The personal pronouns and the relative or interrogative pronoun *who* have three forms, depending on whether the pronoun is used as a subject, an object, or a possessive. Writers frequently encounter a few problems in proper case usage.

The object form of a pronoun is used after a preposition.

Incorrect:	The work was divided between **he** and **I.**
Correct:	The work was divided between **him** and **me.**
Incorrect:	That is the data processor about **who** I have spoken.
Correct:	That is the data processor about **whom** I have spoken.

In written English, *than* is considered a conjunction, not a preposition, and it is followed by the form of the pronoun that would be used in a complete clause, whether or not the verb appears in the construction.

I am more experienced than **she** [is].

I like him better than [I like] **her.**

In general usage, many educated people say "It is me" or "This is her," but Standard English usage prefers the subject form after the linking verb *be*.

It is **I.**

This is **she.**

That is **he.**

It will be **I** who fail.

Although the distinction between *who* and *whom* is disappearing in oral communications, Standard English usage prefers the distinction in writing. *Who* is the standard form when it is the subject of a verb; *whom* is the standard form when it is the object of a preposition or the direct object.

That is the professor **who** taught me chemistry.   (Who is the subject of the verb <u>taught</u>.)

That is the woman **whom** I recommended for promotion.   (Whom is the direct object of <u>recommended</u>.)

To **whom** are you speaking?   (Whom is the object of the preposition <u>to</u>.)

Reflexive And Intensive Pronouns

The reflexive form of a personal pronoun is used to refer back to the subject in an expression where the doer and recipient of an act are the same.

He had only **himself** to blame.

I timed **myself** typing.

The same form is sometimes used as an intensive to make another word more emphatic.

The raise was announced by the president **himself.**

Safety **itself** is crucial.

In certain constructions, writers mistakenly consider *myself* to be more polite than *I* or *me,* but in Standard English, the reflexive forms are not used as substitutes for *I* or *me.*

Faulty: Mr. Jones and **myself** attended the meeting.
Correct: Mr. Jones and **I** attended the meeting.

Faulty: The work was completed by Ms Burns and **myself.**
Correct: The work was completed by Ms Burns and **me.**

EXERCISE

Select the correct pronoun in each of the following sentences.

1. From *(who, whom)* will we receive the instructions?
2. The director of training assigned the project to Jones and *(I, me).*
3. It is *(we, us)* who were to leave early.
4. She was later than *(I, me).*
5. Smith, White, and *(I, myself)* were assigned to conduct the survey.

USAGE GLOSSARY

Many words in the English language are so similar that they cause confusion. Following is a brief glossary of such terms for you to review. (Words that are asterisked are corruptions of correct usage; they are not to be used in formal writing.)

accept, except

Accept means "receive" or "agree to."

Community colleges accept a wide variety of students.

As a preposition, *except* means "other than."

I did all of the work except your report.

As a verb, *except* means "exclude," "omit," "leave out."

If you except Mr. Jones, no other president has owned his own Lear Jet.

advice, advise

Advice is a noun meaning "guidance."

> If I wanted it, I would ask for your advice.

Advise is a verb meaning "counsel," "give advice to," "recommend," or "notify."

> I advise you to exercise stock options.

affect, effect

Affect means "change," "disturb," or "influence."

> The rising cost of fuel has drastically affected the trucking industry.

It can also mean "feign" or "pretend to feel."

> Although she knew she was to be promoted, she affected surprise when notified.

As a verb, *effect* means "bring about," "accomplish," or "perform."

> She effected a perfect word-processed report.

As a noun, *effect* means "result" or "impact."

> Her extra work had no effect on the vice-president.

aggravate, irritate

Aggravate means "make worse."

> The recent charge of police brutality has aggravated racial hostilities.

Irritate means "annoy" or "bring discomfort to."

> His endless twitching began to irritate me.

allude, refer

Allude means "call attention indirectly."

> When the candidate spoke about the Middle East, he alluded to the past administration's involvement in secret arms sales.

Refer means "direct attention specifically."

> He referred to the second paragraph, which stated the fact.

all ready, already

Use *all ready* when *all* refers to things or people.

> At noon the secretaries were all ready to lunch.

Use *already* to mean "by this time" or "by that time."

> I have already typed that report.

all right, *alright

All right means "completely correct," "safe and sound," or "satisfactory."

My answers to the interviewer were all right. (The meaning is that *all* of the answers were *right*.)

Despite a few cuts, I was all right.

Do not use *all right* to mean "satisfactorily" or "well."

*Do not use *alright* anywhere; it is a misspelling of all right.

almost, *most all

Almost means "nearly."

By the time we reached Miami, the tank was almost empty.

*In formal writing, do not use *most all;* use *almost all* or *most*.

Almost all (or most) of the employees were eligible for vacation.

a lot, *alot

A lot of and *lots of* are colloquial and wordy. Use *much* or *many*.

*Do not use *alot* anywhere; it is a misspelling of *a lot*.

among, between

Use *among* with three or more persons, things, or groups.

There are no secrets among my friends.

Use *between* with two persons, two things, or two groups.

Voters must choose between the Democrat and Republican candidates.

amount, number

Use *amount* when discussing uncountable things.

No one knows the amount of good a female President might do.

Use *number* when discussing countable things or persons.

A number of us are thinking of a third candidate.

anyone, any one

Anyone is an indefinite pronoun (like *everyone, everybody*).

Anyone who desires to run for office must file by Wednesday.

Any one means "any one of many."

I could not find any one who could fix my tire.

as, as if, like

Use *as* or *as if* to introduce a clause.

As I drove into the parking lot, my tire blew out.

He looked as if he had worked all night.

Use *like* to mean "similar to."

The logo looked like ours.

assure, ensure, insure

Assure means "state with confidence to."

    I assure you that he will be hired.

Ensure means "make sure" or "guarantee."

    There is no way to ensure that every policy is understood.

Insure means "make a contract for payment in the event of specified loss, injury, or death."

    He insured the package for $100.00.

bad, badly

Bad is an adjective meaning "not good," "sick," or "sorry."

    We paid a lot for the movie, but it was bad.

    In spite of the medicine, I still felt bad.

    She felt bad about losing her job.

Badly is an adverb meaning "not well."

    She starts races well but ends badly.

Used with *want* or *need, badly* means "very much."

    The diver badly wanted a gold award.

    I need some money badly.

Do not use *bad* as an adverb.

    *He swam bad for the first two laps.

    Revised: He swam badly for the first two laps.

censor, censure

As a verb, *censor* means "exercise censorship."

    Some groups want the NEA to censor certain art.

As a noun, *censor* means "one who censors."

    Rap groups should act as censors of their own lyrics.

As a verb, *censure* means "find fault with" or "reprimand."

    The Prime Minister was censured by Parliament.

As a noun, *censure* means "disapproval" or "blame."

    Ridicule often hurts more than censure.

complement, compliment

As a verb, *complement* means "bring to perfect completion."

    His red tie complemented his blue suit.

As a noun *complement* means "something that makes a whole when combined with something else" or "the total number needed."

    Practice is the complement of learning.

    Without a full complement of workers, we cannot complete the task.

As a noun, *compliment* means "expression of praise."

He complimented the appearance of my report.

continual, continuous

Continual means "going on with occasional slight interruption."

At the office there is a continual humming of typewriters.

Continuous means "going on with no interruptions."

We have 24-hour guard service; surveillance is continuous.

council, counsel, consul

Council means "groups of persons who discuss and decide certain matters."

The city council authorized a new park.

As a verb, *counsel* means "advise."

My stockbroker counseled me not to sell my bonds.

As a noun, *counsel* means "advice" or "lawyer."

However, I didn't follow his counsel and sold short.

A *consul* is a government official working in a foreign country to protect the interest of his or her country's citizens there.

If you lose your passport in France, you will need to see the American consul.

criterion, criteria

A *criterion* is a standard by which someone or something is judged.

The criterion for selecting the car was color.

Criteria is the plural of *criterion*.

The criteria for selecting the car were color, factory air conditioning, and horsepower.

data

Data is the plural of Latin *datum* (literally, "given"), meaning "something given"—that is, a piece of information; however, *data* may be treated as singular.

The data stored in my computer is readily accessible.

different from, *different than

*Do not use *than* after different. Use *from*.

His management style is different from mine.

due to, because of

Due to means "resulting from" or "the result of."

The leaking roof was due to broken tiles.

Because of means "as a result of."

Animosities persist in the Middle East because of Iraq's nuclear capability.

eminent, imminent, immanent

Eminent means "distinguished," "prominent."

An eminent astronomer has questioned the "big bang" theory.

Imminent means "about to happen."

The dark sky and distant rumbling indicated the storm was imminent.

Immanent means "inherent," "existing within."

The Trobrianders believe that evil is immanent in stone, sand, and water.

farther, further

Farther means "a greater distance."

The road went farther into the forest.

As an adjective, *further* means "more."

He needs a further education beyond high school.

As a conjunctive adverb, *further* means "in addition, moreover, to a greater extent."

I dislike carrots; further, I detest asparagus.

As a transitive verb, *further* means "promote" or "advance."

How much has the mayor done to further tourism in our town?

few, fewer, little, less

Use *few* or *fewer* with countable nouns.

We have fewer employees than we did a year ago.

Use *little* or *less* with uncountable nouns.

I have less experience than you on the word processor.

frightened, afraid

Frightened is followed by *at* or *by*.

Many people are frightened by thunder.

I was frightened at the thought of gaining weight.

Afraid is followed by *of*.

I am afraid of heights.

good, well

Use *good* as an adjective, but not as an adverb.

The proposal for staggered work hours sounded good to many employees.

Use *well* as an adverb when you mean "in an effective manner," "ably."

He did so well on the project that he was promoted.

Use *well* as an adjective when you mean "in good health."

She hasn't looked well since her operation.

hanged, hung

Hanged means "executed by hanging."

The rustler was hanged at dawn.

Hung means "suspended" or "held oneself up."

I hung my hat on the rack.

I hung on the side of the cliff until I could get a foothold.

hopefully

Use *hopefully* to modify a verb.

She looked hopefully at her boss as he scanned her proposal.

*Do not use *hopefully* when you mean "I hope that," "we hope that," or the like.

Incorrect: Hopefully, the company will make a profit this quarter.

  Revised: The stock holders hope that the company will make a profit this quarter.

immigrate, emigrate

Immigrate means "enter a country in order to live there permanently."

Those Haitians decided to immigrate to Florida.

Emigrate means "leave one country in order to live in another."

My grandfather emigrated from Cuba.

in, into, in to

Use *in* when referring to a direction, location, or position.

Everyone wants to live in the Sun Belt.

Use *into* to mean movement toward the inside.

He dove into the water.

Use *in to* when the two words have separate functions.

I walked in to see the movie.

its, it's

Its is the possessive form of it.

I like this company because of its location and its benefits.

It's means "it is."

It's evident that we are an expanding company.

kind of, sort of

In formal writing, do not use *kind of* or *sort of* to mean "somewhat."

Faulty: When I moved my head, I felt sort of queasy.

Revised: When I moved my head, I felt rather queasy.

lie, lay

Lie (*lie, lying, lies, lied*) means "tell a falsehood."

She sometimes lies about her age.

She has lied about her age for 20 years.

Lie (*lie, lying, lay, lain*) means "rest," "recline," or "stay."

I love to lie in bed late.

Lying on the sand, I watched the water sparkle.

After I lay on the sand for two hours, I was sunburned.

That book has lain on the table for two months.

Lay (*lay, laying, laid*) means "put in a certain position."

A good education lays the foundation for a good life.

The workers were laying tile on the floor.

They laid the tile into stacks.

The hurricane struck after the tile had been laid.

likely, apt, liable

Likely indicates future probability.

It is likely that inflation will continue.

Apt indicates a usual or habitual tendency.

Most cars are apt to rust after a few years.

Liable indicates a risk or adverse possibility.

Apartments which are left unlocked are liable to be burgled.

not very, none too, *not too

In formal writing, use *not very* or *none too* instead of *not too.

Faulty: Not too pleased with the weather, we changed our plans.

Revised: None too pleased with the weather, we changed our plans.

precede, proceed, proceeds, proceedings, procedure

To *precede* is to come before or go before in place or time.

A dead calm often precedes a hurricane.

To *proceed* is to move forward or go on.

The boat could not proceed because the bridge would not open.

When the bridge opened, the boat proceeded.

Proceeds are funds generated by a business deal or a money-raising event.

The proceeds from the garage sale will be used to pay the rent.

Proceedings are formal actions, especially in an official meeting.

The stenographer transcribed the proceedings of the meeting.

A *procedure* is a standardized way of doing something.

The correct procedure for moving a block of type is in the manual.

regardless, *irregardless

Use *regardless,* not *irregardless.*

Faulty: Irregardless of the potential circumstances, Scott decided to bungee jump.

Revised: Regardless of the potential circumstances, Scott decided to bungee jump.

stationary, stationery

Stationary means "not moving."

The typewriter was on a stationary table.

Stationery means "writing paper."

We had to order more stationery from the printing department.

such a

Do not use *such a* as an intensifier unless you add a result clause beginning with *that.*

Weak: The political rally was such a bore.

Better: The political rally was such a bore that I fell asleep.

EXERCISE

Supply the correct word. You may have to change the tense of the verbs.

1. Everyone has _____ the invitation _____ Sam.
   (accept, except)

2. I _____ you to follow your instructor's _____.
   (advise, advice)

3. The malfunctioning air conditioning _____ our tempers.
   The manager _____ a defiant look.
   The _____ of nuclear fallout are under study.
   (affect, effect)

4. Finally, the reports were xeroxed, and we were _____ to begin the board meeting.

    The President had _____ left for lunch when I reported for our interview.

    (already, all ready)

5. It is _____ with me if you use correction tape.

    (alright, all right)

6. The technician looked _____ he were tired.

    (as, as if, like)

7. I _____ you that we can _____ your right to strike.

    (ensure, assure, insure)

8. If there are _____ members, it means _____ work for the secretary.

    (fewer, less)

9. The departmental members work _____ together, and I feel _____ about their cooperation.

    (good, well)

10. Ace Company is an ideal employer because of _____ benefits, and _____ improving _____ stock option program each year.

    (its, it's)

Punctuation and Mechanical Conventions

INTRODUCTION

The professional or technical writer must be conscious and demanding of punctuation and mechanical conventions to prevent vagueness and misreading. Two practices of punctuation are prevalent today. A few businesses and industries adopt an open punctuation system, which favors only essential marks and omits those that can be safely omitted, such as the comma before *and* in a series (screws, nuts, and bolts). The majority of professional writers use standard, or close, punctuation conventions because these conventions promote greater accuracy. This appendix reviews close punctuation conventions.

One of the major characteristics of professional and technical writing is the extensive use of abbreviations, numbers, symbols, and other mechanics. This appendix reviews the general conventions that govern mechanics.

Both the punctuation and mechanical conventions are arranged in alphabetical order to help you edit your writing quickly.

PUNCTUATION

1.0 Apostrophe

1.1 Use an apostrophe to indicate the possessive case of the noun. (If the noun ends in *s,* use an apostrophe after it, but do not use an additional *s.*)

> The company's product
> Jack and Bob's office (joint possession)
> Bill's or Jack's car (individual possession)
> his sister-in-law's law practice
> The Jones' car crashed.

1.2 Use an apostrophe to indicate the possessive case of indefinite pronouns:

> | another's tools | neither's wrench |
> |---|---|
> | anybody's d'esk | one's customers |
> | each one's station | somebody's computer |

Do *not* use an apostrophe to indicate the possessive case of personal pronouns:

his schedule
Ours is the newest model.
The mistake was hers.
Its handle is steel.

1.3 Use an apostrophe to indicate the omission of letters in contractions:

> | I'm | o'clock |
> |---|---|
> | he'll | we're |
> | can't | you're |

Do not confuse *they're* with *their* or *there.*

Their supervisor knows they're there.

Do not confuse *it's* (it is) with *its* (a possessive).

It's demonstrating its graphic function.

1.4 Use an apostrophe to indicate the plural symbols and cited words:

You use too many *and*'s.

The *&*'s are broken on all of the typewriters.

1.5 Do not use an apostrophe to form the plural of a term in all capital letters:

The *CD-ROMs* are by the computer.

The university awarded eight *Ph.D.s* in engineering.

1.6 Do not use an apostrophe to form the plural of letters and numbers unless confusion would result without one:

Your *rs* look like your *ls*.

Two IOUs

the 1980s

nine 1's.

2.0 Brackets

2.1 Use brackets within a quotation to add clarifying words that are not in the original.

Mr. Roberts stated, "They [computers] have revolutionized his business."

2.2 Use brackets within a quotation to enclose the Latin word *sic* ("so," "thus"), which indicates that a misspelling, grammatical error, or wrong word was in the original:

He wrote, "Your [sic] selected to head the committee."

3.0 Colon

3.1 Use a colon after a formal salutation:

Dear Ms Benson:

Good morning:

3.2 Use a colon to introduce a phrase or clause that explains or reinforces a preceding sentence or clause:

Food processing consists of three main steps: selecting the blade, measuring the ingredients, and processing at the appropriate speed.

The position sounds attractive: the salary is high and the opportunities for advancement are excellent.

3.3   Use a colon when a clause contains an anticipatory expression *(the following, as follows, thus, these)* and directs attention to a series of explanatory words, phrases, or clauses:

The requirements for the position are as follows:

1. A master's degree
2. Three years experience
3. Willingness to relocate

3.4   Use a colon to express ratios, to separate hours and minutes, and to indicate other relationships.

3:1	signal:noise
A:B	8:25 P.M.
Acts: 14:7	12:101–104 (volume 12, pages 101–104)

3.5   Use a colon between the main title and subtitle of a book:

Technical Writing: An Easy Guide

4.0   Comma

Refer to Table B.1 for a review of comma and semicolon usage in compound and complex sentences.

4.1   Use a comma to separate independent clauses joined by a coordinating conjunction.

The cursor shows where you are typing, and it moves across the screen as you type.

4.2   Use a comma after an introductory dependent clause:

If you have a two-drive computer system, you place your program diskette in drive A.

4.3   Use a comma after a conjunctive adverb introducing a coordinate clause:

This system is easy to use; however, we suggest that you read the directions carefully.

4.4   Use a comma to separate a nonrestrictive word, phrase, or clause from the rest of the sentence:

You should, however, make copies of your original diskette for safekeeping.

His goal, to become computer literate, is easy.

TABLE B.1   *Comma and Semicolon Review for Compound and Complex Sentence Construction*

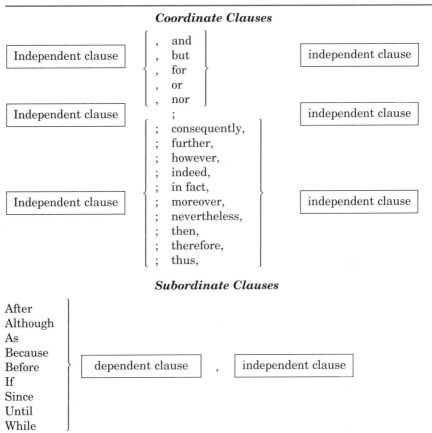

The cursor, which is a blinking square, shows where the entry will appear on the screen.

4.5   Use a comma to separate items in a series:

These instructions will teach you how to create, edit, or proof a file.

The Personnel Department will receive all letters of application, forward them to the appropriate search committee, and handle all correspondence.

The comma is often omitted in company names:

Jones, Smith and Scully

4.6   Use a comma to separate a series of adjectives or adverbs not connected by a conjunction:

Store your thin, sensitive, magnetic disks out of direct sunlight.
The computer blinked haphazardly, noisily.

4.7   Use a comma in a date to separate the day and year:

June 9, 1997
the April 15, 1998, deadline

Do *not* use a comma in the military or international form:

9 June 1987

4.8   Use a comma to separate city and state, state and county, county and state:

Fort Lauderdale, Broward County, Florida, has 27 electronics firms.

4.9   Use a comma to separate titles from names and to set off appositives:

Juan Murillo, M.D., prescribed aspirin.
Julie Koenig, attorney-at-law, tried the case.
K. D. Marshal, Jr., ran for office.
Robert H. Larsen, president, hired a computer guru.
Jerry Noosinow, the Delta pilot, checked the KAL log.

4.10 Use commas to group numbers into units of three in separating thousands, millions, and so forth:

3,845
74,763
9,358,981

4.11 Use a comma after the salutation of an informal letter and after the complimentary close of most letters:

Dear Sally,
Very truly yours,
Cordially,

4.12 Use a comma to separate corporation abbreviations for company names:

Trolleys, Ltd.
Jones and Scully, Inc.
Ace Company, Inc.

4.13   Use a comma after expressions that introduce direct quotations:

President Dan Barker said, "We must double our profits."

If the phrase interrupts the quotation, it is set off by two commas:

"We must," said President Dan Barker, "double our profits."

5.0   Dash

5.1   Use dashes to set off emphatic and abrupt parenthetical expressions:

The idea of this program—it has been tested thoroughly—is to simplify spelling correction.

5.2   Use a dash to mark sharp turns in thought:

He was an arrogant man—with little to be arrogant about.

5.3   Use dashes to separate nonrestrictive material that contains commas from the rest of the sentence:

These manuals—*Guide to Operations, Disk Operating System, Word Proof,* and so on—are protected by copyright.

6.0   Ellipsis

6.1   Use the ellipsis (three spaced periods) to indicate any omission in quoted material:

Martin stressed, "The technical writer . . . must master punctuation." (The words *as well as the professional writer* have been omitted.)

6.2   Use four spaced periods to indicate the ellipsis at the end of a sentence.

The consultant stressed, "Write carefully. . . ." (The words *and edit endlessly* have been omitted.)

7.0   Exclamation Point

7.1   Use an exclamation point at the end of an exclamatory sentence to show emotion or force:

Shred the files!
We will not file bankruptcy!

7.2 Use an exclamation point at the end of an exclamatory phrase:

What a disaster!

8.0 Parentheses

8.1 Use parentheses to enclose an abruptly introduced qualification or definition within a sentence:

You may place your program diskette in drive A and your storage diskette (the one with your file on it) in drive B.

8.2 Use parentheses to enclose a cross-reference within a sentence:

The ellipsis (see Section 6.0) is used in direct quotations.

8.3 Use parentheses to enclose figures or letters to enumerate points:

To use this program, (a) insert your DOS diskette in drive A, (b) turn on your computer, (c) type in the date, and (d) press the Enter key.

9.0 Period

9.1 Use a period to signal the end of declarative or imperative sentences:

Diskettes are sensitive to extremes of temperature.
Do not try to clean diskettes.

9.2 Use a period with certain abbreviations (see pages 681–683 for exceptions):

Dr.	Ph.D.
A.M.	Nov.
P.M.	J. C. Lewis
Jr.	etc.
Mr.	

9.3 Use a period before fractions expressed as decimals, between whole numbers and decimals, and between dollars and cents:

.10	$3.50
3.6	$0.92 (or 98¢ or 93 cents)

9.4 Use a period after number and letter symbols in an outline:

    I.
        A.
        B.
            1.
            2.

10.0 Question Mark

10.1 Use a question mark at the end of an interrogative question:

Do you own a personal computer?

Do *not* use one after an indirect question.

He asked me if I owned a personal computer.

10.2 Use a question mark in parentheses to indicate there is a question about certainty or accuracy:

This is the best (?) computer.

11.0 Quotation Marks

11.1 Use quotation marks to set off direct speech and material quoted from other sources:

Dr. William Haskell writes, "Before 1960 the thing rarer than a marathoner was a health professional trained to care for one. Most doctors," he points out, "forbade post-cardiac patients to do anything more vigorous than walk to the refrigerator."

11.2 Use quotation marks to indicate nonstandard terms, ironic terms, and slang words:

This is a "gimmick."
His "problem" was his genius IQ.
He "got his act together."

11.3 Use quotation marks to indicate titles of articles, essays, short stories, chapters, short poems, songs, television and radio programs, and speeches:

I read the article "Ten Years of Sports Medicine" in *Runner's World.*

Do *not* use quotation marks around quoted material that requires more than four lines in your paper. Display a long quote by indenting it 10 spaces from your regular margins and omitting the quotation marks.

Commas and periods are always placed *inside* the closing quotation mark:

"Yes," Ms Gloss said, "we have a swine flu epidemic."
He said "electrons," but meant "electronics."

Semicolons and colons are placed *outside* the closing quotation mark:

He said, "The trapped air bubble will leave honeycombs";
honeycombs are sections of little indentations.
He said, "The trapped air bubble will leave honeycombs": little
indentations.

Question marks, exclamation points, and dashes are placed inside *or* outside the final quotation mark, depending upon the situation:

He asked, "Is the oscillator connected to the mixer?"
Did he say, "The assembly is constructed of heavy-gauge stainless
steel"?

12.0 Semicolon
Refer to Table B.1 for a review of comma and semicolon usage in compound and complex sentences.

12.1 Use a semicolon between coordinate clauses not connected by a conjunction:

The new system will use low-powered transmitters; it is called a
cellular radio.

12.2 Use a semicolon before a conjunctive adverb including a coordinate clause:

Retort pouches are like cans; however, they do not dent.

12.3 Use a semicolon before a coordinating conjunction introducing a long or loosely related clause:

Niobium, which is used primarily as an alloy, is a metallic element
that resists heat and corrosion and hardens without losing
strength; and it is widely available in Canada and South America.

12.4   Use a semicolon in a series to separate elements containing commas:

J. D. Smyth, member of the board; Carol Winter, president; Glenn Morris, committee chairperson; and I attended the conference.

13.0   Virgule (Slash)

13.1   Use a virgule to indicate appropriate alternatives:

Define and/or use the words in sentences.

13.2   Use a virgule to represent *per* in abbreviations:

17 ft/sec   12 mi/hr

13.3   Use a virgule to separate divisions of a period of time:
the April/May report
the 1996/97 academic year

MECHANICAL CONVENTIONS

1.0   Abbreviations

1.1   Avoid the overuse of abbreviations. (See Table B.2 for a list of common technical abbreviations.)

1.2   Explain an abbreviation the first time you use it:

He has worked for the Department of Transportation (DOT) and the Office of Mental Health (OMH).

1.3   Omit most internal and terminal punctuation in abbreviations of technical measurements:

BTU   lb
psi   ft
DNA   rpm

1.4   If the abbreviation forms another word, use the internal and terminal punctuations to avoid confusion:

in.   A.M.
gal.   No.

TABLE B.2   *Common Technical Abbreviations*

ac	alternating current	**kw**	kilowatt
amp	ampere	**kwh**	kilowatt-hour
A	angstrom	**l**	liter
az	azimuth	**lat**	latitude
bbl	barrel	**lb**	pound
BTU	British Thermal Unit	**lin**	linear
C	Centigrade	**long**	longitude
Cal	calorie	**log**	logarithm
cc	cubic centimeter	**m**	meter
circ	circumference	**max**	maximum
cm	centimeter	**mg**	milligram
cps	cycles per second	**min**	minute
cu ft	cubic foot	**ml**	milliliter
db	decibel	**mm**	millimeter
dc	direct current	**mo**	month
dm	decimeter	**mph**	miles per hour
doz	dozen	**No.**	number
dp	dewpoint	**oct**	octane
F	Fahrenheit	**oz**	ounce
f	farad	**psf**	pounds per square foot
fbm	foot board measure	**psi**	pounds per square inch
fl oz	fluid ounce	**qt**	quart
FM	frequency modulation	**r**	roentgen
fp	foot-pound	**rpm**	revolutions per minute
fpm	feet per minute	**rps**	revolutions per second
freq	frequency	**sec**	second
ft	foot	**sp gr**	specific gravity
g	gram	**sq**	square
gal.	gallon	**t**	ton
gpm	gallons per minute	**temp**	temperature
gr	gram	**tol**	tolerance
hp	horsepower	**ts**	tensile strength
hr	hour	**v**	volt
in.	inch	**va**	volt ampere
iu	international unit	**w**	watt
j	joule	**wk**	week
ke	kinetic energy	**wl**	wavelength
kg	kilogram	**yd**	yard
km	kilometer	**yr**	year

1.5   Use uppercase (capital) letters for acronyms and degree scales:

NASA (National Aeronautics and Space Administration)

VHF (very high frequency)

OEM (original equipment manufacturer)

C (Centigrade)   F (Fahrenheit)

1.6   Use lowercase (small) letters for abbreviations for units of measure:

gph (gallons per hour)

cc (cubic centimeters)

rpm (revolutions per minute)

mph (miles per hour)

bps (bits per second)

1.7   Write plural abbreviations in the same form as the singular:

17	in.
47	lb
5	hr
30	gph
10	cc

2.0   Capitalization

2.1   Use Standard English conventions.

2.2   Begin all sentences with a capital letter.

2.3   Capitalize all proper nouns (proper names, titles that precede proper names, book and chapter titles, languages, days of the week, months, holidays, names of organizations and groups, races and nationalities, historical events, names of structures and vehicles, and so forth):

John Doe	Ace Construction Company
Professor Jane Doe	American Federation of Labor
Introduction to Nursing	Caucasian
French	Jewish
Monday	the Korean War
October	the Statue of Liberty
Labor Day	a Ford Mustang

2.4   Capitalize adjectives derived from proper nouns:

English

Elizabethan

2.5   Capitalize words like *street, avenue, corporation,* and *college* when they accompany a proper name:

Elm Street

Forty-second Avenue

Ace Company, Inc.

Yale University

Broward Community College

2.6 Capitalize *north, east, midwest, near east,* and so on when the word denotes a specific location:

the South

the Midwest

the Near East

101 Northwest First Street

2.7 Capitalize brand names:

Kleenex tissues	Xerox photocopies
Scotch tape	a Frigidaire
a Formica counter	the Astro-Turf field
a Polaroid camera	Sanforized

3.0 Hyphenation

3.1 Use a hyphen between some compound names for family relations:

Hyphenated: brother-in-law's company

One word: my stepmother's portfolio

Two words: my half brother was graduated

3.2 In nontechnical prose, use a hyphen in compound numbers from twenty-one to ninety-nine and in fractions:

thirty-seven cartons

forty-third year

four-fifths of the book

one-eighth inch

3.3 Use a hyphen after the prefixes *all-, ex-, self-,* and before the suffix *-elect:*

all-American

ex-president

self-contained

president-elect

3.4 Use a hyphen in some compound nouns:

kilowatt-hour

dyne-seven

foot-pound

3.5 Use a hyphen in compound adjectives when the latter precedes the word it modifies:

alternating-current motor

closed-circuit television

high-pressure system

easy-to-build model

3.6 Use a hyphen between a number and a unit of measure when they modify a noun:

6-inch ruler

12-volt charge

a 3-week-old prescription

4.0 Italics

4.1 Use italics (underline in handwritten or typed material) to indicate the names of books, magazines, newspapers, and other complete works published separately:

the book *Introduction to Nursing*

the magazine *Newsweek*

the movie *The Right Stuff*

Dante's *Devine Comedy*

Word Proof: A Manual

4.2 Use italics to indicate the names of ships and planes:

the H.M.S. *Ark Royal*

the U.S.S. *Independence*

4.3 Use italics to indicate words, symbols used as words, and foreign words that are not in general English usage:

The prizewinning orchids were *Alleraia* Ocean Spray, *Bloomara* Jim, and *Guantlettara* Noel.

coup d'etat

deus ex machina

The word *thrombosis* is derived from the Greek word *thrombos,* which means "a clot" or "a clump."

Your *9*'s look like *7*'s.

Do not italicize foreign expressions that are established as part of the English language, such as:

a priori	bona fide	habeas corpus	pro tem
ad hoc	carte blanche	laissez faire	resume
ad infinitum	etc.	per annum	status quo
	ex officio	pro rata	

5.0   Numbers

5.1   Handle numbers consistently in any one report.

5.2   Write out single digit numbers from zero through nine when the number modifies a noun:

five disks	two printers
three word processors	nine keyboards

5.3   Use numerals for zero through nine when the number modifies a unit of measure, time, dates, pages, chapters, sections, percentages, money, proportions, tables, and figures:

2 inches	section 9
3-second delay	2 percent
5 gph	a 4% increase
9 years old	$50
2:40 A.M.	$0.05 or .05 cents
June 9, 1996	1:9
9 June 1996	4 to 2 odds
page 7	Figure 2
Chapter 1	Table 6

5.4   In technical prose, use numerals for decimals and fractions:

0.6	1/4 or 0.25
3.341	7/16 in.
3/5 or 0.6	6½ lb

5.5   In technical prose, use numerals for any number greater than nine:

10 psi   237 lb
97 employees   101,400 people

5.6   Write out numbers that are approximations:

a half cup of coffee
a quarter of a mile farther
a fifth of the energy
approximately three times as often

5.7   Place a hyphen after a number of a unit of measure when the unit modifies a noun:

7-in. handle
6-inch-diameter circle
10½ -lb box
27-gal. capacity

5.8   When many numbers, both smaller than and greater than nine, are used in the same section of writing, use numerals:

Buy 4 sheets of 8 x 11½-in. paper, 15 sheets of 8 x 20-in. paper, and 7 manila envelopes.

Exception: If none of the numbers are greater than nine, write them all out:

The office contains eight desks, seven chairs, six file cabinets, and seven typewriters.

5.9   When one number appears immediately after another as a part of the same phrase, avoid confusion by writing out the shorter number:

nine 50-watt bulbs
two 4-inch wrenches
thirteen 20-pound packages
twenty-two 2,500-component circuit boards

5.10   Place a comma in numbers in the thousands:

1,000
17,276
427,928

5.11   Write numbers in the millions in one of two ways:

> 2,700,000   *or*   2.7 million
> 16,000,000   *or*   16 million
> $1,500,000   *or* $1.5 million
> 72,110,427

5.12   Write numbers in the billions, trillions, quadrillions, in one of two ways:

> 15,000,000,000   or   15 billion
> 2,700,000,000   or   2.7 billion
> 47,337,426,104,900

5.13   Do not begin a sentence with a numeral:

> Fifteen inches of rain fell.
> *not*
> 15 inches of rain fell.

6.0   Symbols

6.1   Use symbols sparingly.

6.2   Define symbols in your text. (See Table B.3 for common technical symbols.)

TABLE B.3   *Common Technical Symbols*

Symbol	Word	Symbol	Word
%	percent	=	equals
°	degree	F	Fahrenheit
&	and	C	Centigrade
′	feet	Rx	take (on prescriptions)
″	inches	☉	the Sun, Sunday
$	dollar	£	pound
¢	cents	#	number, or leave space
@	at (12 @ $2.00 each)	Hb	mercury (the element)
+	plus	☿	Mercury (the planet)
−	minus	X	snow
×	times	↑	gas
÷	divide	Ω	ohm
>	greater than	S	Silurian soil or sulfur
‖	parallel		

7.0   Spelling

7.1   Use a dictionary when in doubt about the proper and preferred spelling of a word. (See Table B.4 for a list of frequently misspelled words.)

TABLE B.4   *Frequently Misspelled Words*

accidentally	comparative	heroes	prominent
achievement	competitive	humorous	propaganda
acquaintance	consensus	immediately	psychology
amateur	contemptible	indispensable	pursue
analysis	convenience	irrelevant	questionnaire
anonymous	courageous	irresistible	receive
anxiety	criticism	knowledgeable	rhythm
appreciate	definitely	laboratory	schedule
arctic	descent	leisure	scissors
athletics	desirable	lieutenant	secretary
auxiliary	despair	lighting	seize
awkward	disappear	loneliness	separate
bachelor	discipline	maneuver	sergeant
beggar	efficient	meant	siege
beginning	eighth	medieval	similar
believe	eligible	minimum	sophomore
benefited	equipped	mortgage	souvenir
bookkeeper	exaggerate	necessary	subtle
breath	exercise	ninth	succeed
bulletin	exhausted	noticeable	successful
bureau	existence	occasionally	surprise
business	familiar	occurred	synonym
calendar	fascinating	omitted	thoroughly
campaign	fatigue	opportunity	tragedy
caricature	fiery	parallel	twelfth
catastrophe	foreign	paralysis	unforgettable
cemetery	forty	pastime	unmistakable
colonel	government	possibility	vacuum
coming	guarantee	privilege	vengeance
committee	height	procedure	weird

CHAPTER 1

Allen, Lori, and Dan Voss. *Ethics in Technical Communication: Shades of Gray*. New York: Wiley, 1997.

Allen, Nancy. "Ethics and Visual Rhetorics: Seeing's Not Believing Anymore." *Technical Communication Quarterly* 5.1 (Winter 1996): 87–105.

Anson, Chris M., and L. Lee Forsberg. "Moving Beyond the Academic Community. Transitional States in Professional Writing." *Written Communication* 7 (1990): 200–31.

Bass, John, Manager, Motorola, Inc. Media and Communication Technology Center–Radio Products Group. Personal interview. 12 Mar. 1997.

Beauchamp, Tom L., and Norman E. Bowie. *Ethical Theory and Business*. 2nd ed. Englewood Cliffs, NJ: Prentice Hall, 1983.

Chodorow, N. *The Reproduction of Mothering: Psychoanalysis and the Sociology of Gender*. Berkeley, CA: University of California Press, 1978.

Dagga, Sam. "A Question of Ethics: Lessons from Technical Communicators on the Job." *Technical Communication Quarterly* 6.2 (Spring 1997): 161–78.

Dombrowski, Paul. "Ethics in Action: Ethics in Working With the Federal Government." *ATTW Bulletin* 9.1 (Spring 1999): 3, 6.

Finsberg, Stephen. "Dilbert Tackles Ethics Dilemmas." *Sun–Sentinel* 17 Apr. 1997: D1–2.

Gerson, Sharon J., and Steven M. Gerson. *Technical Writing: Process and Product*. 3rd ed. Upper Saddle River, NJ: Prentice Hall, 2000.

Goodman, Ellen. "Court Must Breathe Real Life Into Cartoon Image of Abortion." *The (Miami) Herald* 25 Apr. 2000: 11B.

Gormen, Christine. "Drugs by Design." *Time* 11 Jan. 1999: 79–82.

Helyar, P.S. "Products Liability: Meeting Legal Standards for Adequate Instructions." *Journal of Technical Writing and Communications* 22.2 (1992): 125–47.

Hiaason, Carl. An Editorial. *The (Miami) Herald* 25 Apr. 2000: 52D.

Kent, Peter. *Making Money in Technical Writing: Turn Your Writing Skills into $100,000 a Year*. New York: Arco, 1998.

Kienzler, Donna S. "Visual Ethics." *Journal of Business Communication* 34.2 (Apr. 1997): 171–87.

Lannon, John M. *Technical Writing*. 7th ed. New York: Addison-Wesley Educational Publishers, Inc., 1997.

Markel, Mike. *Technical Communication: Situations and Strategies*. 5th ed. Boston, MA: Bedford/St. Martin's, 1998.

Ruggiero, Vincent R. *The Art of Critical and Creative Thought*. 4th ed. New York: HarperCollins Publishers, 1995.

Sigma Xi. *Honor in Science*. New Haven, CT: The Scientific Research Society, 1989.

Society for Technical Communications. "STC Ethical Principles for Technical Communications." 2000. 12 June 2000 <http://www.stc–va.org>.

Syms, Brenda R. "Linking Ethics and Language in the Technical Communications Classroom." *Technical Communications Quarterly* 2.3 (Summer 1993): 285–99.

University of Michigan. *Engineering: Master of Science in Technical Information Design and Management*. Ann Arbor, MI: University of Michigan, n.d.

"Writers and Editors, Including Technical Writers." *Occupational Outlook Handbook*. 2000–2001 ed. 15 Apr. 2000. Bureau of Labor Statistics. 20 July 2000 <http://stats.bls.gov/oco/ocos889.htm>.

CHAPTER 2

Corbett, Jan. "From Dialog to Praxis: Crossing Cultural Borders in the Business and Technical Communication Classroom." *Technical Communication Quarterly* 5.4 (Fall 1996): 411–24.

Cross, Mary. "Aristotle and Business Writing: Why We Need to Teach Persuasion." *The Bulletin of the Association for Business Communication* 54.1 (March 1991): 3–6.

Dragga, Sam. "Ethical Intercultural Technical Communication: Looking Through the Lens of Confucian Ethics." *Technical Communication Quarterly* 8 (Fall 1999): 365–81.

Gerson, Sharon J., and Steven M. Gerson. *Technical Writing: Process and Product.* 3rd ed. Upper Saddle River, NJ: Prentice Hall, 2000.

Havelock, Eric. *The Muse Learns to Write: Reflections on Orality and Literacy from Antiquity to the Present.* New Haven, CT: Yale University Press, 1986.

Hunt, Peter. "The Teaching of Technical Communication in Europe: A Report from Britain." *Technical Communication Quarterly* 2.3 (Summer 1993): 319–30.

Lay, Mary M., and William M. Karis, eds. *Collaborative Writing in Industry: Investigation in Theory and Practice.* Amityville, NY: Baywood Publishing Company, Inc., 1991.

Lustig, M. W. and Jolene Koester. *Intercultural Competence: Interpersonal Communication Across Cultures.* Upper Saddle River, NJ: Prentice Hall, 1999.

Markel, Mike. *Technical Communications: Situations and Strategies.* 5th ed. Boston, MA: Bedford/St. Martin's, 1998.

Maylath, Bruce. "The Enigma of International Technical Communication: Measuring Translation Quality." *ATTW Bulletin* (Fall 1998): 4–6.

Pinker, Steven. "Chasing the Jargon." *Time* 13 Nov. 1995: 30–31.

Roundy, N., and D. Mair. "The Composing Process of Technical Writers: A Preliminary Study." *Journal of Advanced Composition* 3 (1982): 89–101.

Thrush, Emily A. "Bridging the Gaps: Technical Communication in an International and Multicultural Society." *Technical Communication Quarterly* 2.3 (Summer 1993): 271:83.

CHAPTER 3

Bikle, David. Lucent Technologies, Manager, Public Relations Letter to the author. 15 Apr. 1997.

"Color Graphics: The Results of Our Tests." *Consumer Reports* Oct. 1996: 58–59.

Tritt, Merrill D, Information Systems Analyst. Spherion Corporation. July 2000.

XACT Development System. *XEPLD Schematic Design Guide.* San Jose, CA: XILINX, Dec. 1994.

CHAPTER 4

Bernhardt, Stephen A. "Seeing the Text." *College Composition and Communication* 37 (1986): 66–78.

Corel WordPerfect Suite 7: Quick Results. Dublin, Ireland: Corel Corporation Limited. 1996.

Kostelnick, Charles. "Supra-textual Design: The Visual Rhetoric of Whole Documents." *Technical Communication Quarterly* 5.1 (Winter 1996): 9–23.

Kramer, Robert, and Stephen A. Bernhardt. "Teaching Text Design." *Technical Communication Quarterly* 5.1 (Winter 1996): 35–60.

Lay, Mary. "The Non-Rhetorical Elements of Design." *Technical Writing Theory and Practice.* New York: Modern Language Association, 1989.

Markel, Mike. "Using Design Principles to Teach Technical Communication." *Journal of Business and Technical Communication* 9 (1995): 206–18.

Schaefer, Heather. Art work provided to the author. 12 Aug. 2000.

CHAPTER 5

Caravello, Patti S. "Judging Quality on theWeb." 2 Apr. 1999. UCLA. 4 July 2000 <http://library.ucla.edu/libraries/url/referec/judging.html>.

Eavenson, Kathleen, Head of Broward County Main Library Community Technology Center. Personal interview. 27 June 2000.

Gorman, Jessica. "Sink the Nukes." *Discover* July 2000: 24.

Harnack, Andrew, and Eugene Kelppinger. *Online! A Reference Guide to Using Internet Sources.* New York: St. Martin's Press, 1997.

Internet Access Company, The. "The Ramp to the Web." 2000. 3 July 2000 <http://corp.tiac.net/support/intro/searching>.

Kirk, Elizabeth. "Evaluating Information Found on the Internet." 4 Jan. 2000. Johns Hopkins University. 3 July 2000 <http://milton.msc.jhu.edu.8001/research/education/net.html>.

Linger, Neil, Reference Librarian, Broward Community College Library/Central Campus. Personal interview. 29 June 2000.

Revkin, Andrew C. "100–year Forecast: Warmer." *The New York Times* 29 June 2000: 8B.

Tillman, Hope N., "Evaluating Quality on the Net." 29 May 2000. Babson College. 3 July 2000 <http://www.tiac.net/users/hope/findqual/html>.

CHAPTER 6

American Psychological Association. "Electronic Reference Format Recommended by the American Psychological Association." 19 Nov. 1999. 11 July 2000 <http://www.apa.org/journals/webref.html>.

American Psychological Association. *Publication Manual of the American Psychological Association,* 4th ed. Washington, DC: American Psychological Association, 1994.

American Chemical Society. "ACS Style Sheet. Footnotes and Bibliography Series." Jan. 1999. Penn State Lehigh Valley. 17 July 2000 <http://www.lehigh.edu/~inhelp/acs.html>.

American Chemical Society. "Citation Style for Internet Services." Jan. 1999. 17 July 2000 <http://www.lv.psu.edu/jkl1/writing/citechem. html>.

Chicago Manual of Style. "Using Chicago Style to Cite and Document Sources." 2000. 11 July 2000 <http://www.bedfordstmartins.com/ online/cite7.html>.

Council of Biology Editors. *The CBE Manual for Authors, Editors, and Publishers.* Bethesda, MD: Council of Biology Editors, Inc. 1994.

Gibaldi, Joseph. *MLA Handbook for Writers of Research Papers.* 5th ed. New York: Modern Language Association of America, 1999.

Grossman, John, ed. *Chicago Manual of Style.* 14th ed. Chicago: University of Chicago Press, 1993.

Lester, James D., and James D. Lester, Jr. *The Essential Guide to Writing Research Papers.* New York: Addison Wesley Longman, Inc., 1999.

Modern Language Association of America. "MLA Documenting Sources from the World Wide Web." 4 Apr. 2000. 11 July 2000 <http:www. mla.org/ste/sources.htm>.

Riger, Stephanie. "Epistemological Debates, Feminist Voices—Science, Social Values, and the Study of Women." *American Psychologist* 57 (June 1992): 78–92.

CHAPTER 7

Bass, John, Manager, Motorola, Inc. Media and Communication Technology Center–Radio Products Group. Personal interview. 12 Mar. 1997.

Dental Products of 3M, *3M Scotchbond™ Etching Gel.* St. Paul, MN: 3M Scotchbond™, Apr., 1998.

Technical Writing Group. *Evaluation and Comment Sheet*. Miami, FL: Racal-Milgo, 1999.

Weiss, Amy. Manager, Motorola, Inc. Integrated Technical Communications Paging Products Group. Personal interview. 7 Apr. 1997.

Weiss, Edmond H. *How to Write a Usable User Manual*. Philadelphia, PA: ISI, 1985.

Whitaker, Ken. *A Guide to Publishing User Manuals*. New York: Wiley, 1995.

CHAPTER 8

Bryson, Bill. *The Mother Tongue: English and How It Got That Way*. New York: William Morrow and Company, Inc., 1990.

Fuhrhop Debra. "A Definition of Volcano." Broward Community College student, Ft. Lauderdale, FL, 1987.

Johns, Robert. "Definition of Microburst." AlliedSignal, Inc., Ft. Lauderdale, FL, 2000.

McCrum, William Cran, and Robert MacNeil. *The Story of English*. New York: Viking Penguin, Inc., 1986.

CHAPTER 9

Braunwald, E., ed. *Heart Disease*. 4th ed. 2 vol. Philadelphia, PA: W.B. Saunders, 1991.

Columbia University College of Physicians and Surgeons. "How the Heart Works." *Complete Home Medical Guide*. Rev. ed. New York: Crown Publishers, Inc. 1989.

Frazier. O. H. "The Human Heart." *Grolier Multimedia Encyclopedia*, 1996. 23 Apr. 1997 <http://gi.grolier.com/encyclopedia>.

"Jargon: The Time Digital Dictionary." *Time* 10 Mar. 1997: 48.

Tierney, L. M. *Current Medical Diagnosis and Treatment*. 34th ed. Stamford, CT: Appleton & Lange, 1997.

CHAPTER 10

Clement, David E. "Human Factors, Instructions, and Warning, and Product Liability." *IEEE Transactions on Professional Communication* 30:3 (1987): 149–56.

Spyridakis, Jan, and Michael J. Wenger. "Writing for Human Performance: Relating Reading Research to Document Design." *Technical Communication* 39.2 (1992): 202–15.

CHAPTER 11

Baker, David, and Dean Weinlaub. "Fossil Fuels." *Sun-Sentinel* 1 May 1994: G6.

"Geography Facts." *The Universal Almanac.* 1994. 8 Mar. 1994. <http://uni.almanac.edu>.

Graves, Heather Brodie, and Roger Graves. "Masters, Slaves, and Infant Mortality: Language Challenges for Technical Editing." *Technical Communication Quarterly* 7:4 (Fall 1998): 389–414.

Krantz, Michael. "A Tube for Tomorrow." *Time* 14 Apr. 1997: 69.

Lemonick, Michael D. "How to Prevent a Meltdown." *Time* April–May 2000: 61–63.

"Making Electricity and Steam Together." *The Lamp of Exxon Corporation* 78.4 (Winter 1996): 4–5.

"Science Explained, The Way Nature Works." *Grolier's Academic American Encyclopedia.* 1997. 7 Mar. 1997 <http://ea.grolier.com>.

CHAPTER 12

Austin, Paul, Vice-president. Handy and Hardy Refining. Letter to the author. 7 April 1995.

Blake, Gary. *Quick Tips for Better Business Writing.* New York: McGraw-Hill, Inc. 1995.

Doup, Liz. "Mind Your E-Manners." *The Sun–Sentinel* 3 July 2000: D1.

Extejt, Marian M. "Teaching Students to Correspond Effectively Electronically." *Business Communications Quarterly* 61.2, (June 1998): 57–67.

Rosen, Jeffrey. "The Eroded Self." *New York Times Magazine,* 30 Apr. 2000: 46–53, 129.

Schiller, Judy. Director, Alumni Relations, Miami University. Letter to the author, 15 June 2000.

Smelzer, Mark. Vice-president, Entertainment, Latitude 90, Inc. <marks@l90.com>, E–mail to the author. 3 Aug. 2000.

CHAPTER 13

AI Resume Service. "Resume Examples." 31 July 2000 <http://resume-service.net/examples.htm>.

Forman, Ellen. "'Judgment Day' Job Interview Survival Guide Advises Applicants to be Prepared for Any Question." *Sun-Sentinel* 31 Mar. 1997: B3.

Forman, Ellen. "Online Bulletin Boards Aiding Job Hunters." *Sun-Sentinel* 12 May 1997: B2.

Grappo. Gary Joseph. "Selling Me, Inc." *Sun-Sentinel* 24 Feb. 1997: B3.

Isaacs, Kim. "Develop a Powerful Resume with Monster's Resume Builder." Monster Career Center. 2000. 31 July 2000 <http://content.monster.com/resume/resources/ newresumebuilder>.

Johns, Robert W. AlliedSignal, Inc. Letter to the author. 13 July 2000.

Kleiman, Carol. "Let the Interviewer Know You're Eager." *Sun-Sentinel* 18 Nov. 1996: B12.

Lorek, L. A. "Cyberways: Let Your Mouse Pound the Pavement When Seeking a Job." *Sun– Sentinel* 31 May 1998: 4F.

Racal-Datacom. "Software Engineering Job Posting." 1 May 1997. 9 June 1997 <http://www.careermosaic.com>.

Tritt, Merrill D. Information Systems Analyst. Spherion Corporation. Aug. 2000.

"Writers and Editors." *Occupational Outlook Handbook*. 2000–2001 ed. 15 Apr. 2000. <http://stats.bls.gov/ocos089.html>.

CHAPTER 14

Johns, Robert, Senior Design Engineer. AlliedSignal, Inc. Evaluation of GAAD Dimmer Driven Circuity report. 9 June 1997.

Johns, Robert. Guardian-American Company. Report for the author. 30 Aug. 2000.

Mills, Gerald. Marine Resources Investigator. Brief Field Report. 6 July 1994.

Ryder Truck Rental, Inc. Travel Expense Form. 3 Mar. 1985.

Sicliri, Ron. Fort Lauderdale Fire Department Request for Leave Form. 12 Apr. 1985.

Venci, Cathy. Broward Community College student incident report. 10 May 1985.

Vordermeier, Kenneth. Vordermeier Company Realtors. Field report. June 1985.

Whitten, Kenneth W., and Kenneth D. Gailey. Experiments in General Chemistry. Ft. Worth, TX: Saunders College Publishing, 1981.

CHAPTER 15

Bowman, Joel P., and Bernadine P. Branchaw. *How To Write Proposals That Produce*. Phoenix, AZ: Oryx, 1992.

Vaughan, David K. "Abstracts and Summaries: Some Clarifying Distinctions." *The Technical Writing Teacher* 18:2 (1991): 132–41.

Johns, Robert. "Abstract for RDB-4B White Paper." AlliedSignal, Inc. Aug. 10, 2000.

Warren, David. "Shipboard Damage Control." *Naval Engineers Journal* Jan. 1992: 63.

Smith, Charles E. Coopers & Lybrand. "Reimbursable Expense Policy and Monitoring and Control System Proposal." May 1985.

CHAPTER 16

Galindo, Y, K. McLachlan, and Z. Kasloff. "Microscopic Study of Smooth Silver-plated Retention Pins in Amalgam." *Journal of Dental Research* Feb. 1980:124–28.

Pearsall, Tom, Cynthia Chapman, and Melinda Kreth. "Interchange: What Are the Differences Between 'Scientific' and 'Technical' Writing?" *ATTW Bulletin* 7:1 (Fall 1996): 6–7.

Sobel, Davis. "A Reporter At Large among Planets." *The New Yorker* 9 Dec. 1996: 47–52.

"Submission Guidelines." 1999. *Technical Communication Quarterly*. 1999. 18 Aug. 2000 <http://english.ttu.edu/attwtest/ ATTWpubs TCQsubmissions.asp>.

Venci, Cathy. "Hospice Versus Hospital: An Examination of Contrasting Approaches in Patient Care." Broward Community College student paper. May 1987.

CHAPTER 17

Alvear, Jose. "Meeting and idea-sharing Online." *Internet World* Aug. 1997: 33–34.

CHAPTER 18

Greengard, Samuel. "Picture Perfect." *U.S. Airways Magazine* Aug. 1997: 8–15.

Grice, Roger A., and Lenore S. Ridgway. "Presenting Technical Information in Hypermedia Format: Benefits and Pitfalls." *Technical Communication Quarterly* 4:1 (1995): 35–36.

Maddison, Gordon. Instructor of English. Broward Community College. Interview with the author. 1993.

CHAPTER 19

"America Online's Web Pages Made Easy." CD–ROM. America Online, Inc. 2000.

Gerson, Sharon J., and Steven M. Gerson. *Technical Writing: Process and Product*. 3rd ed. Upper Saddle River, NJ: Prentice Hall, 2000.

Markel, Mike. *Technical Communication: Situations and Strategies*. 5th ed. Boston, MA: Bedford/St. Martin's, 1998.

Tritt, Merrill D. Information Systems Analyst. Spherion Corporation. Aug. 2000.

INDEX